Solid State Chemistry and Physics

VOLUME 2

Solid State Chemistry and Physics

AN INTRODUCTION

VOLUME 2

Edited by Paul F. Weller
Department of Chemistry
State University College
Fredonia, New York

MARCEL DEKKER, INC. New York 1974

MARCEL DEKKER, INC.
305 East 45th Street, New York, New York 10017

LIBRARY OF CONGRESS CATALOG CARD NUMBER: 73-85302
ISBN: 0-8247-6026-3

Printed in the United States of America

PREFACE

During the last decade the field of solid-state science — or materials science — has become increasingly interdisciplinary. Physicists and metallurgists have been joined by chemists in solving many problems relating to computer and communications technology, microcircuitry, lasers, photography, and so on. And now the importance of electrical and magnetic effects in biological processes has involved many biologists in the study and application of solid state principles.

The present text is intended for students of these interdisciplinary fields. We have made an effort to present the fundamental principles and practices of the solid state in a language and a form that is understandable not only to those acquainted with solid-state physics, but also to chemists and biologists or anyone else sincerely interested in the properties of solids. The concepts and topics, along with the necessary mathematical background, covered in a typical undergraduate physical chemistry course are assumed requirements for the entire book. In a few specific cases somewhat more sophisticated mathematical treatments are used and described. We have attempted to prepare the reader for these more detailed sections by dividing the text into four major parts, with Part I serving as a relatively non-mathematical introduction to the concepts used throughout the rest of the text.

The introductory chapter of Part I treats many of the concepts and properties covered in more detail in later sections of the book. Many intuitive, nonmathematical physical pictures and analogies are used to introduce necessary principles. Crystal structures are covered in Chapter 2 without the use of sophisticated mathematics, and the very important topic of chemical bonds and their roles in solids is developed in Chapter 3. Then the effects and interactions of various imperfections in solids are treated in considerable detail. Rather strong emphasis has been placed on imperfections throughout the text. Their presence can have pronounced effects on the properties of materials, properties that are of critical importance to present and future technology and, possibly, even to fundamental life processes.

Part II builds on the concepts introduced in Part I and covers many of the properties of solids — electrical, optical, magnetic, etc. — in more rigorous detail. Since the study and correct understanding of many of the properties described in Part II require highly pure single crystals or crystals with carefully controlled impurity contents, Part III considers many aspects of sample preparation, purification, and single-crystal growth. In Part IV the principles developed in Parts I through III are applied to the cases of polymers and biological processes.

While this text is certainly not an exhaustive treatment of all topics important to the solid-state scientist, it is hoped that the presentation will whet the appetite and stir the interest for further investigation by individuals with various backgrounds and training. We believe that the coverage is sufficient in scope and depth to provide a firm foundation in the principles of the solid state, from which a more detailed study of the entire field or of a given specialized area can be launched.

Paul F. Weller

LIST OF CONTRIBUTORS

R. F. BREBRICK,* Lincoln Laboratory, Massachusetts Institute of Technology, Lexington, Massachusetts

EDWARD KOSTINER,** Department of Chemistry, Cornell University, Ithaca, New York

JEROME S. PRENER, General Electric Research and Development Center, Schenectady, New York

A. REISMAN, IBM Thomas J. Watson Research Center, Yorktown Heights, New York

GABOR A. SOMORJAI, Department of Chemistry, University of California, Berkeley, California

H. TI TIEN, Department of Biophysics, Michigan State University, East Lansing, Michigan

BERNHARD WUNDERLICH, Department of Chemistry, Rensselaer Polytechnic Institute, Troy, New York

*Present address: Marquette University, Milwaukee, Wisconsin

**Present address: Department of Chemistry and Institute of Materials Science, University of Connecticut, Storrs, Connecticut

CONTENTS

PART III

PURIFICATION AND CRYSTAL GROWTH

CONTENTS OF VOLUME 1

Solid State Chemistry and Physics

VOLUME 2

PART II (continued)

PHYSICAL PROPERTIES AND IMPERFECTIONS

Chapter 8

POINT DEFECTS IN SOLIDS

Jerome S. Prener

General Electric Research and Development Center
Schenectady, New York

I. INTRODUCTION

The concept of a crystalline solid, as an orderly array of stationary groups of identical atoms extending indefinitely in three dimensions, is an idealized one. Not only are such crystals unattainable, as we shall see later, but they would be useless for technical applications and of little interest for many research studies in solid state chemistry and physics. Any deviation from the ideal structure of a perfect crystal we may assign to the presence of defects or imperfections of one type or another. The role of many of these defects in determining the optical, electrical, magnetic, mechanical, and chemical properties of solids is the subject of the latter part of this book (Chapters 4 through 14). Because defects do determine many of the technologically important properties of solids, an ever increasing amount of attention is being devoted to their study. Our task in this chapter is to answer such questions as:

1. What sorts of defects do we encounter in solids and how are they formed?

2. How does the stoichiometry of a solid depend on the concentration of defects?

3. How do the concentrations of defects depend on experimentally controllable variables?

To a solid state scientist, the answers to such questions are of prime importance, since they offer a means of controlling those properties of crystalline solids related to defects and of designing new materials with desirable properties.

In this chapter we will not give either a rigorous treatment or an exhaustive survey of the field of defects in solids. We intend, rather, to familiarize the reader with the terms and concepts used in the study of defects or imperfections in solids. Much of the experimental detail useful for studying defects in solids will be found in Part II of this book. In Parts III and IV will be found a number of examples of specific solids. The concepts of defect chemistry have made it possible to understand many of the chemical and physical properties of these solids.

For those who are interested in pursuing further the subject of defects in solids, there are, fortunately, a number of excellent specialized books available. Particularly recommended are:

1. F.A. Kröger, The Chemistry of Imperfect Crystals, North-Holland Publ., Amsterdam, The Netherlands, 1964. This is an excellently written and very complete treatment of the subject of defects. It contains a very large number of references to the original literature.
2. R.A. Swalin, Thermodynamics of Solids, Wiley, New York, 1962. This is a very fine textbook dealing with the application of thermodynamics to solid phases. Three chapters are devoted specifically to defects in solids.
3. H.G. Van Buren, Imperfections in Crystals, North-Holland Publ., Amsterdam, The Netherlands, 1960. This book treats the subject of dislocations in solids quite extensively.
4. J.H. Shulman and W.D. Compton, Color Centers in Solids, Pergamon, New York, 1962. This is an excellent review of an important class of defects that have been studied in great detail.

II. POINT DEFECTS IN SOLIDS

A. The Perfect and the Real Crystal

If we consider a simple ionic solid, say NaCl, then our picture of the ideal or perfect crystal would consist of an ordered array of exactly the same Na^+ and Cl^- ions, each at specified positions in the face centered cubic (fcc) lattice, this array extending indefinitely in three dimensions. Now it is obvious that any real crystal of NaCl must be of finite size. Further, the constituent ions, Na^+ and Cl^-, are not in fixed positions but are constantly vibrating. The ions themselves are not all the same but consist of isotopes of different masses. It is also quite evident that it is impossible to rid this solid of the last traces of impurities (see Section IV.B). Already we have introduced into our picture of a real crystal a number of defects: surfaces, lattice vibrations, and impurities (including isotopes of the constituent ions). The effect of surfaces upon the properties of solids is the subject of Chapter 10, and we shall say no more about these. Lattice vibrations, also referred to as phonons, constitute an important defect in solids since they are the agents whereby the lattice reaches a state of thermal equilibrium. The energy distribution, or spectrum of the phonons, is important in problems involving optical absorption, heat capacity, electrical resistivity, x-ray diffraction, line broadening, luminescence, and many others. The changes in the phonon spectrum accompanying the formation of atomic defects in solids contribute to the entropy of the solid, as we shall see later. We will not treat phonons here. The reader interested in learning more about these defects should consult a text on solid state physics [1].

In addition to the several defects already mentioned, there are others which are present to a larger or smaller extent in all crystals. Present in all crystals are dislocations, and in some crystals stacking faults. For discussions and descriptions of these defects, the reader is referred to the book by Van Buren mentioned previously. Our primary concerns in this chapter will be atomic defects, also called point defects or imperfections. Although these point defects occur in all classes of solids (Chapter 1), we shall consider mainly semiconducting and insulating crystalline compounds.

B. Impurities

We have already mentioned the impurity defect. It is experimentally impossible to remove the last traces of impurities no matter how carefully and diligently one purifies the solid (see Section IV. B). The element Ge has been obtained with a total electrically active impurity concentration of less than 10^{11} cm^{-3} and is one of the purest solids available. In Chapter 11 various purification schemes are described as well as methods for the controlled introduction of impurities into solids. The effects of even trace amounts of impurities upon many of the physical and chemical properties of solids can be tremendous and of great technological importance. These effects are discussed in the latter part of this book starting with Chapter 4.

Let us consider again NaCl and replace a small number of the Na^+ ions by Tl^+ ions [2]. For example, the fraction of Na^+ ions replaced by Tl^+ ions may well be in the range from 10^{-3} to 10^{-6}. The thallous ion not only substitutes for the sodium ion, but has the same charge. This type of impurity we symbolize by writing Tl_{Na}^{x}. The superscript x is used to indicate a zero effective charge. This means that relative to the Na^+ ion which it replaces, the Tl^+ ion impurity exhibits no net charge. The subscript indicates the atom which the impurity replaces in the structure. Similarly Br^- at Cl^- sites are written as Br_{Cl}^{x}. Such defects, having zero effective charge, are called neutral defects. Other examples of neutral impurity defects are Cu_{Zn}^{x}, Al_{Zn}^{x}, and Cl_{S}^{x} in ZnS, N_{C}^{x} in SiC, and B_{Si}^{x} in Si. Another impurity whose properties have been studied in alkali halides is Ca [3]. For the Ca^{2+} ion at Na^+ sites we write $Ca_{Na}^{\bullet}$. The Ca^{2+} ion has a single positive charge relative to the Na^+ ion which it replaces in the structure.

We indicate this single positive effective charge by a $^{\cdot}$ as a superscript. Similarly O^{2-} ions at Cl^- ion sites are written as O'_{Cl}, the ' symbolizing a single negative effective charge.* Such defects having nonzero effective charges are referred to as charged defects. Other examples of charged impurity defects are: Cu'_{Zn}, $Al^{\cdot}_{Zn}$, and $Cl^{\cdot}_{S}$ in ZnS, $N^{\cdot}_{C}$ in SiC, and B'_{Si} in Si.

So far, we have considered impurities that substitute, in small numbers, for the major constituents of the solid. These impurities, both neutral and charged, are called substitutional impurities or defects, It is also possible for certain small atoms, such as H, Li, C, N, B, and Cu, to reside, as impurities, in the interstices of the crystal structure. These interstitial impurities or defects may also be neutral or charged. Thus H atoms at interstitial sites in ZnO [5] are written as H^{x}_{i} whereas Li^+ ions at interstitial sites in ZnO [6] are written as $Li^{\cdot}_{i}$. We note that the interstitial site in a lattice has no charge associated with it, so that the effective charge and the true charge of interstitial defects are equal.

Since charged defects introduce extra charges into a crystal (these being equal to their effective charges) they must be compensated by some other defect of opposite effective charge. The crystal as a whole must be electrically neutral. These compensating defects may be other impurities. In ZnS phosphors used for television screens the impurities responsible for the fluorescence are equal numbers of Ag'_{Zn} and $Cl^{\cdot}_{S}$ (about 10^{-4} atom fraction of each). The Ag impurity has an effective negative charge if we think of a Ag^+ ion as substituting for a Zn^{2+} ion. The Cl impurity has an effective positive charge since Cl^- substitutes for S^{2-}.

---

*The symbols ^{x}, $^{\cdot}$, and for the effective charges of point defects are those adopted by Kröger [4]. In the literature the symbols 0, +, and - are also used, but since these are also used for the real ionic charges, they are best avoided to prevent confusion. It should be pointed out that the effective charges are not dependent on any ionic model for bonding. We shall discuss this point in Section II. G.

An impurity atom may, in principle, occupy all the sites available in a given structure, that is, both the various substitutional and the various interstitial sites. Which site is actually occupied by a given impurity in a given structure is determined largely by the energy it takes to incorporate it at that given site. This energy depends on attractive forces (bond formation, electrostatic attraction, etc.) and on repulsion forces (size effects). In ionic compounds (KCl, CdF_2, ZnS, Fe_2O_3, etc.) the constituents differ in electronegativity, and substitutional impurities tend to substitute for those atoms to which they are closest in chemical behavior. Thus metallic impurities substitute for metal atoms of the compound and nonmetallic impurities for the nonmetallic constituents of the compound. Ca and Tl, for example, substitute for K in KCl, but Br and O substitute for the halide. In the complex compound $Ca_5(PO_4)_3F$, used as a phosphor in fluorescent lamps, we may find the impurity defects Mn^{x}_{Ca}, $Sb^{\cdot}_{Ca}$, $(OH)^{x}_{F}$, O'_{F}, and $(MnO_4)^{x}_{PO_4}$, among others [7]. For those compounds in which the electronegativity difference between the constituents is small, size effects may become important (see Ref. [8] for a discussion of electronegativity). Thus, Si as an impurity occupies Sb sites in InSb [9], but both Ga and As sites in GaAs [10, 11].

For interstitial defects, size effects are predominant. As already mentioned, only small atoms are found to occupy interstitial sites. Thus, such anions as Cl^-, Br^-, O^{2-}, and S^{2-} do not occur as interstitial defects although the smaller F^- ion does.

C. Native Defects

Charge compensation may also take place by way of native defects, also called intrinsic defects. If a crystal of NaCl is grown from a melt containing a small amount of $CaCl_2$ impurity, one finds in the crystal $Ca^{\cdot}_{Na}$ defects, as already mentioned, and an equal number of missing Na^+ ions or Na^+ ion vacancies [3], the symbol for these being V'_{Na}. The negative effective charge arises from the fact that the removal of the positive Na^+ ion has left locally a region of net negative charge in the crystal. Similarly, Cl^- ion vacancies in NaCl and Zn^{2+} and S^{2-} vacancies in ZnS are written as $V_{Cl}^{\bullet}$, V_{Zn}'', and $V_S^{\bullet\bullet}$, respectively. As another example, one may introduce just Cl as an impurity into ZnS instead of AgCl as described (for example, by firing the ZnS in HCl). One then obtains a fluorescent material of a different color in which the defects are $Cl_S^{\bullet}$, each two of which are compensated by one V_{Zn}'' [12]. Native defects may also exist in the form of interstitials. Thus, an important defect in CaF_2 and CdF_2 is F_i' [13, 14], and in CdTe it is believed that Cd_i^x defects exist [15]. In Section IV. F the conditions under which compensation of charged impurity defects by native defects becomes important will be discussed.

Another type of native imperfection is the misplaced atom. Thus, in the compound Bi_2Te_3 a small number of Bi atoms are found on Te sites and vice versa. This type of defect is also found in some spinel compounds. In the spinel $MgFe_2O_4$, some of the Mg^{2+} ions may occupy sites normally occupied only be Fe^{3+} in the ideal crystal. Such a defect is written as Mg_{Fe}'. Further details on this type of substitutional disorder can be found in Chapter 18 of the book by Kröger (see also Section III. B).

D. Electrons and Holes

In Chapter 3 the electronic ground state of perfect crystals was discussed. The electrons are in states whose energies group themselves into a series of energy bands separated by energy gaps. For nonmetallic solids, and it is these with which we are concerned, the group of states with the highest energies which can be fully occupied by electrons form the valence band. This is separated from a group of unoccupied states forming a band of still higher energies called the conduction band.* The separation between these two, in energy, is called the band gap energy.

In perfect crystals containing no point defects, at a temperature of 0 °K, all the electrons reside in states of lowest energy; the states of the valence band are completely occupied and there are no electrons in conduction band states. At any temperature greater than 0 °K, some electrons will be thermally excited to conduction band states. Some of the states in the valence band formerly occupied by electrons are now empty and behave as mobile effective positive charges (see Chapter 4). These mobile charges are referred to as holes. It takes an amount of energy equal to the band gap energy E_G to generate one electron-hole pair. The reverse transition, in which a conduction band electron (also called a free electron) returns to an empty valence band state, can also take place. This process is referred to as electron-hole recombination (Chapter 7). At equilibrium the generation and recombination rates of electron-hole pairs are equal. The result is that, in a given solid at a temperature T greater than 0 °K, there will be a certain equilibrium concentration n (number of electrons per cm^3 of solid) of electrons in conduction band states. The symbol for these electrons is written as e'. There will also be a certain equilibrium concentration p of holes in valence band states, and these are written as $h^{\bullet}$. The role played by these electronic defects (i. e., conduction band electrons and valence band holes) in determining many of the physical properties of solids is discussed in all the chapters in the latter part of this book. Of particular interest and importance is the role of electronic defects in the electrical conduction process in solids, as described in Chapter 4, and in optical absorption and emission processes, as described in Chapter 7.

---

*The compounds containing transition metals may contain only partly filled bands. These substances are discussed in some detail in Chapter 4 and will not be considered here.

In this chapter we consider the interaction of electrons and holes with point defects.

In a solid free of atomic point defects it is necessary that $n = p$. These electron-hole pairs assure charge neutrality, since conduction electrons and valence band holes have opposite effective charges. The values of n and p at any temperature T and pressure P will depend mainly on the band gap energy, as we will show in Section IV. They will be larger for smaller band gap materials, and they increase with temperature in any given material. In any real crystal containing various point defects, however, n does not necessarily have to equal p, due to the interaction of these charge carriers with the defects.

E. Effective Charge States of Defects

In a solid containing atomic point defects there may exist localized electronic states having discrete energy levels within the forbidden gap. These arise from a disruption, by the point defects, of the periodic array of the atoms making up the perfect crystalline solid. When the perturbation of the periodic potential (Chapter 3) by the defects is large enough, localized states result. Since these states are associated with the presence of particular defects, the occupancy or nonoccupancy of these states by electrons can change the effective charge of the defects. In other words, the same defect may exist with more than one effective charge. We have already alluded to this possibility when we mentioned such impurity defects as Al_{Zn}^{x} and $Al_{Zn}^{\cdot}$ in ZnS and B_{Si}^{x} and $B_{Si}^{\prime}$ in the element Si (Section II. B).

Let us consider a binary compound such as ZnS containing the impurity $Cu_{Zn}^{\prime}$ in small concentrations. We may describe the conversion of the negative effective charge state of this defect to one of zero effective charge state by the equation*

---

*In equations representing reactions, the energy terms E_i will be written either on the left-hand side or the right-hand side so as to make them positive quantities.

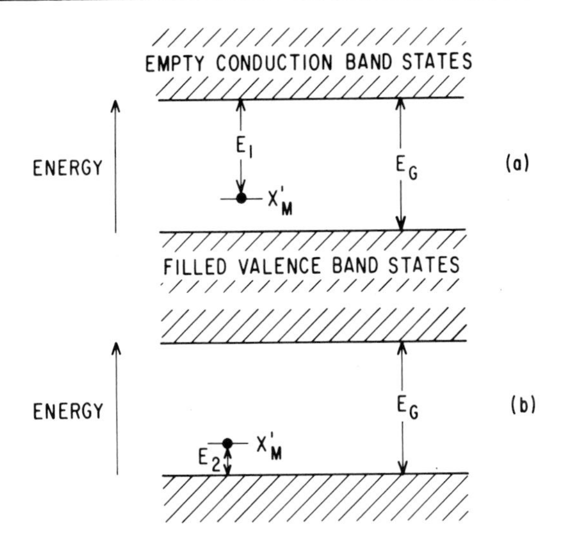

FIG. 1. (a) Energy level of the localized ground electronic state associated with the impurity defect X'_M in the binary compound MX. E_1 is the ionization energy of the electron in the state and E_G is the band gap energy. (b) Same as (a). E_2 is the binding energy of a hole to X'_M.

$$Cu'_{Zn} + \text{energy } E_1 \rightarrow Cu^x_{Zn} + e' \quad . \tag{1}$$

The electron in a localized state associated with substitutional Cu impurity can be ionized, by thermal energy, to a conduction band state.* The smallest energy required for this process is E_1. We say that the energy E_1 locates the energy level of the ground electronic state of the defect Cu'_{Zn} with respect to the bottom of the conduction band of ZnS. This is shown for the defect X'_M in the compound MN in Fig. 1(a). We note that the host crystal, ZnS, was not specifically included in writing Eq. (1). The particular host crystal must, however, be specified somewhere, since the energy E_1 will vary with different materials.

The same change in the effective charge state of Cu_{Zn} in ZnS can

*Since in this chapter we are interested in equilibrium states of solids, we consider only thermal processes. Electrons can also be excited and ionized by optical processes, as described in Chapter 7.

be described in another way. In the previous section we saw that an electron-hole pair can be created by the process

$$E_G = e' + h^{\cdot} \quad .$$

Here, E_G is the band gap energy of ZnS, which is again not written out specifically. If we now add Eq. (1) to the above equation we find that

$$Cu^{x}_{Zn} + E_2 \rightarrow Cu'_{Zn} + h^{\cdot} \quad , \tag{2}$$

with $E_2 = E_G - E_1$. Physically, Eq. (2) is interpreted in the following way. At any temperature above 0°K an electron in the valence band can be thermally excited to the localized state associated with the copper defect, thus converting Cu^{x}_{Zn} into Cu'_{Zn}. Or using the concept of holes [Eq. (2)] we say that a hole trapped at Cu^{x}_{Zn} may be ionized to a valence band state. The minimum energy required for this process is given by E_2, and we say that E_2 locates the energy level of the ground electronic state associated with the defect Cu'_{Zn} with respect to the top of the valence band of ZnS. The energy E_2 is also referred to as the binding energy of a hole to Cu'_{Zn}. This is shown in Fig. 1(b). We see from Figs. 1(a) and 1(b) that $E_1 + E_2 = E_G$, the band gap energy. The reverse of the ionization processes given by Eqs. (1) and (2) are free electron-localized hole and free hole-localized electron recombination processes, respectively. In fact, at temperatures above 0°K, these ionization and recombination processes result in equilibrium concentrations of Cu'_{Zn}, Cu^{x}_{Zn}, e', and $h^{\cdot}$ defects. What the factors are which determine these concentrations in any given situation is the subject of Section IV. In semiconductors, neutral defects, from which a hole can be ionized [Eq. (2)] or which can accept an electron [reverse of Eq. (1)], are called singly ionizable acceptors. Another example of an impurity acceptor in ZnS is P^{x}_{S}. Acceptors play an important role in the conduction processes in semiconductors, as described in Chapter 4. It is also possible for some defects to behave as double acceptors. For example, we might consider the processes

$$A^{x} + E_3 \rightarrow A' + h^{\cdot} \tag{3}$$

and

$$A' + E_4 \rightarrow A'' + h^{\cdot} \quad . \tag{4}$$

It is possible that the energies of the ground electronic states of both A' and A'', E_3 and E_4, lie within the forbidden gap. We might expect E_4 to be larger than E_3 (i.e., the A'' level lies higher above the top of the valence band than the A' level), since it is more difficult to ionize the second hole than the first.

At this point, we might ask, in reference to Eqs. (1) and (2): Is a localized state associated with the defect Cu_{Zn}^{x}, and where does its energy lie with respect to the conduction band? In other words, we want the energy E_5 of the process

$$Cu_{Zn}^{x} + E_5 \rightarrow Cu_{Zn}^{\cdot} + e' \quad . \tag{5}$$

It appears likely that, in ZnS, this energy is greater than the band gap energy, so that the defect $Cu_{Zn}^{\cdot}$ does not exist in an equilibrium state. The only effective charge states of the substitutional Cu impurity in ZnS of concern to us are therefore Cu_{Zn}^{x} and Cu_{Zn}'. In Section II.G we shall discuss the relationship between these effective charge states and the ionic states Cu^{+} and Cu^{2+}, familiar to chemists. Although Cu in ZnS appears to exist only in two charge states, this is not true for Ag in KCl. As shown by Delbec and co-workers [16], Ag_K can have the charge states Ag_K^{x}, Ag_K', and $Ag_K^{\cdot}$. The reactions involving the changes in the charge states may be written as

$$Ag_K^{x} + E_6 \rightarrow Ag_K^{\cdot} + e' \tag{6}$$

and

$$Ag_K^{x} + E_7 \rightarrow Ag_K' + h^{\cdot} \quad . \tag{7}$$

In Eq. (6), the energy E_6 locates the level of Ag_K^{x} with respect to the bottom of the conduction band, and in Eq. (7), the energy E_7 locates the level of Ag_K' with respect to the top of the valence band [compare

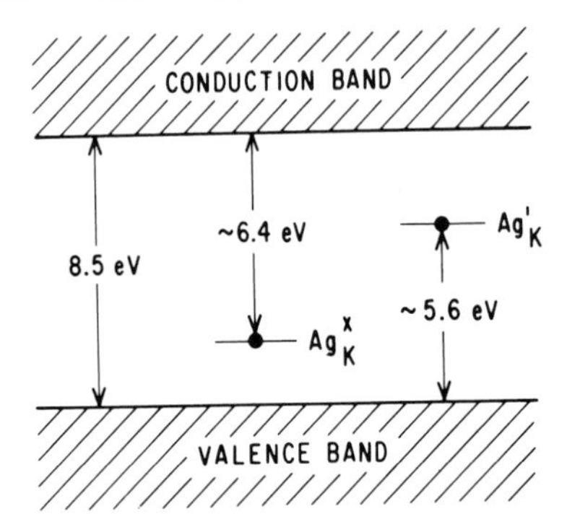

FIG. 2. The energy level scheme of the localized electronic states arising from the defects Ag_K^x and $Ag_K^{'}$ in KCl [16].

with Eqs. (1) and (2)]. These two levels arising from Ag_K in KCl are shown in Fig. 2. Now we note from Eq. (7) that Ag_K acts as an acceptor impurity in KCl, since a hole can be ionized from the neutral defect Ag_K^x. In Eq. (6) we see that an electron can also be ionized from the neutral defect Ag_K^x [note this difference between Eq. (7) and Eq. 1)]. In this situation we say that Ag_K can behave like a donor in KCl. A defect that can behave both as an acceptor and as a donor is called an amphoteric defect.

As another example of a donor, we consider the defect Al_{Zn}^x in ZnS. It too can be converted to a different charge state by the ionization process

$$Al_{Zn}^x + E_8 \rightarrow Al_{Zn}^{\cdot} + e' \tag{8}$$

or

$$Al_{Zn}^{\cdot} + E_9 \rightarrow Al_{Zn}^x + h^{\cdot} \quad . \tag{9}$$

Again we say that E_8 locates the energy of the ground electronic state of the defect Al_{Zn}^x with respect to the bottom of the conduction band in ZnS. It is also referred to as the binding energy of an electron to $Al_{Zn}^{\cdot}$. The energy E_9 locates the ground electronic state of the defect Al_{Zn}^x with respect to the top of the valence band. This is shown in Fig. 3 where, obviously, $E_8 + E_9 = E_G$. Such neutral defects from which an

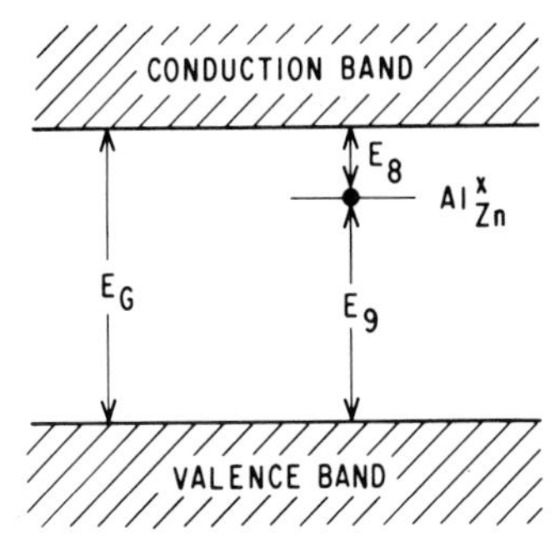

FIG. 3. The energy level of the state arising from the defect Al^{X}_{Zn} in ZnS. The meaning of the quantities E_8 and E_9 is given in the text.

electron can be lost by ionization or which can capture a hole are, as we just saw, called singly ionizable donors. Another example of a donor impurity in ZnS is Br_S. Again, double donors are possible if the energies E_{10} and E_{11} are less than the band gap energy:

$$D^{X} + E_{10} \rightarrow D^{\cdot} + e' \tag{10}$$

$$D^{\cdot} + E_{11} \rightarrow D^{\cdot\cdot} + e' \quad . \tag{11}$$

The energies E_{10} and E_{11} locate the levels of the defects D^{X} and $D^{\cdot}$ with respect to the bottom of the conduction band. As with Cu, we may inquire as to the existence of other effective charge states of Al_{Zn} in ZnS such as Al'_{Zn} or $Al^{\cdot\cdot}_{Zn}$. There is in fact no experimental evidence for the existence of these defects.

Interstitial impurities can also give rise to localized states which can be occupied or empty of electrons and in semiconductors behave as donors or acceptors. Since most interstitial impurities are small, electropositive elements, such as H, Cu, and Li which can take part in ionization reactions like

$$Li^{X}_{i} + E_{12} \rightarrow Li^{\cdot}_{i} + e' \quad , \tag{12}$$

behave as donors.

There are examples in several semiconductors of the same impurity acting as both a donor and acceptor in the same compound, because it

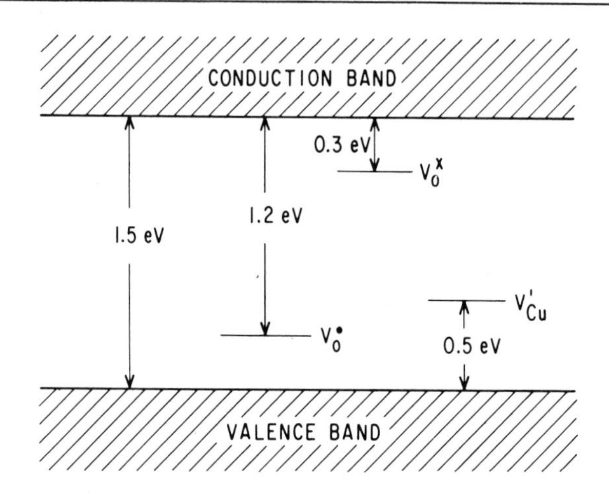

FIG. 4. Energy level of localized states arising from native defects in Cu_2O at 1000°C [18].

can occupy two different sites in the same structure. As examples we might cite Cu in PbS [17] and Si in GaAs [8]. In PbS, Cu_{Pb} is an acceptor, but Cu_i is a donor. In GaAs, Si_{Ga} is a donor but Si_{As} is an acceptor. As with Ag_K in KCl, already described, such impurities are also called amphoteric impurities.

Just as impurities can give rise to localized states with discrete energy levels in the forbidden gap, so can native defects. Vacancies and native interstitials can undergo changes in their effective charge states and in semiconductors may behave as donors or acceptors. For reasons to be discussed in Section II. G, we generally find that vacancies of the electropositive (metallic) constituent behave as acceptors and those of the electronegative (nonmetallic) constituent behave as donors. Native metallic interstitials generally behave as donors. For example, studies of Cu_2O [18] indicate that the native defects V_{Cu} and V_O play an important role in determining the electrical and optical properties of this compound. At elevated temperatures, the following reactions involving these defects can take place:

$$V_{Cu}^{x} + E_{13} \rightarrow V_{Cu}' + h^{\cdot} \quad , \tag{13}$$

$$V_{O}^{x} + E_{14} \rightarrow V_{O}^{\cdot} + e' \quad , \tag{14}$$

$$V_{O}^{\cdot} + E_{15} \rightarrow V_{O}^{\cdot\cdot} + e' \quad . \tag{15}$$

Approximate energy levels at 1000 °C of the defects V_{Cu}', V_{O}^{x}, and $V_{O}^{\cdot}$ in Cu_2O as deduced by Bloem [18] are shown in Fig. 4. It appears that V_{Cu} behaves as a singly ionizable acceptor and V_O as a doubly ionizable (or simply a double) donor.

In CdTe, Cd_i is a prominent native defect. According to de Nobel [15], it behaves as a donor, although recent work has cast some doubt on this. For donor behavior,

$$Cd_i^{x} + E_{16} \rightarrow Cd_i^{\cdot} + e' \tag{16}$$

with the energy $E_{16} = 0.02$ eV at room temperature. Also, in this compound, V_{Cd} apparently acts as a double acceptor. Finally we might mention that a much studied native defect in the alkali halides is the so-called F-center. This defect is discussed in detail in Chapter 7. Here we would merely like to indicate that in the chlorides it is the defect symbolized by V_{Cl}^{x} and is formed from a Cl^- ion vacancy, ($V_{Cl}^{\cdot}$), by the capture of an electron.

In Section IV, we show how the control of the concentrations of the various native defects may be achieved.

F. Association of Defects

So far, we have discussed various isolated defects, substitutional and interstitial, impurity and native, neutral and charged. If these various defects are distributed at random throughout the crystal, there will be a certain probability that two or more of them may occupy neighboring sites in the structure. That is, they may form associates (pairs, triplets, etc.). At low defect concentrations – and this situation is the one of interest to us at present* – the number of such near neighbor associates will be small. Shortly we shall estimate how small this concentration is. However, if there are forces between defects, the concentration of associates will be quite different. The concentration will be higher than the random concentration if the forces are those of attraction, and smaller if they are those of repulsion.

The most important attractive force between defects is the coulombic attraction between a defect having an effective positive charge and one having an effective negative charge. Thus, for example, the impurity defect $Ca_K^{\bullet}$ and the charge compensating native defect V_K' in KCl will be attracted coulombically. We can show this by a reaction:

$$Ca_K^{\bullet} + V_K' \rightarrow (Ca_K V_K)^x + E_{17} \quad . \qquad (17)$$

The interaction energy** can be estimated by

$$E_{17} = q^2/\epsilon r = 14.4/\epsilon r \quad (eV) \quad . \qquad (18)$$

In Eq. (18), q is the electronic charge, r, in Å, is the distance between the defects (i.e., an interatomic spacing), and ϵ the static dielectric constant of the particular solid. This equation gives only an approximate value of the interaction energy, since repulsive forces, covalent bonding effects between near neighbor defects, and rearrangements of the atoms near the associated imperfection have not been taken into account. Also, the use of the static dielectric constant is not

---

*In Section V we consider compounds with very large defect concentrations.

**In many cases we speak of ionization energy, interaction energy, and so on, in solids. Actually, for processes like those given in Eq. (1) or (17), there may be a minute change in volume associated with the given reaction. For reactions occurring at constant pressure, there will be a pressure-volume work term in addition to the energy term, and we should speak of the enthalpy change rather than the energy change in a reaction. Except for experiments carried out at extremely high pressures, the pressure-volume term is negligible compared to the energy term. Thus, the reaction enthalpy and energy can be considered to be the same (see also Section IV.A).

completely correct at the small interatomic distances between associated defects.

Values of the interaction energy have been determined experimentally from conductivity and photochemical effects for a number of associated defects. In view of the approximate nature of Eq. (18), the agreement between the experimental values and those calculated from the equation are not too bad. A few results are given in Table 1.

TABLE 1

Experimental and Theoretical Values of the Interaction Energy of Associated Defects

Solid	Association reaction	E [from Eq. (18)] (eV)	E(exptl) (eV)	Ref.
KCl	$Ca^{\bullet}_{K} + V'_{K} = (Ca_{K}V_{K})^{x}$	0.69	0.52	[3]
NaCl	$Ca^{\bullet}_{Na} + V'_{Na} = (Ca_{Na}V_{Na})^{x}$	0.60	0.67	[19]
NaCl	$Mn^{\bullet}_{Na} + V'_{Na} = (M_{Na}V_{Na})^{x}$	0.60	0.70	[20]
AgBr	$Cd^{\bullet}_{Ag} + V'_{Ag} = (Cd_{Ag}V_{Ag})^{x}$	0.22	0.16	[21]

Association can also occur between a substitutional impurity and a vacancy, between vacancies, between impurities and interstitials, and so on. Some further examples are given by the reactions

$$V^{\bullet}_{Cl} + V'_{Ag} \rightarrow (V_{Ag}V_{Cl})^{x} \quad \text{in AgCl [22]},$$

$$Sm^{\bullet}_{Cd} + F'_{i} \rightarrow (Sm_{Cd}F_{i})^{x} \quad \text{in } CdF_2 \text{ [23]},$$

$$Al^{\bullet}_{Zn} + V''_{Zn} \rightarrow (Al_{Zn}V_{Zn})' \quad \text{in ZnS [12]},$$

$$V^{\bullet}_{F} + O'_{F} \rightarrow (V_{F}O_{F})^{x} \quad \text{in } Ca_5(PO_4)_3F \text{ [24, 25]}.$$

The last compound, $Ca_5(PO_4)_3F$, offers some interesting examples of an association of three defects. The F^- ions in the structure all lie on one-dimensional chains, so that three defects can be observed as an associate with different geometric arrangements [24, 25]:

$$2V_F^{\bullet} + O_F' \rightarrow (V_F O_F V_F)^{\bullet}$$

and

$$2V_F^{\bullet} + O_F' \rightarrow (O_F V_F V_F)^{\bullet} \quad .$$

We will now consider the problem of determining the number of near neighbor associated pairs in a compound containing given concentrations of defects with opposite effective charges. Although various solutions of this problem have been described [26-30], the simplest, and most frequently used, approach is based on the laws of mass action and appears to give results not too different from more sophisticated treatments. Using Eq. (17) as an example, we can write a mass action equation:

$$\frac{[(Ca_K V_K)^x]}{[Ca_K^{\bullet}][V_K']} = \frac{K_{ass}}{[L_K]} \quad . \tag{19}$$

The symbols [ ] indicate concentrations of defects in number per cm^3 and $[L_K]$ is the number of K sites per cm^3. The equilibrium constant K_{ass} can be shown [26] to be equal to

$$K_{ass} = Zf \exp(q^2/\epsilon rkT) \quad . \tag{20}$$

In Eq. (20), Z is the number of equivalent sites that can be occupied by V_K' at the smallest distance in the crystal to $Ca_K^{\bullet}$. For KCl, Z = 12. The quantity f results from the change in vibrational entropy upon association, which is caused by the change in the phonon spectrum of the crystal due to the association of the defects. Values of f varying from 0.1 to 3×10^{-3} for various materials have been found experimentally. This indicates that there is an increase in vibrational entropy upon forming the defect. If f is taken as unity, as is frequently done, the calculated number of associated pairs will be somewhat too high. For a total concentration of $Ca = [Ca]_{total}$ atoms per cm^3 in the crystal of KCl we have $[Ca_K^{\bullet}] + [(Ca_K V_K)^x] = [Ca]_{total}$. Assuming that these are the only defects present, charge compensation requires that $[V_K'] = [Ca_K^{\bullet}]$. Introducing the fractions

$$\beta_{ass} = \frac{[(Ca_K V_K)^x]}{[Ca]_{total}} , \qquad f_{Ca} = \frac{[Ca]_{total}}{[L_K]}$$

$$\beta_{Ca_K^{\bullet}} = \frac{[Ca_K^{\bullet}]}{[Ca]_{total}} = \beta_{V_K'} = 1 - \beta_{ass} ,$$

we have from Eq. (19) that

$$\beta_{ass}/(1 - \beta_{ass})^2 = f_{Ca} \times K_{ass} . \tag{20a}$$

We can see from Eq. (20a) that, for low values of $[Ca]_{total}$, low values of the interaction energy, or high temperatures, $\beta_{ass} \ll 1$. Thus, most of the defects are isolated and randomly distributed. As the concentration of Ca increases, the temperature decreases, or as the interaction energy increases, the concentration of associated pairs increases. By setting $f = 1$ and $T = \infty$ (i. e., setting the interaction energy equal to zero), we can get the number of associates in a random distribution:

$$\beta_{ass}/(1 - \beta_{ass})^2 = 12 f_{Ca} .$$

For a typical value of $f_{Ca} = 10^{-4}$, we see that $\beta_{ass} = 12 \times 10^{-4}$. At a temperature of $T = 600\,°K$, $f = 1$ and $q^2/\epsilon r = 0.69$ eV, we find that $\beta_{ass} = 0.96$.

A remark concerning the temperature T is in order. In principle, of course, one can calculate the equilibrium concentration of associated pairs at any temperature, but in practice equilibrium may not be reached at low temperatures. If a crystal of KCl is grown at the freezing point of KCl (1049 °K) with a calcium content, $f_{Ca} = 10^{-4}$, and then allowed to cool to a temperature of 600 °K, the concentration of associated pairs will be smaller than that calculated above. At the high temperature the number of associated pairs is small and association takes place during cooling, by the diffusion of the isolated defects. Diffusion rates are, however, very temperature sensitive (Chapter 9), and a temperature will be reached (which is greater than 600°K) below which the diffusion rate will be too small for equilibrium to be established in a reasonable time. Thus, we might expect that the concentration of associated pairs will depend on whether the crystal is cooled slowly or quenched rapidly from some high temperature. Frequently rapid-quenching experiments are carried out in an attempt to freeze in some high temperature configuration of defects, which are then measured at room or lower temperatures. How successful these quenching experiments are depends on the particular system under study.

We have, in Section II. E, discussed briefly the discrete energy levels associated with dilute concentrations of isolated point defects. In general, we expect that the binding energy of the isolated defects for holes and electrons are decreased markedly during association. This will be discussed in the next section.

The isolated atomic defects described in Sections II. B, C, and E are called primary defects, whereas associates formed between primary defects of opposite effective charge, as already described, are called secondary defects. Secondary defects can also form due to attractive forces other than coulombic ones, such as those that result from covalent bonding and those from elastic interactions in which lattice strains are relieved by association. In Fig. 5, for example, are shown some of the primary and secondary defects founds in alkali halides. Such a variety of secondary defects undoubtedly exist in other compounds as well but have not been studied to any extent. In this chapter, we emphasize the role of primary defects and simple associates and disregard any complex secondary defects for several reasons:

(a) Very little is known about complex secondary defects in solids.

(b) At the high temperatures required for reaching the equilibrium

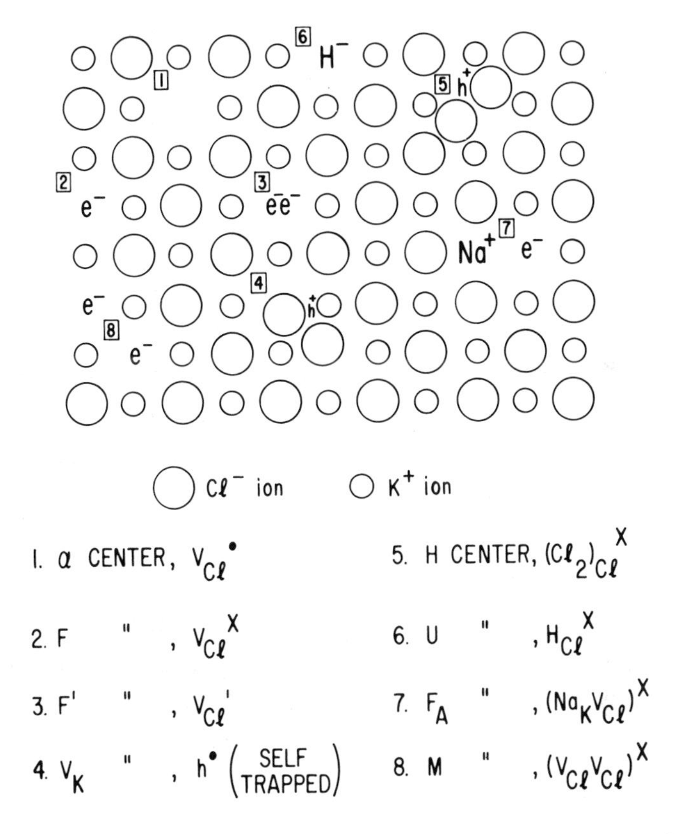

FIG. 5. Schematic models of some primary and secondary defects in alkali halides.

state of a solid, we might expect many of the complex secondary defects to be dissociated into their primary defects.

(c) In many cases physical properties of solids dependent on the presence of defects can be understood in terms of models involving primary defects and simple associates between oppositely charged defects.

We do not mean to imply, however, that our knowledge of even primary defects is complete. The elucidation of the kind of defects actually present in a solid prepared and treated in a specified manner is an extremely difficult problem. In many cases reasonable guesses have been made only to be proven wrong at some later date. Much of Part II of this book is concerned with the various physical methods used to study defects. The reader interested in going into greater detail regarding the complexities in imperfect solids can consult the books mentioned in the Introduction.

G. Bonding and Point Defects

In this section we will consider the relationship between the type of bonding in a solid and the properties of point defects. We will also discuss further the donor and acceptor behavior of defects in semiconducting solids. We shall see that the qualitative behavior of impurities and native defects is not dependent on the nature of the bonding between the constituent atoms of the solid. Impurity defects present some interesting features and are considered first.

As a specific example, let us look at impurities in a binary compound like ZnS and consider first that ZnS is an ionic solid composed of Zn^{2+} ions and S^{2-} ions. Both of these have filled electronic orbitals; the electron configurations being for Zn^{2+} $[Ar]3d^{10}$ and for S^{2-}, $[Ne]3s^23p^6$ (the elements in the square brackets indicate the filled orbitals of the Ar and Ne configurations). For an impurity substituting for the Zn^{2+} ions let us take one with three valence electrons. Such an impurity might be Al, Ga, Sc, or the rare earths. Using Al as an example, we can write for the substitution reaction

$$(Zn^{2+}[Ar]3d^{10})_{ZnS} + Al[Ne]3s^23p \rightarrow Zn[Ar]3d^{10}4s^2 + (Al^{3+}[Ne] + e^-)_{ZnS}. \quad (21)$$

In writing Eq. (21), we have made use of the principle that the ions shall have closed shell configurations. The transition elements and the rare earths are exceptions in that stable ions with partially filled d and f orbitals are to be found. This leads to the possibility of having transition and rare earth ion impurities in solids in various oxidation states. We shall discuss these a little later in this section. We see that the electron balance in Eq. (21) requires that there be one electron more in the solid after the substitution. The defect ($Al^{3+} + e^-$) has the same charge as the Zn^{2+} ion which it replaces and, in the effective charge symbolism, is represented as Al_{Zn}^{x}. The question might now be asked as to the possible states for the extra electron.

One such state of the electron is a localized bound state. The electron can be considered as moving in a "H-like" orbit about the excess fixed unit positive charge of the Al^{3+} ion, much the same as the electron in the H atom moves about the positive proton. The rest of the lattice is taken into account in the simplest model [31] by considering that the system ($Al^{3+} + e^-$) is imbedded in a medium with the static dielectric constant of ZnS and that the electron has an effective mass m^* rather than the free mass m_0 (Chapters 3 and 4).

If the foregoing description is a good approximation to the state of the electron, then we can write from quantum mechanics an expression for the ionization energy of this "H-like" system:

$$E_i = \frac{13.6}{\epsilon^2}\left(\frac{m^*}{m_0}\right) \text{ eV} , \tag{22}$$

where ϵ is the static dielectric constant of the solid. This ionization energy is the energy required for the reaction, in ZnS,

$$Al_{Zn}^{x} + E_i \rightarrow Al_{Zn}^{\bullet} + e' \tag{23}$$

where e' represents, as before, a conduction electron. This equation is the same as Eq. (8) with $E_i = E_8$. Al_{Zn} is a donor impurity and E_i represents the energy level of Al_{Zn}^{x} with respect to the bottom of the conduction band. We can also say that the Al^{3+} ion impurity is charge compensated by a bound electron, the binding energy being E_i. The species Al_{Zn}^{x} are called un-ionized donors, whereas the species $Al_{Zn}^{\bullet}$ are termed ionized donors.

Donors having energy levels below the conduction band as given by Eq. (22) are called shallow donors, and shallow donors have been studied in many compounds as well as in semiconducting elements such as Si and Ge (Chapter 4). The Bohr radius r_B of the "H-like" orbit of the bound electron is given by

$$r_B = 0.53\epsilon(m/m^*)\ \text{Å} \tag{24}$$

Treating the $(Al^{3+} + e^-)$ defect as a "H-atom-like" system does not take into account the specific nature of the impurity. Actually some of the electron density lies in the region of the impurity, so that the potential peculiar to each different impurity will affect the binding energy of the electron. The effect is to make this energy somewhat larger than that given by Eq. (22), and in fact various shallow donors in the same material have slightly different energy levels below the conduction band. The Bohr radius of the bound electron is also somewhat smaller than that given by Eq. (24). For some donor defects the potential peculiar to the defect is dominant, and the binding energy is much larger than that given by Eq. (22). Such donors are said to be deep donors and to give rise to deep energy levels. That is, the energy levels of deep donors lie further below the conduction band edge than that given by Eq. (22). The actual binding energy of an electron to any given donor impurity in a solid must be determined from experiment (Chapter 4).

At this point, it should be emphasized that the system we have been discussing, $(Al^{3+} + e^-)$, represented by the effective charge symbol Al_{Zn}^x, is not equivalent electronically to an Al^{2+} ion at Zn^{2+} sites although the defect symbol would be the same. The electronic configuration of Al^{2+} would be [Ne]3s and the 3s electron would have quite a different binding energy. There is, in fact, no evidence for the existence of Al^{2+} ions, as such, in ZnS. We shall come back to this point shortly in discussing the properties of the rare earths in some fluorides. The point is, therefore, that the effective charge symbols for defects do not give any information as to the electronic state of the system.

Having shown that Al impurities in ZnS can have donor properties, the reader can readily write equations similar to Eq. (21) to show that the halogens at sulfur sites also have donor properties. The extension to other compounds and impurities is usually straightforward.

The discussion in the preceding paragraphs was based on a completely ionic model of ZnS. Let us see now whether using a covalent model for the bonding in ZnS will lead to the same conclusions regarding the behavior of Al in ZnS. In a covalent model, both Zn and S can be considered to have a hybrid sp^3 configuration [32]. This configuration allows both Zn and S to form four tetrahedral bonds to neighboring atoms. The formation of ZnS can be written as

$$Zn[Ar]3d^{10}4s^2 + S[Ne]3s^2 3p^4 \rightarrow Zn^{2-}[Ar]3d^{10}4s4p^3 + S^{2+}[Ne]3s3p^3 . \quad (24a)$$

Each of the four sp^3 orbitals are accupied by an electron, and the bonds are formed by the overlap of the sp^3 orbitals of Zn and S. Thus, each tetrahedral covalent bond contains two electrons. Actually, the electron distribution will be strongly enhanced in the region of the S^{2+}, and this is equivalent, physically, to an ionic contribution to the bonding. The species Zn^{2-} and S^{2+} are, therefore, merely a convenient basis for the discussion of a covalent bonding model of ZnS and are not to be considered as ions. Furthermore, we must not really picture the electrons in the bonds as localized in that region. In Chapter 3 it was indicated that the electronic states of the bonding electron actually extend throughout the crystal. In the region between the atoms along the tetrahedral bonding directions there is an increase in electron density. The electronic states of the valence band are those of the least tightly bound electrons in the system, namely, those involved in the Zn to S bonds. Into this crystal of ZnS, we now introduce an Al impurity at a Zn site. According to the covalent model, this substitution may be written as

$$Al[Ne]3s^2 3p + (Zn^{2-}[Ar]3d^{10}4s4p^3)_{ZnS} \rightarrow (Al^-[Ne]3s3p^3 + e^-)_{ZnS} + Zn[Ar]3d^{10}4s^2 . \quad (25)$$

In writing Eq. (25), we have used the principle that all bonds are completed in the substitution. Again, we note the presence of an extra electron in the system and that (Al^- + e^-), having the same charge as Zn^{2-}, can be written as Al^x_{Zn}. The electron, as before, can be in a bound state due to the extra effective positive charge of Al^- at Zn^{2-} sites. The results are, therefore, the same as those based on the ionic model. The effective charge symbol Al_{Zn}^{x} thus contains no information as to the bonding in the solid. Since one need not consider the nature of the bonding in the solid for the thermodynamic treatment of defects to be discussed in Section IV, the effective charge symbolism is very useful. Inasmuch as both the ionic and covalent models of bonding in solids lead to the same results regarding the role played by the defects, we will henceforth use the ionic symbolism when the electronic structure of defects is of interest and use the effective charge symbolism in all other cases.

Up to now, in this section, we have considered one possible state of the electron in the defect ($Al^{3+} + e^{-}$), namely, a bound "H-like" state. It is also possible for the electron to reside in one of the conduction band states. In this case we will have Al^{3+} ions charge compensated, not by a bound electron but by a conduction electron. The defects, in effective charge symbols, are then $Al_{Zn}^{\cdot}$ and e', as we have already seen. This will be the case for donor impurities in semiconductors at sufficiently high temperatures (Section IV).

A third possible state for the extra electron accompanying the incorporation of donor-like impurities will now be considered. The compound CdF_2 with the fluorite structure [Fig. 6(a)] can be grown readily as large single crystals containing up to several atomic percent of rare earth impurities. From optical, chemical, and electrical measurements it has been shown that the rare earths substitute for Cd^{2+} ions as tripositive ions RE^{3+} and are charge compensated by interstitial F^{-} ions [14, 23, 33-35]. The interstitial F^{-} ion and the RE^{3+} ion form an associated pair [Fig. 6(b)]. In other words, the defects are $RE_{Cd}^{\cdot}$ and F_i', associated into the neutral defect $(RE_{Cd}F_i)^{x}$. We will concentrate on two ions, Sm^{3+} and Eu^{3+}, to be specific. The CdF_2 containing these ions as impurities are colorless highly insulating crystals. If these crystals are now heated for a few minutes at 500°C in Cd vapor, a remarkable change is observed [34, 35]. The Sm^{3+}-doped crystal is colored a deep blue and becomes a

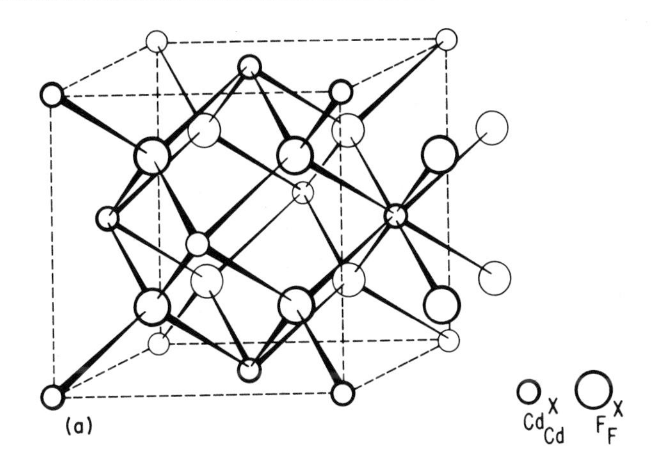

FIG. 6(a). Fluorite structure of CdF_2.

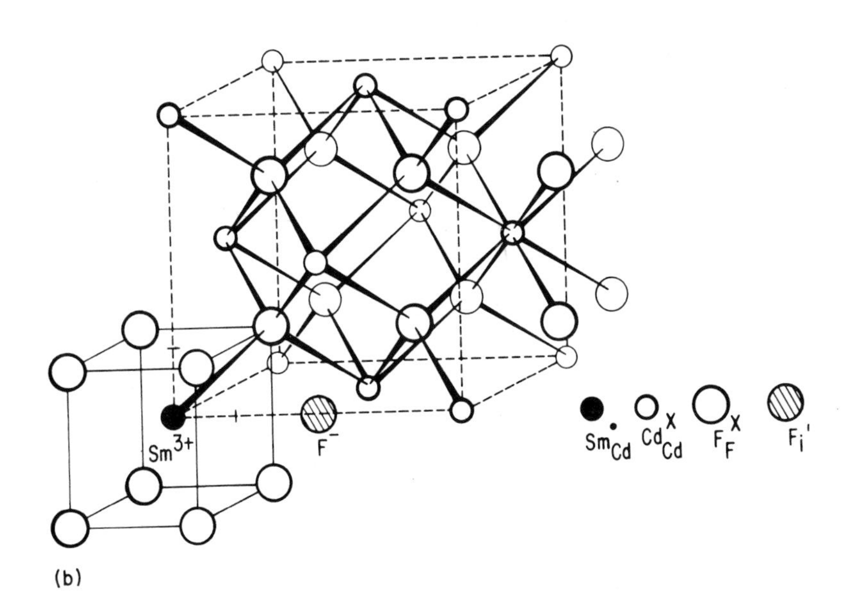

FIG. 6(b). $Sm_{Cd}^{\cdot}$ and F_i' defects in CdF_2.

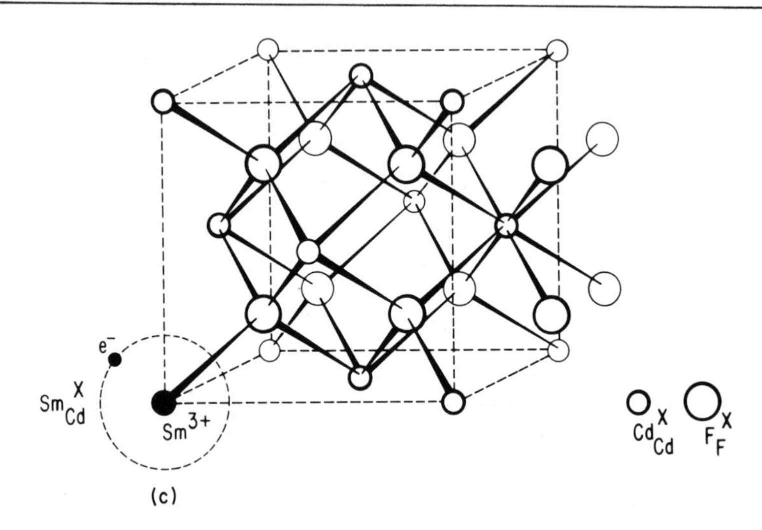

FIG. 6 (c). Model of Sm_{Cd}^{X} defect.

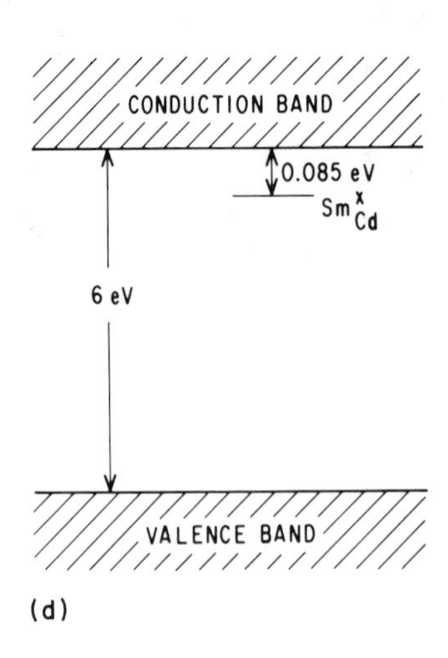

FIG. 6 (d). Energy level of state arising from Sm_{Cd}^{X} in CdF_2.

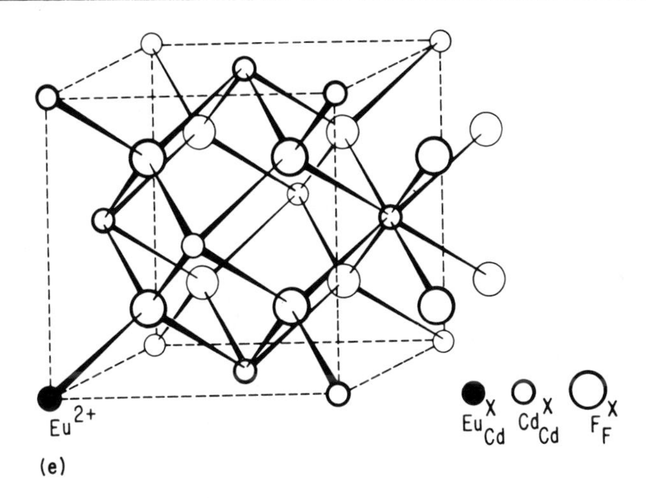

FIG. 6 (e). Model of Eu^{X}_{Cd} defect.

semiconductor with low electrical resistivity.* The Eu^{3+}-doped crystal, however, remains colorless and highly insulating but develops a strong optical absorption band in the near ultraviolet. The behavior of these materials upon heating in Cd vapor has been explained in the following manner. At 500°C, the interstitial F^- ion defects are highly mobile and diffuse rapidly to the crystal surface where they are annihilated by the reaction

$$2F^-(\text{interstitial}) + Cd(\text{vapor}) \rightarrow CdF_2(\text{solid}) + 2 \text{ electrons} . \quad (26)$$

The two electrons diffuse into the crystal where they compensate the charge of two RE^{3+} ions. The compensating electron can be bound in a "H-like" state to the $Sm^{\cdot}_{Cd}$ giving the donor defect Sm^{X}_{Cd} [Fig. 6(c)]. The donor level has been found to lie 0.085 eV below the bottom of the conduction band [Fig. 6(d)], and the conductivity is due to the ionization, at room temperature, of some of the Sm^{X}_{Cd} to give conduction band electrons:

*The same effect is observed with most of the rare earth ions as well as with Y^{3+}.

$$Sm^{x}_{Cd} + 0.085 \text{ eV} \rightarrow Sm^{\cdot}_{Cd} + e' \quad . \tag{27}$$

The behavior of Eu in CdF_2 is, however, different, for the electronic structure of the Eu^{x}_{Cd} defect that is formed is not the same as that of the Sm^{x}_{Cd} defect. The electronic configuration of Eu^{3+} (i. e., $Eu^{\cdot}_{Cd}$) is $[Xe]4f^6$. The electrons introduced into the system, when the CdF_2 crystal containing Eu is heated in Cd, go into the 4f orbitals yielding $Eu^{2+}[Xe]4f^7$ [Fig. 6(e)]. This occurs with Eu but not with Sm because the $4f^7$ configuration (half-filled 4f shell) is a very stable configuration and yields a state of lower energy than the bound "H-like" state. Again we see that the effective charge symbol (Eu^{x}_{Cd}, Sm^{x}_{Cd}) gives no information as to the electronic configuration of the defect. Although Eu^{x}_{Cd} yields a deep donor level, it can be ionized just as Sm^{x}_{Cd} [Eq. (27)], except that the required temperatures are higher [35]. In both cases, after ionization, one is left with tripositive rare earth ions and conduction electrons.

Examples like the one just cited, in which the impurities can exist in different oxidation states, are quite common. The transition metals in particular, when present as impurities in solids, can take on different oxidation states depending on such factors as the presence of other impurities and the heat treatment of the solid in oxidizing or reducing atmospheres. For example, vanadium defects in Al_2O_3 can exist in three different charge states. Each of the ionic states V^{2+} (V'_{Al}), V^{3+} (V^{x}_{Al}), and V^{4+} ($V^{\cdot}_{Al}$) give rise to characteristic electron paramagnetic resonance spectra [36]. Another example involves the impurity Mn in β-Al_2O_3. When heated in an oxygen atmosphere, Mn^{4+} ions are found (i. e., $Mn^{\cdot}_{Al}$). In a reducing atmosphere, such as H_2, Mn^{2+} ions are found (i. e., Mn'_{Al}) [37]. Similarly, for AgCl containing Cu impurities, the valency of Cu changes from Cu^{2+} ($Cu^{\cdot}_{Ag}$) at high Cl_2 pressures to Cu^{+} (Cu^{x}_{Ag}) at low Cl_2 pressures [38]. In MgO containing small amounts of Mn it was found that Mn^{2+} (Mn^{x}_{Mg}) defects predominate. When Li^{+} ions are added as an additional impurity, however, it was found that Mn^{4+} ($Mn^{\cdot\cdot}_{Mg}$) defects predominate. This is due to the fact that the double positive effective charge of $Mn_{Mg}^{\bullet\bullet}$ is compensated by two Li_{Mg}' defects without the necessity of creating Mg ion vacancies (V_{Mg}'') [39]. At sufficiently high oxygen pressure, however, the defects $Mn_{Mg}^{\bullet\bullet}$ can be formed even without the Li_{Mg}'. A quantitative treatment illustrating the effect of the atmosphere on the effective charge state of impurity defects will be given in Section IV. F.

Another possibility for the extra electron introduced into a solid with a donor impurity is for it to transfer to a state of lower energy which may be provided by acceptor impurities or native acceptor defects. For example, if ZnS is doped (i. e., impurities added) with small equal concentrations of both Al and Ag, the defects formed are $Al_{Zn}^{\bullet}$ and Ag_{Zn}'. Or under suitable conditions doping with Al alone will lead to the defects $Al_{Zn}^{\bullet}$ and V_{Zn}'', there being one-half as many of the latter defects as the former. In either case, we say that the semiconductor is compensated by equal numbers of donors and acceptors. Exactly why this electron transfer can take place from a donor to an acceptor will be discussed shortly.

We have described so far four possible states for extra electrons accompanying the introduction of donor impurities into solids. Still another possibility exists when the constituents of the solid themselves are ions of variable valency. Such solids are the compounds of the transition and rare earth elements. When Ti is introduced as an impurity into Fe_2O_3 we may write, for the substitution reaction,

$$(Fe^{3+}[Ar]3d^5)_{Fe_2O_3} + Ti[Ar]3d^24s^2 \rightarrow (Ti^{4+}[Ar] + e^-)_{Fe_2O_3} + Fe[Ar]3d^64s^2. \quad (28)$$

Once again electron balance requires that there be an extra electron in the system per Ti^{4+} ion introduced, and Ti in Fe_2O_3 acts as a donor impurity. In this system, as with many transition metal compounds, the electron may be crudely considered as bound to an Fe^{3+} ion in the vicinity of the Ti^{4+} defect, thereby converting it to a Fe^{2+} ion. A useful model consists, in effective charge symbols, of the defects $Ti_{Fe}^{\bullet}$ and Fe_{Fe}'. This bound electron can partake in electrical conduction

TABLE 2

Models for Donor Defects

Effective charge symbol	Ionic symbol	Description
D_M^x	$(D^{3+} + e^-)$	Unionized donors; electron in a localized bound donor state
D_M^x	D^{2+}	Change in oxidation state of D; electron bound in atomic orbital of D
$D_M^{\cdot} + e'$	$D^{3+} + e^-$	Ionized donor; electron in conduction band state
$D_M^{\cdot} + A'$	$D^{3+} + A^+$	Donors compensated by acceptors which may be impurities or native defects; electron in a bound localized acceptor state

processes by a "hopping mechanism." The bound electron moves from one Fe^{2+} ion to a neighboring Fe^{3+} ion, in effect interchanging the role of these ions. Pictorially, we show this conduction process as

$$\cdots Fe^{3+}Fe^{3+}Fe^{2+}Fe^{3+}Fe^{3+}\cdots \rightarrow \cdots \underset{\text{electron motion}}{\underrightarrow{Fe^{3+}Fe^{3+}Fe^{3+}Fe^{2+}Fe^{3+}}}\cdots$$

While useful for descriptive purposes, this simple ionic picture of transition metal compounds is certainly not exact. The more correct model involving conduction in a partly occupied narrow d band formed by the overlap of the d orbitals of the transition metal atoms is described in Chapter 3. We will not consider such materials in this chapter (see Section V), our primary concern being with ionic solids composed of ions with filled shells or covalent solids in which all the bonds are complete.

In Table 2 are summarized models for the possible defects formed by the incorporation, at M sites, of small concentrations of a donor-type impurity D into the compound MX composed of M^{2+} and X^{2-} ions.

Having discussed donor impurities at some length, we will now consider, briefly, acceptor impurities. As an example, let us look at the behavior of P at Te sites in CdTe. As before, we write down a substitution reaction using an ionic model for the bonding in CdTe. In writing down the substitution reaction we again require that the ions be formed with closed shell configurations. Just as with donors, using a covalent model or any intermediate model for the bonding in the solid would lead to the same results. The substitution reaction is

$$(Te^{2-}[Kr]4d^{10}5s^2 5p^6)_{CdTe} + e^- + P[Ne]3s^2 3p^3 \rightarrow$$
$$(P^{3-}[Ne]3s^2 3p^6)_{CdTe} + Te[Kr]4d^{10}5s^2 5p^4 \quad . \tag{29}$$

In Eq. (29), electron balance requires that an extra electron be added to the system along with the P impurity. This extra electron may be provided in a number of ways.

First, under suitable conditions, the electron may be provided by the filled valence band of the CdTe, leaving a mobile hole in the band. We then write P'_{Te} and $h^{\bullet}$ for the defects formed. The solid is then said to contain ionized acceptors and holes. Just as a mobile electron could at low temperatures be bound to a donor, so we can consider the possibility of the mobile hole being bound to the acceptor (Section IV), giving the defect P^{x}_{Te}. The acceptor is then said to be un-ionized. The binding energy is given by an equation analogous to Eq. (22), except that m^* is the effective mass of the hole (Chapters 3 and 4). This bound hole can be ionized into the valence band by the reaction

$$P^{x}_{Te} + E_{30} \rightarrow P'_{Te} + h^{\bullet} \quad . \tag{30}$$

This is similar to Eq. (2) so that E_{30} locates the level of P'_{Te} with respect to the top of the valence band at CdTe. If E_{30} is given by Eq. (22) we speak of shallow acceptors, and such defects are known. In many cases the binding energy of a hole is greater than the "hydrogenic" value; these defects are termed deep acceptors.

Just as with donors, the symbol P_{Te}^{x} does not necessarily imply a P^{2-} ion. In some cases, however, a change in valency of the impurity on capturing a hole from the valence band is a valid description of the state of the acceptor. An example of this is Cu in ZnS. Cu having one valence electron less than Zn acts as an acceptor impurity in ZnS. Studies of the optical spectra in the infrared of the Cu_{Zn}^{x} state of the defect indicate that the state of the Cu can be best described as a Cu^{2+} ion with an electronic configuration $[Ar]3d^9$ [40]. Since in the ionic model the ionized Cu defect (Cu_{Zn}') would have an electronic configuration $Cu^+[Ar]3d^{10}$, we can describe the hole as being bound in a 3d orbital of Cu (i.e., the empty 3d orbital of Cu^{2+} is equivalent to a bound hole). The case of Cu^{2+} in ZnS is thus analogous to that of Eu^{3+} in CdF_2. Although the hole is trapped in the 3d orbital of Cu, it can still be ionized into the valence band by the reaction given in Eq. (2). The value of E_2 has been found to be 1.4 eV [41]. Such examples of defects with variable oxidation states are, as previously indicated, common among the transition and rare earth elements.

Instead of the required electron in Eq. (29) being supplied by the valence band of the solid it may be supplied by the simultaneous incorporation of a donor defect, either impurity or native. For example, In can be added along with P to CdTe during crystal growth. By adding Eq. (29) to one like Eq. (21) we find that the substitution reactions become

$$CdTe + In[Kr]4d^{10}5s^2 5p^1 + P[Ne]3s^2 3p^3 \rightarrow$$

$$(In^{3+}[Kr]4d^{10} + P^{3-}[Ne]3s^2 3p^6)_{CdTe} + Cd + Te \quad . \tag{31}$$

The ionic species In^{3+} at Cd^{2+} sites and P^{3-} at Te^{2-} sites are, in effective charge symbols, $In_{Cd}^{\cdot}$ and P_{Te}'. Such a transfer of electrons from the un-ionized donor In_{Cd}^{x} to the un-ionized acceptor P_{Te}^{x} to give the ionized species $In_{Cd}^{\cdot}$ and P_{Te}' can only take place if the localized donor level lies above the localized acceptor level in energy. In other words, the level associated with In_{Cd}^{x} must be above the level associ-

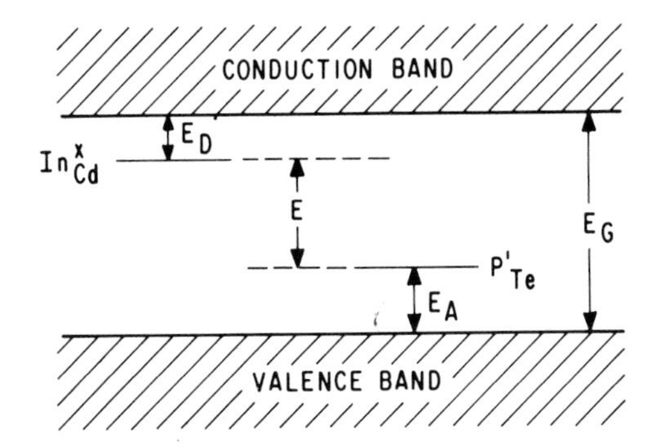

FIG. 7. Model of compensated CdTe containing equal amounts of the ionized impurity species $In^{\cdot}_{Cd}$ and P'_{Te}. Note, however, that the energy levels are those of In^{x}_{Cd} and P'_{Te}.

ated with P'_{Te}. Referring to Fig. 7, we can write a series of reactions:

$$In^{x}_{Cd} + E_D \rightarrow In^{\cdot}_{Cd} + e' ,$$

$$P^{x}_{Te} + E_A \rightarrow P'_{Te} + h^{\cdot} ,$$

$$e' + h' \rightarrow E_G .$$

Adding these we get

$$In^{x}_{Cd} + P^{x}_{Te} \rightarrow In^{\cdot}_{Cd} + P'_{Te} + E_G - (E_D + E_A) . \tag{32}$$

The quantity $E_G - (E_D + E_A)$ is seen, from Fig. 7, to be the energy difference between the donor and acceptor levels. When this energy is positive the reaction given by Eq. (32) will occur, and the ionized species will result. Under these conditions, if the concentrations of donor and acceptor impurities are equal, the CdTe is said to be <u>compensated</u>, as described previously.

Finally, if the constituents of the solids themselves are capable of existing in more than one oxidation state (mainly transition or rare earth elements), then the system, acceptor plus hole, has been described, in an ionic model, as consisting of a hole bound to one of the constituent atoms, thereby changing its oxidation state. As an example, we consider Li^+ ion impurities in NiO. The substitution reaction is

$$Li[He]2s + (Ni^{2+}[Ar]3d^8)_{NiO} \rightarrow (Li^+[He]_{NiO} + h^+)_{NiO} + Ni[Ar]3d^8 4s^2 . \quad (33)$$

This positive hole has been described as residing in one of the 3d orbitals of Ni in NiO yielding the species $Ni^{3+}[Ar]3d^7$. Thus, the defects in effective charge symbols would be Li'_{Ni} and $Ni^{\bullet}_{Ni}$. As discussed, this ionic picture, while perhaps useful for descriptive purposes, is actually not a very good description of the state of the hole (see Chapter 4).

To summarize, Table 3 lists models for an acceptor defect A, at M sites, in a binary compound MX composed of M^{2+} and X^{2-} ions. This table is similar to Table 2 for donor defects.

TABLE 3

Models for Acceptor Defects

Effective charge symbol	Ionic symbol	Description
A^{x}_{M}	$(A^+ + h^+)$	Un-ionized acceptor; hole in a localized bound acceptor state
A^{x}_{M}	A^{2+}	Change in oxidation state of A; hole bound in atomic orbital of A
$A'_{M} + h^{\bullet}$	$A^+ + h^+$	Ionized acceptor; hole in valence band state
$A'_{M} + D^{\bullet}$	$A^+ + D^{3+}$	Acceptors compensated by donors which may be impurities or native defects; hole bound in a localized donor state

We now turn to the consideration of vacancies and ask whether the donor or acceptor properties of vacancies in compound semiconductors can be predicted. We also want to see if the behavior depends on whether the solid is ionically or covalently bonded. The qualitative properties of a native defect such as a sulfur vacancy in ZnS are, fortunately, independent of the bonding in the compound. If we consider first that ZnS is an ionic compound composed of Zn^{2+} and S^{2-} ions, the formation of a neutral sulfur vacancy may be represented by

$$(S^{2-})_{ZnS} \rightarrow (V_S + 2e^-)_{ZnS} + S(gas) \quad . \tag{34}$$

The two electrons left behind in the solid by the removal of a neutral sulfur atom to the gas phase can be considered as being trapped in the vicinity of the vacancy to yield the neutral defect, written as V_S^x. The electrons can be removed one at a time into the conduction band by thermal ionization. These processes are written as

$$V_S^x + E_{35} \rightarrow V_S^{\bullet} + e' \tag{35}$$

and

$$V_S^{\bullet} + E_{36} \rightarrow V_S^{\bullet\bullet} + e' \quad . \tag{36}$$

Thus, a sulfur vacancy can exist in three charge states, V_S^x, $V_S^{\bullet}$, and $V_S^{\bullet\bullet}$, and can be expected to act as a double donor. The energy E_{35} locates the level of V_S^x with respect to the bottom of the conduction band and E_{36} that of $V_S^{\bullet}$. If we now consider the other extreme of covalent bonding, ZnS is made up of Zn^{2-}[Ar]$3d^{10}4s4p^3$ and S^{2+}[Ne]$3s3p^3$, each with a tetrahedral sp^3 configuration, giving rise to the four covalent bonds of the zincblende structure. There are eight bonding electrons, four from the Zn (or S) and one each from the four covalently bonded neighbors. Removal of a sulfur atom [Ne]$3s^2 3p^4$ to the gas phase removes six of the eight bonding electrons, thus leaving two electrons in the vicinity of the vacancy. Since a neutral sulfur atom was removed the defect left must be neutral with respect to the lattice, V_S^x. Again, these two electrons can be ionized thermally one at a time as shown in Eqs. (35) and (36). Thus, either an ionic or covalent bonding model for ZnS indicates the double donor behavior for a sulfur vacancy. A zinc vacancy, by similar arguments, is expected to behave as a double acceptor. For example, on a covalent model the removal of a neutral Zn atom $[Ar]3d^{10}4s^2$ to the gas phase forms a V_{Zn}^{x} defect and removes two of the eight bonding electrons. There are then two holes left in the group of eight bonding electrons, and these holes can be considered as trapped at the zinc vacancy. In summary, we expect that vacancies of the metallic component of a binary compound will show acceptor behavior whereas vacancies of the nonmetallic component will show donor behavior.

Before closing this section, we can consider what happens to the localized levels of defects when they associate into pairs. Consider a donor D and an acceptor A, either impurity or native. The donor level is given by the energy E_{37}:

$$D^{x} + E_{37} \rightarrow D^{\cdot} + e' \quad , \tag{37}$$

whereas the acceptor level is given by E_{38}:

$$A^{x} + E_{38} \rightarrow A' + h^{\cdot} \quad . \tag{38}$$

We shall, for simplicity, assume that these are both hydrogenic levels, so that E_{37} and E_{38} are given by Eq. (22), and that (m^*/m_0), the effective mass ratio for electrons is equal to that for holes. By combining Eq. (22) with Eq. (24) we can also write for the energies

$$E = 7.21/\epsilon r_B \ \text{eV} \quad . \tag{39}$$

The association of $D^{\cdot}$ and A' can be written as

$$D^{\cdot} + A' \rightarrow (DA)^{x} + E_{40} \tag{40}$$

where E_{40} is given by Eq. (18):

$$E_{40} = 14.4/\epsilon r \ \text{eV} \quad . \tag{41}$$

By adding Eqs. (37) and (38) we get

$$D^{x} + A^{x} + E_{42} \rightarrow D^{\cdot} + A' + e' + h^{\cdot} \tag{42}$$

with

$$E_{42} = E_{37} + E_{38} = 14.4/\epsilon r_B \quad . \tag{43}$$

The separation, in energy, between the donor and acceptor levels is given by

$$E_G - E_{42} = E_G - 14.4/\epsilon r_B \quad . \tag{44}$$

Now, if we consider that the donors and acceptors are associated, we must add together Eqs. (42) and (40) to give

$$D^x + A^x + E_{45} \rightarrow (DA)^x + e' + h^{\cdot} \tag{45}$$

where

$$E_{45} = E_{42} - E_{40} = \frac{14.4}{\epsilon}\left(\frac{1}{r_B} - \frac{1}{r}\right) \quad . \tag{46}$$

The separation in energy between the donor and acceptor levels of the associated pair is now

$$E_G - E_{45} = E_G - \frac{14.4}{\epsilon}\left(\frac{1}{r_B} - \frac{1}{r}\right) \quad . \tag{47}$$

For a nearest neighbor pair, generally $r < r_B$, so that the separation between the levels of the associated pair is greater than the band gap. In other words, the localized levels in the forbidden energy gap disappear and the associated pair cannot be expected to bind either an electron or hole.* When the donor and acceptor are farther apart in the structure, r becomes larger, and very shallow localized levels appear. They become deeper as r increases. These considerations have been used in explaining the very interesting many-line fluorescence spectra seen in GaP doped with donors and acceptors [42].

*Actually, at small values of r (of the order of an interatomic distance), Eq. (47) is not completely valid. The levels may not disappear completely but may become very shallow. Electrons and holes may be bound at very low temperatures.

III. STOICHIOMETRY OF SOLIDS

A. Stoichiometry and Point Defects

The ideas of Dalton regarding a fixed composition and a rational formula or stoichiometry as definitive criteria for chemical compounds have been part of chemistry lore for many years. It was not until the late 1920's and the early 1930's that the work of Schenck and Dingmann [43] on the Fe-O system and the studies of Biltz and Juza [44] on the dissociation equilibria of binary compounds indicated the widespread existence of many ionic and covalent compounds with variable compositions over limited ranges. In fact, ferrous oxide, the oxide of the very typical +2 oxidation state of iron, was found not to exist at all with the composition FeO. This composition lay outside the very large existence field of the wüstite (i. e., ferrous oxide) phase. Studies [45] have shown that the wüstite phase in fact could be represented by a composition range FeO_{1+x} with x ranging from 0.09 to 0.19 at 900 °C (Fig. 8).

In their classical paper on the statistical thermodynamics of real crystals, Wagner and Schottky [46] showed that, in principle, at any temperature above absolute zero every solid, crystalline compound can exist as a single phase over a range of composition. The solid with the ideal stoichiometry, as deduced from simple valence rules, had no special thermodynamic status, although the extent to which different compounds could exist with off-stoichiometric compositions varied widely. In general, one can distinguish two broad classes of solids, although there is, of course, no sharp dividing line between them. On the one hand are the large group of compounds, particularly those of the transition and lanthanide elements, which, from chemical analysis, equilibrium vapor pressure studies, and structural determination by x-ray diffraction, have been found to exist as stable homogeneous phases over a considerable range of composition. Most frequently, but not always, the range embraces some rational stoichiometry, and the solid of this

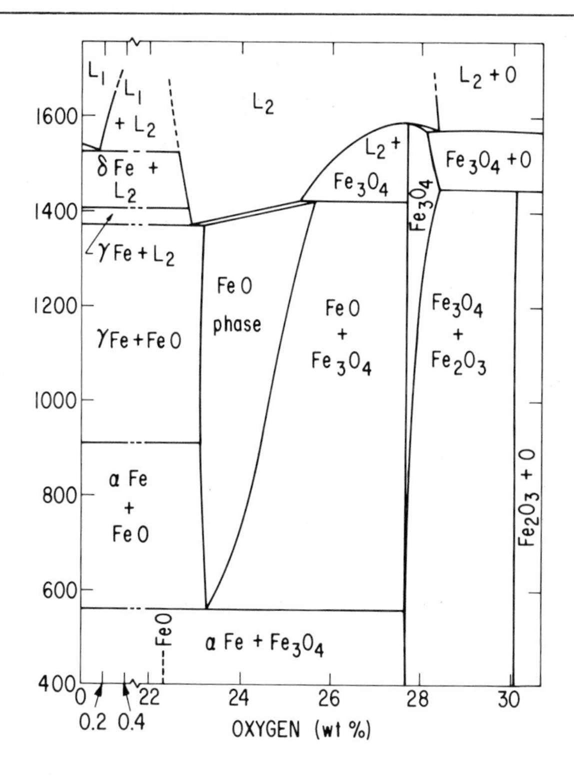

FIG. 8. Phase diagram of Fe-O system at a total pressure of 1 atm. Note that the field of stability of "FeO" does not include the 1:1 composition [45].

composition is termed the stoichiometric compound. The other compositions within the range are then referred to as nonstoichiometric compounds or compounds exhibiting stoichiometric deviations. To this class belong FeO_{1+x} as well as FeS_{1+x}, the oxides of titanium, cerium, and uranium, the sulfides of chromium, the hydrides of palladium, and many, many others. These substances, exhibiting large deviations from stoichiometry, are discussed further in Section V.

The other class of compounds, such as NaCl, PbS, ZnO, ZnS, CdTe, $BaTiO_3$, $Ca_5(PO_4)_6F_2$, and CdF_2, generally exist as stable phases over only very small composition regions ranging about the stoichiometric composition as determined by simple valence rules. The

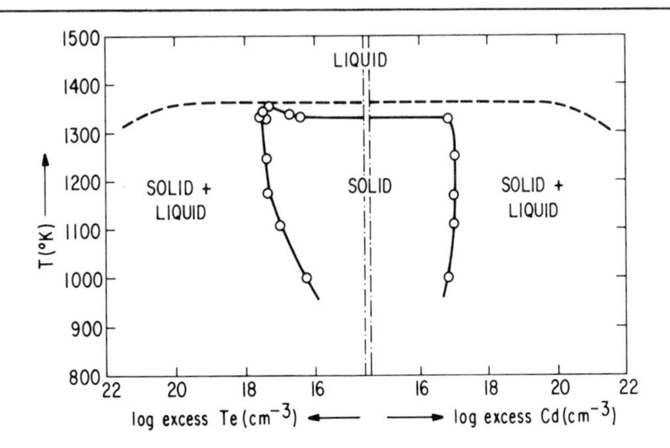

FIG. 9. Phase diagram of the Cd-Te system near the region of 1:1 composition. Note that the maximum melting point does not come at the 1:1 composition but at a composition containing excess Te [15].

compositional variations are generally so small as to be beyond the limits of study by conventional analytical, chemical, and x-ray diffraction techniques. Rather, these compounds with small stoichiometric deviations are best studied by optical, electrical, and magnetic measurements, as described in several chapters. It is with these compounds that we shall be primarily concerned in this section.

Stoichiometric deviations in impurity-free compounds result from lattice disorder arising from the presence of native defects; vacancies, interstitials, and misplaced atoms, charged and neutral. Thus, in CdTe [15] it is believed that the predominant primary native defects are cadmium interstitials and cadmium vacancies. An excess number of one over the other gives either Cd-rich or Te-rich CdTe. The maximum deviations from the ideal 1:1 stoichiometry in CdTe have been found to be about 10^{17} excess Te or Cd atoms per cm^3, compared with 1.5×10^{22} Cd or Te lattice sites per cm^3. The ideal 1:1 stoichiometry in CdTe lies within the existence region of the CdTe phase, but even the stoichiometric compound has a finite concentration of interstitials and vacancies. In Fig. 9 a portion of the phase diagram of the CdTe system is shown. The existence of the solid CdTe phase over a small but finite concentration range can be seen. The data also show that the maximum melting point of CdTe at about 1350°K does not occur at the 1:1 composition, but for one containing excess Te in the form of Cd vacancies.

B. Stoichiometry and Structural Disorder

We shall, for simplicity, consider a pure binary compound composed of A and B atoms and having a structure in which the ratio of the concentrations (in number per cm^3) of the B and A lattice sites is given by

$$r_L = [L_B]/[L_A] = b/a \quad . \tag{48}$$

The site concentrations $[L_A]$ and $[L_B]$ are readily calculated from the number of sites of each type per unit cell of the structure and the volume of the unit cell. The stoichiometric composition is represented by the chemical formula A_aB_b. In any real solid, there are present various primary and secondary native defects as well as electronic defects. In principle, all of the different types of primary defects can be present at the same time, but due to differences in the energy of formation, two types of primary native defects will generally dominate. The presence of such dominant defect pairs results in various types of <u>lattice</u> or <u>structural disorder</u>.

Under suitable conditions (see Section IV), the compound can, in fact, be stoichiometric and have the atomic ratio B:A = b:a . In general, however, the ratio of B:A will be more or less different from this, and the compound will be nonstoichiometric. Its composition will be given by $A_aB_{b(1+\delta)}$, where δ can be positive or negative depending on whether B or A are in excess over the stoichiometric composition. In this solid the ratio of the total concentration of B and A atoms is:

$$r_C = \frac{[B]}{[A]} = b\frac{(1 + \delta)}{a} \quad . \tag{49}$$

The deviation from stoichiometry is conveniently given by the difference between Eqs. (48) and (49):

$$\Delta = (r_C - r_L) = \frac{b}{a}\delta \quad . \tag{50}$$

We want to relate Δ to the concentrations of primary native defects for several possible types of structural disorder.

(1) The two dominant primary defects may be V_A and V_B, both neutral and charged. The solid then exhibits Schottky disorder, thus termed since Schottky first recognized the importance of such disorder in solids. This type of disorder is found in a large number of solids, such as the alkali halides, BaO, CdS, and PbS. In Schottky disorder, $[A] = [A_A^x]$ and $[B] = [B_B^x]$. Since the A and B sites are occupied either by the respective B and A atoms or by vacancies, the site relations for Schottky disorder are

$$\begin{aligned} [L_B] &= [B] + [V_B] \quad , \\ [L_A] &= [A] + [V_A] \quad , \end{aligned} \tag{51}$$

$[V_B]$ and $[V_A]$ being the total vacancy concentrations. From Eqs. (49), (50), and (51) we find that

$$\Delta = \frac{[L_B] - [V_B]}{[L_A] - [V_A]} - \frac{b}{a} \quad . \tag{52}$$

For the special case of the stoichiometric composition, $\Delta = 0$, and $b[V_A] = a[V_B]$. The degree of Schottky disorder, as measured by the vacancy concentrations, can be large for some compounds and small for others. In Section IV those factors which determine the degree of disorder in a compound will be indicated.

Where the vacancies exist in several effective charge states, and there are electronic defects present, a charge neutrality relation must also hold. For example, one might have for some particular compound the defects $[V_A^x]$, $[V_A']$, $[V_A'']$, e', $[V_B^x]$, $[V_B^\bullet]$, and $p^\bullet$. The charge neutrality relation is

$$[V_A'] + 2[V_A''] + e' = [V_B^\bullet] + p^\bullet \quad . \tag{53}$$

This charge neutrality relation and the ratio of site relations [from

Eqs. (48) and (51)] are important in the thermodynamic treatment of defects in solids given in Section IV.

(2) The predominant defects may be V_A and A_i or V_B and B_i, in which case the solid exhibits Frenkel disorder [47]. This type of disorder is found in the silver halides (V_{Ag} and Ag_i), CaF_2 (V_F and F_i), and probably in CdTe (V_{Cd} and Cd_i), as well as others. Considerations similar to those given above yield, for the case of Frenkel disorder involving the A atoms,

$$\Delta = \frac{[L_B]}{[L_A] - [V_A] + [A_i]} - \frac{b}{a} \quad . \tag{54}$$

For the stoichiometric composition, $[V_A] = [A_i]$. Again, if charged defects, atomic and electronic, are present then a charge neutrality relation must be obeyed.

(3) The two predominant defects may be A_i and B_i. Disorder arising from such a pair is termed <u>interstitial disorder</u>. To date, no examples of this type of disorder have been found.

(4) The predominant defects may be the misplaced atoms, A_B and B_A, resulting in a solid with <u>substitutional</u> or <u>antistructure disorder</u>:

$$\Delta = \frac{[L_B] - [A_B] + [B_A]}{[L_A] - [B_A] + [A_B]} - \frac{a}{b} \quad . \tag{55}$$

Antistructure disorder is expected to occur only for compounds in which the electronegativity difference between the constituents is not too large. It, therefore, occurs primarily in intermetallic compounds such as Bi_2Te_3, Mg_2Sn, and possibly CdSb. An exception to this occurs in some ternary spinels where there are two different sites in the structure, both occupied by different ions. In the spinels, whose general formula is AB_2O_4, the A and B atoms have a valency (ν_A and ν_B) such that

$$\nu_A + 2\nu_B = 8 \quad . \tag{56}$$

For example, in $MgAl_2O_4$, $\nu_A = 2$ and $\nu_B = 3$, but in Mg_2TiO_4, $\nu_A = 4$ and $\nu_B = 2$. The structure has two types of sites. There are eight sites per unit cell (containing 32 O^{2-} anions) in which cations are surrounded by four oxygens (tetrahedral sites) and sixteen sites per unit cell in which the cation is surrounded by six oxygens (octahedral sites). In the completely ordered structure there are two possibilities:

(a) Normal spinel $A_t(B_o)_2O_4$ where t and o stand for tetrahedral and octahedral occupancy, and

(b) Inverted spinel $B_tB_oA_oO_4$.

Antistructure or substitutional disorder can occur when some of the ions normally occupying t sites appear on o sites and vice versa. An example of this has already been given in Section II. D. $MgFe_2O_4$ has the ideal converted spinel structure $Fe_t^{3+}Fe_o^{3+}Mg_o^{2+}O_4$. At the high temperatures required for the preparation of this compound, from MgO and Fe_2O_3, it has been found that some of the Mg^{2+} ions are present on t sites and a corresponding additional number of Fe^{3+} ions are present on the o sites.

(5) The predominant defects may be V_A and A_B or V_B and B_A. No certain examples of this type of disorder are known at present.

(6) The predominant defects may be A_i and B_A or B_i and A_B. Again, no examples of solids with this disorder are known at present.

Before leaving the description of possible structural disorder in compounds, we should briefly mention the existence of certain compounds with disordered sublattices. AgI, below 145.8 °C, crystallizes in the hexagonal wurtzite structure (β form) in which the I atoms are close packed. The Ag ions occupy, in an ordered manner, four of the eight tetrahedral interstitial sites in this close-packed arrangement. Above 145.8 °C the α modification is the stable form of AgI. The iodine atoms take on a bcc arrangement. They form a rigid network between which the Ag ions move freely in a completely disordered arrangement. They can occupy, randomly, any space within the iodine network large enough to accommodate them. Another type of sublattice disorder is found in α-Ag_2Te, α-Cu_2Se, and α-Cu_2S. In these the anions together with half the metal ions form a cubic zincblende structure in the interstices of which the rest of the metal ions are statistically distributed. Still another type of disorder, electronic in origin, is found in Fe_3O_4 with the spinel structure already described. The t sites are occupied by Fe^{3+} ions whereas the o sites are occupied by Fe^{2+} and Fe^{3+} ions randomly distributed. Since Fe^{2+} differs from Fe^{3+} by one electron, this type of disorder can be looked upon as arising from the random distribution of N electrons over 2N sites. The electrons are free to move within the Fe octahedral sublattice, thus giving Fe_3O_4 a high electrical conductivity (see Chapter 4).

IV. CONTROL OF POINT DEFECTS IN SOLIDS WITH SMALL DEFECT CONCENTRATIONS

A. The Phase Rule

In the previous three sections, we have discussed electronic defects as well as the various types of atomic defects in solids, their association, their donor and acceptor behavior in semiconductors, and the structural disorder and stoichiometric deviations that arise from the presence of the native defects. We will now consider, using the methods of statistical thermodynamics, the factors that control the concentrations of defects in a solid. Basic to the application of thermodynamics to the problem of defects in solids is the Phase Rule.

The Phase Rule is a thermodynamic law, first enunciated by Josiah Willard Gibbs, relating the number of phases, the number of components, and the number of degrees of freedom of a system in a state of equilibrium. We are interested in applying the Phase Rule to a single solid phase consisting of one or more chemical components. Thus, the solid Ge is a one-component system, CaF_2 is a two-component system, Al_2O_3 containing Cr^{3+} impurities on Al^{3+} (i.e., Cr_{Al}^{x}) sites is a three-component system, and so on. The Phase Rule as applied to our case is

$$f = C + 1 \quad . \tag{57}$$

Here f, the number of degrees of freedom, is the number of intensive variables which have to be specified in order to determine the state of the system, and therefore its properties, completely. An intensive variable is one like temperature, pressure, concentration of defects, chemical potential of the components, and density, which does not depend on the amount of solid material under consideration. The mass, number of defects, total energy content, volume, etc., are extensive variables depending, as they do, on the amount of solid present.

For a one-component system such as the element Si, Eq. (57) states that $f = 2$. The two intensive variables most readily accessible to experimental control are temperature and pressure. Thus, if the temperature and pressure are fixed, the Phase Rule tells us that the state of the Si is determined completely. Since the state of Si and its properties will depend on the concentration of native defects in various possible charge states, on the concentration of free electrons and holes, on the concentration of native defect associates, and so on, these are therefore determined completely at any given temperature and pressure. It must be remembered that only those defects that are a consequence of an equilibrium state of the system are determined. Thus, dislocations, stacking faults, grain boundaries, and other nonequilibrium defects present in solids cannot be treated by thermodynamic methods.

A few remarks about the temperature and pressure variables are in order. Because of kinetic factors, such as diffusion and surface reaction rates, equilibrium states of solids are reached much faster at higher temperatures than at lower ones. Thus, experimentally, in studies of the effect of external intensive variables upon defect concentrations in a solid, high temperatures are used to reach an equilibrium state in a reasonable time. Measurement of the electrical properties, optical properties, magnetic properties, and so on, which give information on the type and concentration of defects, are generally, but not always, carried out at room temperature or below. In order to freeze in the high temperature equilibrium state, the solids are frequently quenched as rapidly as possible. The assumption is then made that, since the temperature is lowered so rapidly from the high equilibration temperature, the high temperature state is indeed frozen in with regard to the kind and concentration of atomic defects. Electrons and holes, however, have high mobilities even at low temperatures, and these are not frozen in at a high temperature configuration but can redistribute themselves among the various accessible states in accord with the lower temperature to which the crystal is quenched or measured. We shall, a little later (Section IV. E), give a specific example of such behavior of solids on cooling.

With regard to the total hydrostatic pressure P on the solid phase, this is generally not an important variable for experiments carried out at ordinary pressures. Defect equilibria in solids, as we shall see shortly, are characterized by equilibrium constants K(T, P), which are functions of the temperature and the pressure. According to thermodynamics, the pressure dependence of K is given by

$$\left(\frac{d \ln K}{dP}\right)_T = -\frac{v_M}{RT}\left(\frac{\Delta v_D}{v_M}\right) , \tag{58}$$

where v_M is the molar volume of the solid and Δv_D is the volume change associated with the production of one mole (i. e., 6.02×10^{23}) of defects. Integration of Eq. (58) yields

$$K_2/K_1 = \exp[(-v_M/RT)(\Delta v_D/v_M)\Delta P)] . \tag{59}$$

Using some typical large values, $v_M = 50\ cm^3$, $(\Delta v_D/v_M) = 1.6$, at a temperature of 1000°K we get

$$K_2/K_1 = \exp(-0.001\,\Delta P) . \tag{60}$$

Taking $P_1 = 0$ and $P_2 = 100$ atm, $K_2/K_1 = 0.9$. Thus, between 0 and 100 atm, equilibrium constants change by only 10 percent. In most cases the pressure dependence is even smaller. Unless, therefore, the range of pressures is larger than 100 atm, we can disregard the

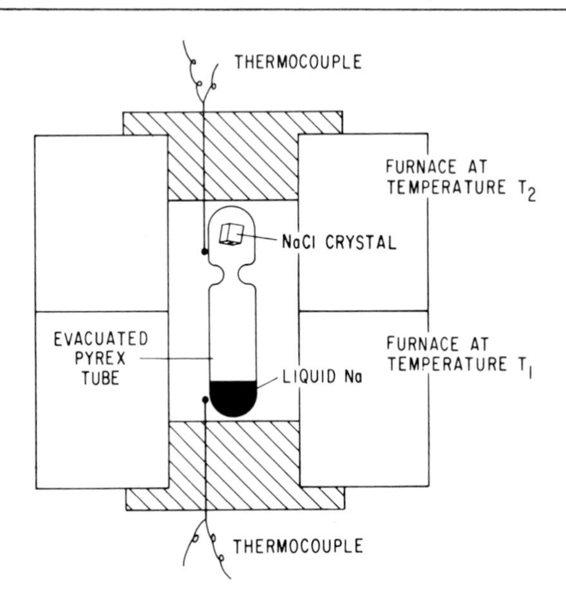

FIG. 10. Experimental setup for heating a crystal of NaCl at a temperature T_2 in a controlled pressure of Na vapor. The Na reservoir is kept at a lower temperature T_1.

hydrostatic pressure as an important variable in discussing the thermodynamics of solids. In the element Si, for example, the state of Si and its properties are determined by the temperature and only to a negligible extent by the pressure.

For a binary compound such as NaCl, Eq. (57) gives $f = 3$. To fix the state and hence the properties of the solid, we must fix the temperature T, the pressure P (although again small differences in P will not affect the equilibrium state), and one other intensive variable of the solid. Most conveniently, this is taken as the chemical potential [48] of either the Na or the Cl. The reason for choosing this variable arises from the fact that the chemical potentials can be fixed by exposing the compound, at the temperature T, to either Na vapor or Cl_2 gas at a specific partial pressure p_{Na} or p_{Cl_2}.* If the gaseous and solid phases

*The effect of small changes in p_{Na} and p_{Cl_2} can be very large. This effect is due to changes in the chemical potential of Na or Cl_2 and is not due to a pressure volume term as in Eq. (59). We already have shown that this is very small and can be disregarded.

are in equilibrium with each other, then thermodynamics tells us that the chemical potentials of every chemical component are equal in both phases. The chemical potentials of Na and Cl in the gas phase are given by [48]:

$$\mu_{Na}(gas) = \mu^0_{Na} + kT \ln (p_{Na}) \quad , \tag{61}$$

$$\mu_{Cl}^{(gas)} = \tfrac{1}{2}\mu_{Cl_2}(gas) = \tfrac{1}{2}\mu^0_{Cl_2} + \tfrac{1}{2} kT \ln (p_{Cl_2}) \quad . \tag{62}$$

The relation $\mu_{Cl} = \frac{1}{2}\mu_{Cl_2}$ derives from the gas phase dissociative equilibrium $Cl \rightleftarrows \frac{1}{2} Cl_2$. The μ^0 are constants that depend only on temperature, so that the μ's are fixed at a given T and p_{Na} or p_{Cl_2}. At equilibrium,

$$\mu_{Na}(solid) = \mu_{Na}(gas) = \mu^0 + kT \ln (p_{Na}) \quad , \tag{63}$$

and a similar relationship holds for Cl_2. One way of carrying out this experiment of fixing T and $\mu_{Na}(solid)$ is shown in Fig. 10. The value of p_{Na} is fixed by the temperature T_1 of the Na reservoir (it is the vapor pressure of Na at temperature T_1) and T is fixed by the temperature in the region of the sample of NaCl (i. e., $T = T_2$). For Cl_2, which is a gas, we can simply fill the tube with an amount of gas which, at the temperature T, will exert the required pressure p_{Cl_2} to fix the chemical potential of Cl_2.

Another type of binary system might be the semiconducting element Si containing a small amount of As impurity. Again, two intensive variables must be fixed (neglecting the total pressure). By fixing the temperature T and the chemical potential of As (by exposing the Si for example to a given pressure of As or AsH_3 vapor, the state of the system is fixed. Thus, not only are the concentration native defects and free and bound electrons and holes fixed, but also the concentration of As impurities in the various possible charge states.

In ternary systems, for example, ZnS containing small concentrations of Cu impurity, three intensive variables must be fixed (again neglecting the total pressure) to completely specify the state and properties of the system. These are conveniently the temperature T, and two other intensive variables. These might be the chemical potential of sulfur fixed by a S_2 vapor over the solid at a fixed pressure, and the concentration of Cu in the solid fixed by growing ZnS crystals containing a given concentration of Cu impurity. Instead of the last intensive variable being the Cu concentration, the chemical potential of Cu may be chosen. We can embed the ZnS crystal in Cu_2S powder and depend on the very rapid diffusion of Cu in ZnS to reach an equilibrium state quickly. The state of the Cu_2S powder will be fixed by the values of T and p_{S_2}, and so the chemical potential of Cu in Cu_2S will be fixed. At equilibrium the ZnS and Cu_2S phases will be in a state of heterogeneous equilibrium, so that the chemical potential of the Cu in the ZnS will also be fixed.

The application of the Phase Rule to more complex examples is very straightforward. We must now turn our attention to more quantitative questions. How, for example, do the concentrations of the various atomic and electronic defects vary with changes in the values of the intensive variables?

B. Thermodynamics and Defects

The statement has been made that, at any temperature above absolute zero, there will be defects present in every solid. Using as a simple example of a solid an element M in which only neutral vacancies V_M^x exist, we shall show why this is so. For such a solid, f = 2, and $[V_M^x]$ should be a function of T and P only. The thermodynamic function we will use to describe the state of the solid is the Gibbs free energy G. From thermodynamics, the equilibrium state will be determined by the condition that dG = 0 for any process occurring in a closed system (i. e., no mass change in the system) at constant temperature and pressure [49]. As already pointed out, however, the equilibrium state of solids depends only to a negligible extent on the pressure. Now G, being an extensive property, will be expected to depend on T and P and on the values of the composition variables $n(M_M^x)$ and $n(V_M^x)$, where the n's stand for the number of M atoms on M sites and the number of neutral vacancies on M sites in the piece of solid under consideration.* G is given by [49]

$$G = \mu_1 n_2 + \mu_2 n_2 \quad , \tag{64}$$

where $n_1 = n(M_M^x)$ and $n_2 = n(V_M^x)$ and the μ's are chemical potentials defined by

$$\mu_i = (\partial G/\partial n_i)_{T,\, P,\, n_{j \neq i}} \quad . \tag{65}$$

The condition for equilibrium is

$$(\partial G)_{T,\, P,\, \text{closed system}} = 0 = \mu_1\, dn_1 + \mu_2 dn_2 \tag{66}$$

The requirement that the system be closed is met by

$$d[n(M)] = d[n(M_M^x)] = dn_1 = 0 \quad . \tag{67}$$

Therefore, Eq. (66) becomes

$$\mu_2 dn_2 = 0 \quad . \tag{68}$$

However, dn_2 being any arbitrary variation in n_2 is not zero. Thus, the condition for equilibrium is contained in the statement that

$$\mu_2 = 0 \quad . \tag{69}$$

To proceed further, we must express G as a function of T, P, n_1, and n_2, evaluate μ_2 from Eq. (65), and set it equal to zero.

We can first write a reaction occurring at T and P:

$$(M_M^x)_{n_1} \rightarrow (M_M^x)_{n_1}(V_M^x)_{n_2} \quad , \tag{70}$$

with $\Delta G = G - G^*$. G^* is the free energy of the solid containing n_1 atoms but no vacancies, whereas G is the free enerty of the imperfect

---

*Note that the quantities $[V_M^x]$, $[M_M^x]$, etc., used previously are $n(V_M^x)$, $n(M_M^x)$ divided by the volume **V** of the solid.

solid containing n_1 atoms and n_2 vacancies. The creation of each single vacancy by such a process will be associated with an enthalpy (or, to a good approximation, energy) change $\Delta\bar{H}(V_M^x)$, provided that the concentrations of the vacancies are sufficiently low so that there is no interaction energy between the vacancies. For low vacancy concentrations, the $\Delta\bar{H}$ values are independent of concentration, and the vacancies will be randomly distributed over available sites. The creation of each single vacancy will also be associated with a vibrational entropy change $\Delta\bar{S}_\nu(V_M^x)$ due to the changes in the phonon spectrum of the solid upon creation of the vacancy. Finally, and most important, is the change in configurational entropy, ΔS_c, associated with the reaction in Eq. (70). This term arises from the fact that there are a very large number of geometrical configurations (i.e., arrangements of n_2 vacancies and n_1 atoms over the $n_1 + n_2$ sites) in the imperfect solid corresponding to the same total enthalpy, whereas in the perfect solid there is only one configuration, all the n_1 atoms being on the n_1 lattice sites. The quantity ΔS_c is given by [50]

$$\Delta S_c = k \ln W - k \ln W^* = k \ln W \quad , \tag{71}$$

where W is the number of possible configurations in the disordered solid and $W^* = 1$. W can be evaluated from statistical considerations. The number of ways of arranging n_1 atoms and n_2 vacancies on $n_1 + n_2$ sites is [50]

$$W = (n_1 + n_2)!/n_1!n_2! \quad . \tag{72}$$

The $n_1!$ and $n_2!$ factors in the denominator arise from the fact that permutations of the atoms or vacancies among themselves do not lead to a new distinguishable configuration; one vacancy or one atom is indistinguishable from any other one. Therefore, for Eq. (71),

$$\Delta S_c = k \ln (n_1 + n_2)!/n_1!n_2! \quad . \tag{73}$$

The expression for ΔG in Eq. (70) is therefore:

$$\Delta G = G - G^* = n_2 \Delta \bar{H}(V_M^x) - n_2 T \Delta \bar{S}_\nu(V_M^x) - kT \ln (n_1 + n_2)!/n_1!n_2! \ . \quad (74)$$

Using Sterling's approximation,

$$\ln N! = N \ln N - N \quad , \quad (75)$$

when N is a very large number, we get, for G,

$$G = G^* + n_2[\Delta \bar{H}(v_M^x) - T \Delta \bar{S}_\nu(V_M^x)] + kT \ln \left[n_1 \ln \frac{n_1}{n_1 + n_2} + n_2 \ln \frac{n_2}{n_1 + n_2} \right]. \quad (76)$$

The required chemical potential μ_2 is therefore

$$\mu_2 = [\Delta \bar{H}(V_M^x) - T \Delta \bar{S}_\nu(V_M^x)] + kT \ln \left(\frac{n_2}{n_1 + n_2} \right) \ . \quad (77)$$

Letting $\mu_2{}^0 = [\Delta \bar{H}(V_M^x) - T \Delta \bar{S}_\nu(V_M^x)]$ and substituting for n_1 and n_2, $n(M_M^x)$ and $n(V_M^x)$, we get

$$\mu_2 = \mu_2{}^0 + kT \ln \left(\frac{n(V_M^x)}{n(V_M^x) + n(M_M^x)} \right) \ . \quad (78)$$

Finally, the equilibrium condition $\mu_2 = 0$ from Eq. (69) yields, after dividing the n's by the volume $\mathbf{V}$,

$$[V_M^x] = ([V_M^x] + [M_M^x]) \exp(-\mu_2{}^0/kT) \ . \quad (79)$$

At low vacancy concentrations, $[V_M^x] + [M_M^x] = [L_M]$ is practically equal to $[M_M^x]$, given by

$$[M_M^x] = \rho \mathscr{L}/A \quad , \quad (80)$$

where ρ is the density of M, A is its atomic weight, and $\mathscr{L}$ is Avogadro's number. Equation (79) gives the required relation between the concentration of vacancies and the two intensive variables T and P. The pressure dependence comes from the enthalpy term, which is $\Delta \bar{E}(V_M^x) + P \Delta \bar{\mathbf{V}}(V_M^x)$. Since $P \Delta \bar{\mathbf{V}}$ is much smaller than $\Delta \bar{E}$, the pressure dependence is negligible at ordinary pressures. The vibrational entropy term $T \Delta \bar{S}_\nu(V_M^x)$ is also smaller than the enthalpy term. The quantity $[L_M] \exp(-\mu_2{}^0/kT)$ is frequently symbolized by K(T, P), an

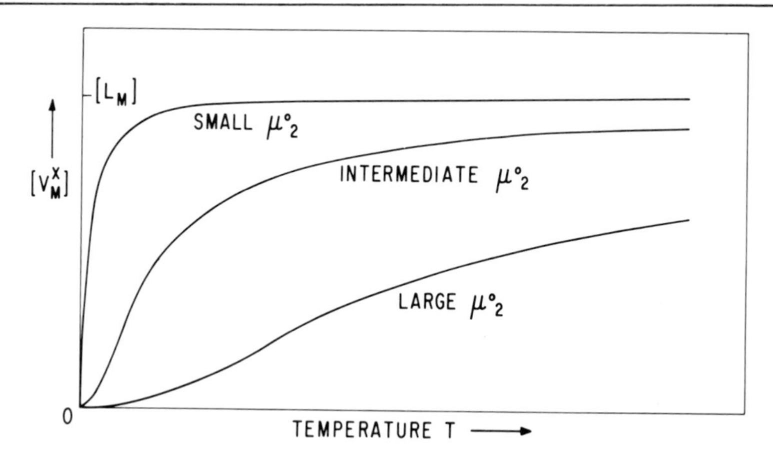

FIG. 11. Plot of $[VM^{\times}]$ against T [Eq. (79)] for various values of $\mu_2{}^0$.

equilibrium constant for the reaction given in Eq. (70):

$$[V_M^{\times}] = K(T, P) \quad . \tag{81}$$

The value of the equilibrium constants for defect production in solids must be determined from experiment. If the defect concentrations at various temperatures are found from suitable experiments, then equilibria equations like Eq. (81) can be used to determine the values of K.

There are several features of interest in the result. First, at $T = 0$, $[V_M^{\times}] = 0$, but it is finite at any $T > 0$. This is what we set out to show. Secondly, at any T, the larger the enthalpy (or energy) required to form a vacancy, the smaller is $[V_M^{\times}]$. In general, the degree of disorder of any type in a solid will depend on the energy required to form the particular primary defects. Finally, as T increases, the disorder as given by $[V_M^{\times}]$ increases rapidly and tends to approach $[L_M]$. Of course at some value far below this, the crystal will melt. These various relations are shown in Fig. 11.

The disorder in a pure solid arises from the increase in configurational entropy with increasing disorder. The same considerations apply to impurities. A pure solid in contact with a container will always tend to dissolve small amounts of impurities, even though it may take energy to replace an atom of the solid by the impurity atom. Those impurities which have lower solution energies will dissolve to the greatest extent. Entropy therefore works against us in the purification of solids.

C. Stoichiometry in Pure Compounds

Having shown that lattice disorder as well as impurity content is a consequence of the increase in configurational entropy of a solid with increasing disorder, some real solids will now be considered. In this section we will consider the factors that control the stoichiometry of a pure binary compound. By a pure compound, we mean, of course, one in which the inevitable concentrations of impurities is much lower than those of the native defects. For such a binary compound $f = 3$, so that the concentration of the various native defects and hence the stoichiometry of the compound depend on the three intensive variables of the solid, T, P, and the chemical potential μ of one of the constituents of the compound.

Bloem [51] studied the compound PbS equilibrated at high temperatures (1000 to 1200°K) under various sulfur pressures (present in the gas phase primarily as S_2 molecules). He presented evidence that the disorder in PbS was of the Schottky type and that the Pb and S vacancies could be present both as neutral (V_{Pb}^{x}, V_S^{x}) defects and as singly charged (V_{Pb}', $V_S^{\bullet}$) defects. In addition, conduction band electrons e' and valence band holes $h^{\bullet}$ were present, leading to electrical conductivity. We shall present a thermodynamic treatment of PbS and show how the concentrations of the defects are expected to vary with changes in the important intensive variables T and μ_S (or μ_{Pb}).

We start with the six defects and the two compound constituents in a piece of PbS: $n(Pb_{Pb}^{x})$, $n(S_S^{x})$, $n(V_{Pb}^{x})$, $n(V_S^{x})$, $n(V_{Pb}')$, $n(S_S^{\bullet})$, $n(e')$, and $n(h^{\bullet})$. The free energy is a function of these numbers as well as of T and P. However, in a solid these numbers are not independent of each other, being connected by the lattice-site and charge-neutrality relations for PbS:

$$n(Pb_{Pb}^{x}) + n(V_{Pb}^{x}) + n(V'_{Pb}) = n(S_{S}^{x}) + n(V_{S}^{x}) + n(V_{S}^{\bullet}) \tag{82}$$

and

$$n(V'_{Pb}) + n(e') = n(S_{S}^{\bullet}) + n(h^{\bullet}) \quad . \tag{83}$$

As a consequence of these two restrictions, the free energy G can be written as a function of the six compositions variables $N_1, \ldots, N_6$:

$$G = \sum_{i=1}^{6} \mu_i N_i \quad . \tag{84}$$

The relation between the n's and the N's is purely arbitrary so long as the conditions in Eqs. (82) and (83) are fulfilled. Thus, we may arbitrarily write:

$$n(Pb_{Pb}^{x}) = N_1 , \quad n(V_{Pb}^{x}) = N_2 , \quad n(V'_{Pb}) = N_3 ,$$

$$n(S_{S}^{x}) = N_4 , \quad n(V_{S}^{x}) = N_5 , \quad n(e') = N_6 ,$$

whence

$$n(V_{S}^{\bullet}) = N_1 + N_2 + N_3 - N_4 - N_5$$

and

$$n(h^{\bullet}) = - N_1 - N_2 + N_4 + N_5 + N_6 \quad .$$

Just as before, the condition for equilibrium is given by

$$(dG)_{T,\,P,\,\text{closed system}} = 0 = \sum \mu_i \, dN_i \quad , \tag{85}$$

and the condition for a closed system is met by

$$dN_1 = dN_4 = 0 \quad . \tag{86}$$

Since the remaining variations dN_i are to be arbitrary, Eq. (85) requires that:

$$\mu_2 = \mu_3 = \mu_5 = \mu_6 = 0 \quad . \tag{87}$$

The chemical potential μ_i can be written in terms of the various n's by using the relation

$$\mu_i = \partial G/\partial N_i = \sum_j [\partial G/\partial n(D_j)\ \partial n(D_j)/\partial N_i] \quad . \tag{88}$$

Following Kroger [4] we call the quantities $\partial G/\partial n(D_j) = \xi(D_j)$ <u>virtual chemical potentials</u>. Then

$$\mu_1 = \xi(Pb_{Pb}^{x}) + \xi(V_S^{\bullet}) - \xi(h^{\bullet}) \quad ,$$

$$\mu_2 = \xi(V_{Pb}^{x}) + \xi(V_S^{\bullet}) - \xi(h^{\bullet}) \quad , \tag{89}$$

$$\mu_3 = \xi(V_{Pb}^{\prime}) + \xi(V_S^{\bullet}) \quad ,$$

and so on. Using statistical methods analogous to those in Section IV.B, we find that for low defect concentrations each $\xi(D_j)$ is of the form

$$\xi(D_j) = \xi^0(D_j) + kT \ln \frac{n(D_j)}{n(L_j)} \quad . \tag{90}$$

The quantities $n(L_j)$ are the numbers of Pb or S sites in the piece of solid under consideration when the D_j are vacancies or atoms of Pb or S. When D_j represents the defects e' or $h^{\bullet}$ then the $n(L_j)$ are, respectively, the effective numbers of states in the conduction band or valence band. For simple parabolic bands, these are given by

$$n(L_{h^{\bullet}}) = n(L_{e'}) = 2(2\pi m^* kT/h^2)^{\frac{3}{2}}\, V \quad , \tag{91}$$

where m^* is the effective mass of the hole or the electron. From Eqs. (89) and (90), expressions for the various chemical potentials μ_i can be readily written down. For ease of interpretation we shall use certain linear combinations of the equilibrium conditions given in Eq. (87). After some simple manipulations, one arrives at the equations representing the equilibrium state of PbS:

$$\mu_2 + \mu_5 = 0 ; \quad [V^{\times}_{Pb}][V^{\times}_{S}] = [L]^2 \exp\{-[\xi^0(V^{\times}_{Pb}) + \xi^0(V^{\times}_{S})]/kT\} ,$$

$$\mu_6 = 0 ; \quad n \cdot p = N_c N_v \exp\{-[\xi^0(e') + \xi^0(h^{\cdot})]/kT\} ,$$

$$\mu_6 - \mu_5 = 0 ; \quad [V^{\cdot}_{S}]n/[V^{\times}_{S}] = N_c \exp\{-[\xi^0(V^{\cdot}_{S}) + \xi^0(e') - \xi^0(V^{\times}_{S})]/kT\} , \tag{92}$$

$$\mu_3 - \mu_2 = 0 ; \quad [V'_{Pb}]p/[V^{\times}_{Pb}] = N_v \exp\{-[\xi^0(V'_{Pb}) + \xi^0(h^{\cdot}) - \xi^0(V^{\times}_{Pb})]/kT$$

In Eqs. (92), $[V^{\times}_{Pb}] = n(V^{\times}_{Pb})/\mathbf{V}$, etc., $[L] = n(L_{Pb})/\mathbf{V} = n(L_S)/\mathbf{V}$; N_v and $N_c = n(L_{h^{\cdot}})/\mathbf{V}$, and $n(L_{e'})/\mathbf{V}$. The terms involving the ξ^0 appearing in the exponentials are easy to interpret.

The equation $\mu_2 + \mu_5 = 0$ refers to the equilibrium reaction in which a pair of neutral Schottky defects are formed:

$$Pb_N S_N \rightleftarrows Pb_N(V^{\times}_{Pb})S_N(V^{\times}_{S})$$

or, subtracting $Pb_N S_N$ from both sides,

$$0 \rightleftarrows (V^{\times}_{Pb})(V^{\times}_{S}) \quad . \tag{93}$$

The quantity $\xi^0(V^{\times}_{Pb}) + \xi^0(V^{\times}_{S})$ is then the standard free energy change (i.e., the change in free energy other than the configurational entropy term) for the reaction (93):

$$\xi^0(V^{\times}_{Pb}) + \xi^0(V^{\times}_{S}) = \Delta\bar{H}(V^{\times}_{Pb}) + \Delta\bar{H}(V^{\times}_{S}) - T[\Delta\bar{S}_{\nu}(V^{\times}_{Pb}) + \Delta\bar{S}_{\nu}(V^{\times}_{S})] . \tag{94}$$

This is close to E_S, the energy required to produce a Schottky pair, if we neglect the $P\Delta\bar{\mathbf{V}}$ term in $\Delta\bar{H}$ and the small contribution of the change in vibrational entropy. Similarly, $\mu_6 = 0$ refers to the equilibrium reaction

$$0 \rightleftarrows e' + h^{\cdot} \quad . \tag{95}$$

This reaction has already been described in Sections II.D and II.E. Disregarding the very small pressure volume terms in the enthalpy and the negligible vibrational entropy change,

$$\xi^0(e') + \xi^0(h^\cdot) = E_G \quad , \tag{96}$$

the band gap energy. Again, $\mu_6 - \mu_5 = 0$ refers to the equilibrium reaction reaction

$$V_S^x \rightleftarrows V_S^\cdot + e' \tag{97}$$

and represents the ionization of the sulfur vacancy donor as described in Section II. E. To the same approximation as Eq. (96),

$$\xi^0(V_S^\cdot) + \xi^0(e') - \xi^0(V_S^x) = E_D \quad , \tag{98}$$

where E_D, the donor ionization energy, locates the level of V_S^x below the bottom of the conduction band. Finally, $\mu_3 - \mu_2 = 0$ refers to the equilibrium reaction

$$V_{Pb}^x \rightleftarrows V'_{Pb} + h^\cdot \tag{99}$$

and represents the ionization of the lead vacancy acceptor as described in Section II. E:

$$\xi^0(V'_{Pb}) + \xi^0(h^\cdot) - \xi^0(V_{Pb}^x) = E_A \quad , \tag{100}$$

where E_A, the acceptor ionization energy, locates the level of V'_{Pb} above the top of the valence band.

We now consider the chemical potentials μ_1 and μ_4. The free energy can be written as in Eq. (84) and also in terms of the actual chemical composition of the PbS:

$$G = \mu_{Pb}\, n(Pb_{Pb}^x) + \mu_S\, n(S_S^x) \quad . \tag{101}$$

Comparing this to Eq. (84), in which $N_1 = n(Pb_{Pb}^x)$ and $N_4 = n(S_S^x)$, we find that $\mu_1 = \mu_{Pb}$ and $\mu_4 = \mu_S$. With some simple manipulation, using the relations given by Eqs. (87), (89), and (90),

$$\begin{aligned} \mu_{Pb} &= \xi(Pb_{Pb}^x) - \xi(V_{Pb}^x) = \xi^0(Pb_{Pb}^x) - \xi^0(V_{Pb}^x) + kT \ln \frac{n(Pb_{Pb}^x)}{n(V_{Pb}^x)} \quad , \\ \mu_S &= \xi(S_X^x) - \xi(V_S^x) = \xi^0(S_S^x) - \xi^0(V_S^x) + kT \ln \frac{n(S_S^x)}{n(V_S^x)} \quad . \end{aligned} \tag{102}$$

As indicated in Section IV.A, μ_{Pb} or μ_S can be fixed by equilibrating the solid PbS with a vapor of Pb or S_2. Then,

$$\mu_{Pb} = \mu^0_{Pb}(\text{gas}) + kT \ln(p_{Pb}) \quad ,$$
$$\mu_S = \tfrac{1}{2}\mu^0_{S_2}(\text{gas}) + kT \ln(p_{S_2})^{\frac{1}{2}} \quad . \tag{103}$$

The quantities $\mu_{Pb}(\text{gas})$ and $\mu_{S_2}(\text{gas})$ are found in texts on statistical thermodynamics [48]. Combining these with Eq. (102) we get, after dividing by the volume **V** and recalling that for low defect concentrations $[Pb^x_{Pb}] = [S^x_S] \approx [L]$,*

$$p_{Pb}[V^x_{Pb}] = [L] \exp\{-[\mu^0_{Pb}(\text{gas}) + \xi^0(V^x_{Pb}) - \xi^0(Pb^x_{Pb})]/kT\} \quad ,$$
$$p^{\frac{1}{2}}_{S_2}[V^x_S] = [L] \exp\{-[\tfrac{1}{2}\mu^0_{S_2}(\text{gas}) + \xi^0(V^x_S) - \xi^0(S^x_S)]/kT\} \quad . \tag{104}$$

Equations (104) describe the equilibria:

$$Pb^x_{Pb}(\text{solid}) \rightleftarrows V^x_{Pb}(\text{solid}) + Pb(\text{gas}) \quad ,$$
$$S^x_S\ (\text{solid}) \rightleftarrows V^x_S\ (\text{solid}) + \tfrac{1}{2}S_2(\text{gas}) \quad , \tag{105}$$

and the quantities in the exponentials are the standard free energy changes for the reactions. The four relations given by Eq. (92), either one of the relations given by Eq. (102), and the charge neutrality relation in Eq. (83) give a total of six equations. From these, the concentrations of the six defects, $[V^x_{Pb}]$, $[V^x_S]$, $[V'_{Pb}]$, $[V^{\bullet}_S]$, n, and p, are determined as functions of T, P, and μ_S or μ_{Pb}, in accord with the requirements of the Phase Rule. We will now consider the methods used for solving

*Strictly speaking, [L], the concentration in cm^{-3} of Pb or S sites in PbS, appearing in Eqs. (104) is not a constant. With changes in defect concentrations there are very small changes in the unit cell volume of PbS, which will change the value of [L] slightly. However, for small defect concentrations this is a negligible change, and [L] can be taken to be a constant.

these equations and thereby determine how the stoichiometry of PbS can be changed by varying the S_2 pressure. This was studied experimentally by Bloem [51].

D. Mass Action Laws

If in Eqs. (92) and (104) we let the right sides of the equations be represented by equilibrium constants $K_i(T, P)$, we get the set of equations

$$[V_{Pb}^{x}][V_S^{x}] = K_S ,$$

$$n \cdot p = K_i ,$$

$$[V_S^{\cdot}]n/[V_S^{x}] = K_D , \quad (106)$$

$$[V_{Pb}^{\prime}]p/[V_{Pb}^{x}] = K_A ,$$

$$p_{S_2}^{\frac{1}{2}}[V_S^{x}] = K_g .$$

It is sometimes convenient to combine Eqs. (106):

$$[V_S^{\cdot}][V_{Pb}^{\prime}] = K_A K_D K_S / K_i = K_S^{\prime} , \quad (107)$$

where $K_S^{\prime} = [L]^2 \exp[-(\Delta\bar{G}^0/kT)]$ and $\Delta\bar{G}^0$ is the standard free energy change for the process of producing a singly charged Schottky defect pair. Neglecting the small contributions of the pressure-volume terms and the thermal entropy,

$$\Delta\bar{G}^0 = E_A + E_D + E_S - E_G .$$

These equations can be set down directly by applying the laws of mass action to the equilibria given in Eqs. (93), (95), (97), (99), and (105). The thermodynamic treatment given here is the justification for applying mass action laws to defect equilibria in solids. It is extremely difficult to get exact solutions of these equations plus the charge neutrality relation for the concentrations of the various defects as a function of the temperature and p_{S_2}. The method described by Kröger and Vink [52] gives very good approximate solutions, which furthermore have the great advantage of being physically simple to interpret.

The charge neutrality relation is [Eq. (83) divided by the volume V]

$$[V'_{Pb}] + n = [V^{\bullet}_{S}] + p \quad .$$

Looking at the mass action equations we see that, at very high sulfur pressures, $[V^{x}_{S}]$ is small. Consequently $[V^{x}_{Pb}]$ is large, $[V'_{Pb}]$ and p will be large, and n will be small. We therefore expect a region of high sulfur pressure in which the charge neutrality relation can be very well approximated by $[V'_{Pb}] = p$. With this, the solutions of Eqs. (106) are readily found:

$$\begin{aligned}
[V^{x}_{S}] &= K_g\, p_{S_2}^{-\frac{1}{2}} \quad , \\
[V^{x}_{Pb}] &= K_S K_g^{-1}\, p_{S_2}^{\frac{1}{2}} \quad , \\
[V^{\bullet}_{S}] &= K_D (K_A K_S)^{\frac{1}{2}} K_i^{-1} K_g^{\frac{1}{2}}\, p^{-\frac{1}{4}} \quad , \\
[V'_{Pb}] &= p = (K_A K_S)^{\frac{1}{2}} K_g^{-\frac{1}{2}}\, p_{S_2}^{\frac{1}{4}} \quad , \\
n &= K_i (K_A K_S)^{-\frac{1}{2}} K_g^{\frac{1}{2}}\, p_{S_2}^{-\frac{1}{4}} \quad .
\end{aligned} \tag{108}$$

As the S pressure decreases, $[V^{x}_{Pb}]$, $[V'_{Pb}]$, and p decrease, whereas $[V^{x}_{S}]$, $[V^{\bullet}_{S}]$, and n increase. Therefore, at the other extreme of very low sulfur pressure, the charge neutrality relation becomes, to a good approximation, $[V^{\bullet}_{S}] = n$. The solutions of Eqs. (106) are then

$$\begin{aligned}
[V^{\bullet}_{S}] &= n = K_D^{\frac{1}{2}} K_g^{\frac{1}{2}}\, p_{S_2}^{-\frac{1}{4}} \quad , \\
[V'_{Pb}] &= K_A K_S K_D^{\frac{1}{2}} K_i^{-1} K_g^{-\frac{1}{2}}\, p_{S_2}^{\frac{1}{4}} \quad , \\
p &= K_i K_D^{-\frac{1}{2}} K_g^{-\frac{1}{2}}\, p_{S_2}^{\frac{1}{4}} \quad .
\end{aligned} \tag{109}$$

The concentrations of V_S^x and V_{Pb}^x have the same dependence on sulfur pressure as in Eq. (108) since they do not appear in the charge neutrality relation.

In the region of intermediate sulfur pressures, two possibilities exist for the approximate charge neutrality relation, either $n = p$ or $[V'_{Pb}] = [V^{\bullet}_S]$. In the first case $n = p = K_i^{\frac{1}{2}}$ and in the second $[V'_{Pb}] = [V^{\bullet}_S] = [K'_S]^{\frac{1}{2}}$, so that the neutrality condition depends on whether K_i is less than or greater than K'_S, where

$$K_i = N_v N_c \exp(-E_G/kT) \quad \text{and} \quad K'_S = [L]^2 \exp(E_G - E_A - E_D - E_S)/kT.$$

For typical insulators with large band gaps and deep donor and acceptor levels, $K_i/K'_S < 1$, whereas for typical semiconductors with small band gaps, $K_i/K'_S > 1$. For PbS it was found that K_i was somewhat larger than K'_S (see Section IV. E), and the appropriate neutrality condition for the intermediate range of sulfur pressures is, therefore, $n = p$. The solutions to Eqs. (106) become

$$
\begin{aligned}
[V^{\bullet}_S] &= K_D K_i^{-\frac{1}{2}} K_g p_{S_2}^{-\frac{1}{2}} , \\
[V'_{Pb}] &= K_A K_S K_i^{-\frac{1}{2}} K_g^{-1} p_{S_2}^{\frac{1}{2}} , \\
n = p &= K_i^{\frac{1}{2}} .
\end{aligned}
\tag{110}
$$

In Fig. 12, the logarithm of the concentration of the various defects are plotted against the log p_{S_2} at a single temperature T. In this way straight lines are obtained with slopes of 0, ± 1/4, and ± 1/2. These approximate solutions differ from the true solutions only near the boundaries between the three regions, as shown by the thin lines. The values of p_{S_2} at the boundaries between the regions are easily determined. For example, the boundary between regions I and II can be obtained by setting $[V^{\bullet}_S]$ for region I [Eq. (109)] equal to $[V^{\bullet}_S]$ for region II [Eq. (110)]. In this way, one finds that, at this boundary,

$$(p_{S_2})_{I,II} = (K_D K_g / K_i)^2 . \tag{111}$$

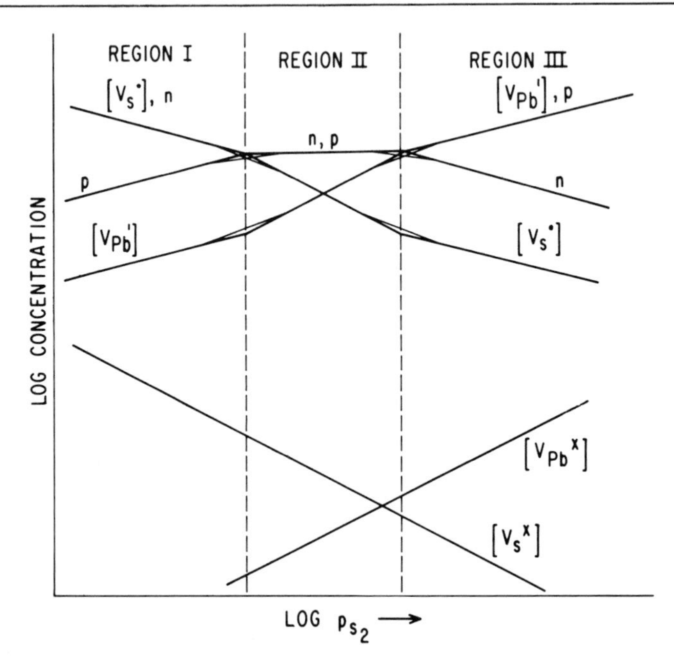

FIG. 12. Plot of the log of the concentrations of the various native defects in PbS against log p_{S_2} at some high temperature T.

In the same way, the boundary between regions II and III lies at

$$(p_{S_2})_{II,III} = (K_D K_g / K'_S)^2 \quad . \tag{112}$$

The width of the intermediate sulfur pressure region II is therefore

$$\log(p_{S_2})_{II,III} - \log(p_{S_2})_{I,II} = 2[\log K_i - \log K'_S] \quad . \tag{113}$$

Experimental observation of the narrowness of this region, as described in the next section, enabled Bloem to show that, for PbS, K_i was only slightly larger than K'_S.

The dependence on temperature of the degree of disorder in PbS as measured by the products $[V_S^x][V_{Pb}^x]$, $[V_S^{\cdot}][V_{Pb}^{\cdot}]$, and $n \cdot p$ is easily obtained. Each of these products is a function of T of the form $\exp(-A/kT)$. Thus, as the temperature decreases, the degree of disorder decreases, and at $T = 0\,°K$, if equilibrium could be maintained, the disorder would vanish.

The stoichiometry of PbS can be determined from the solutions for the three regions of sulfur pressure. In region I, the ratio of $[V_S^{\cdot}]/[V_S^{x}] = (K_D/[V_S^{x}])^{\frac{1}{2}}$, with $K_D = N_c \exp(-E_D/kT)$. From Hall measurements (Chapter 4) at room temperature, Bloem determined that both E_D and E_A were about 0.01 eV. At 1000 °K, $\exp(-E_D/kT) = 0.89$ and $N_c = 9.60 \times 10^{14} T^{\frac{3}{2}} = 0.3 \times 10^{20}\ cm^{-3}$ (taking the effective electron mass in PbS equal to 0.34 times the free electron mass). This gives a value of $K_D = 0.27 \times 10^{20}\ cm^{-3}$, so that $[V_S^{\cdot}] > [V_S^{x}]$ unless $[V_S^{x}]$ becomes larger than $0.27 \times 10^{20}\ cm^{-3}$. For PbS, $[L] = 1.9 \times 10^{22}\ cm^{-3}$, and $[V_S^{x}]/[L]$ would have to be larger than 1.4×10^{-3}. The vacancy concentration in PbS is much smaller than this, so that the sulfur vacancies essentially are all present as $[V_S^{\cdot}]$ (i. e., the S vacancy, which acts as a donor defect, is ionized). Since $[V_S^{\cdot}] >> [V_S^{x}]$ in region I, it remains so in all the regions. In the same way the Pb vacancies, which act as acceptors, are ionized and are almost entirely present as [Pb']. Even at room temperature, $N_c = 0.5 \times 10^{19}$, and the vacancies are still mostly ionized in all ranges. From Eq. (52), the stoichiometric deviation Δ is therefore given by

$$\Delta = \frac{1.9 \times 10^{22} - [V_S^{\cdot}]}{1.9 \times 10^{22} - [V'_{Pb}]} - 1 \approx \frac{[V'_{Pb}] - [V_S^{\cdot}]}{1.9 \times 10^{22}} \tag{114}$$

since $[V'_{Pb}] << [L]$. In region I, where $[V_S^{\cdot}] >> [V'_{Pb}]$, $\Delta = -[V_S^{\cdot}]/1.9 \times 10^{22}$, and from Eqs. (109) we find that

$$\Delta = -K_D^{\frac{1}{2}} K_g^{\frac{1}{2}} p_{S_2}^{-\frac{1}{4}}/1.9 \times 10^{22} \quad . \tag{115}$$

In region III, where $[V'_{Pb}] >> [V_S^{\cdot}]$, $\Delta = [V'_{Pb}]/1.9 \times 10^{22}$, and from Eqs. (108) we find that

$$\Delta = (K_A K_S)^{\frac{1}{2}} K_g^{-\frac{1}{2}} p_{S_2}^{\frac{1}{4}} \quad . \tag{116}$$

Thus, in region I, PbS is rich in Pb, and in region III it is rich in S. There will be a single value of p_{S_2} in region II where $[V_S^{\cdot}] = [V'_{Pb}]$ exactly. From the charge neutrality relation, n = p exactly at this point. Using the set of equations (106), we find that this value of p_{S_2} is given by

$$p_{S_2} = K_D^2 K_g^2 (K_S' K_i)^{-1} \quad . \tag{117}$$

From this result we find that the exact stoichiometric composition PbS comes in the <u>midpoint</u> of region II (Fig. 12).* To the left and right of the midpoint, PbS is, respectively, rich in Pb and S, and equations analogous to Eqs. (115) and (116) can be easily obtained.

In conclusion, the stoichiometry of a binary compound like PbS can be controlled by fixing the temperature T and the pressure of one of its components in the gas phase. In the next section we shall describe the measurements of the electron and hole concentrations in PbS. From these measurements, estimates can be made of the values of the various equilibrium constants which characterize the defect equilibria in PbS.

E. Determination of Equilibrium Constants

The results obtained in the previous section describe the state of PbS at the high temperatures required to bring the solid into a state of equilibrium. The results can, therefore, be verified only by measurements carried out at these temperatures. Most usually, however, experimental measurements are carried out at room or lower temperatures after the solid has been quenched as rapidly as possible from the high temperature. As already described in Section IV. A, the simple assumption is usually made that the total number of native and impurity defects are frozen in at their high temperature values, due to the fact that these can only be created or destroyed by diffusion processes which are strongly temperature dependent. Electrons and holes, however, have

---

*Note that when $[V_S^{\bullet}] = [V_{Pb}']$, each is equal to $(K_S')^{\frac{1}{2}}$, whereas when $[V_{Pb}^{x}] = [V_S^{x}]$, each is equal to $(K_S)^{\frac{1}{2}}$. The latter condition corresponds to a sulfur pressure given by $K_g^2 K_S^{-1}$. This treatment only holds for the case in which the vacancies are largely ionized.

high mobilities, so that they will redistribute themselves over all of the available states in accord with the low temperature equilibria. For example, if there are native defects which can behave as donors and acceptors, these will be ionized at the high temperatures, giving rise to free electrons and holes. At sufficiently low temperatures, these defects can capture the charge carrier. In this way, not only is the concentration of charge carriers decreased, but there can be a change in the effective charge states of the defects. Thus, although one assumes that the total concentration of defects remains constant upon quenching, the ratios of the concentrations of defects in various charge states may change markedly.

In PbS, as we have said, even at room temperature we can consider all the vacancies as ionized, due to the small values of E_A and E_D. Thus, at room temperature, we still have to consider only the defects $V_S^{\bullet}$, V_{Pb}', n, and p. The state of PbS at room temperature can be determined by an examination of the neutrality condition:

$$[V_S^{\bullet}] + p = [V_{Pb}'] + n \quad . \tag{118}$$

In region I (see Fig. 12) we take $[V_S^{\bullet}]$ at room temperature to be equal to its high temperature value, and the neutrality condition requires that, to a very good approximation, $[V_S^{\bullet}] = n$. The same is true in region III, where at room temperature $[V_{Pb}'] = p$ with $[V_{Pb}']$ taken equal to its high temperature value. In region II, however, we do not have the neutrality condition $n = p$ as at high temperatures. At the midpoint of region II, $[V_S^{\bullet}] = [V_{Pb}']$ exactly at room temperature, so that Eq. (118) requires that $n = p = K_i^{\frac{1}{2}}$. Since K_i, being equal to $N_c N_v \exp(-E_G/kT)$, is very much smaller at room temperature than at the high temperatures, n and p will have low values. To the left of the midpoint (i. e., toward lower values of p_{S_2}), the approximate room temperature neutrality condition will be $[V_S^{\bullet}] = n$, and to the right of the midpoint it will be $[V_{Pb}'] = p$. The room temperature concentration of the charged defects plotted against the values of p_{S_2} at which the PbS was equilibrated at some high temperature T is shown in Fig. 13.

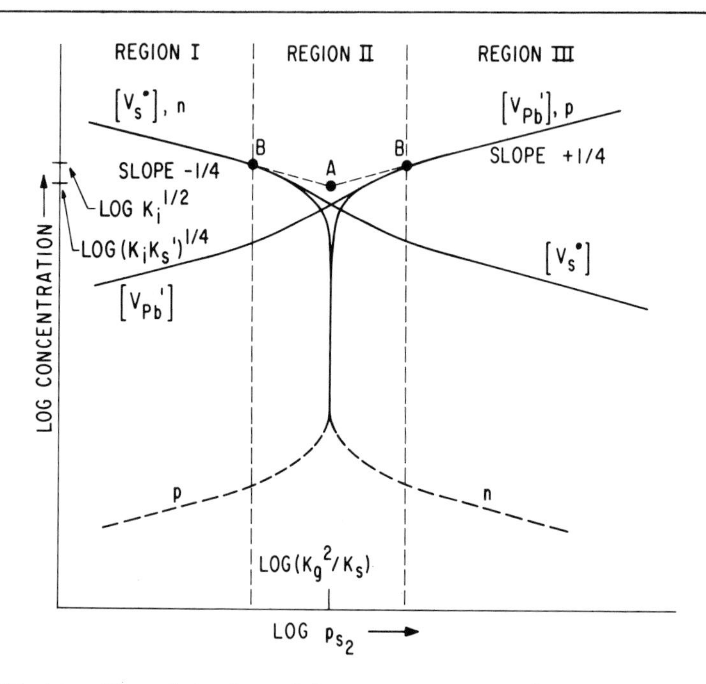

FIG. 13. Plot of the log of the concentration of the charged defects in PbS against log p_{S_2} after quenching to room temperature.

Since Bloem obtained the concentrations of electrons and holes at room temperature from Hall measurements (Chapter 4), these could be compared directly with the values predicted by theory. The results of some of his measurements at several temperatures are given in Fig. 14.

The resemblance between the experimental results and the predicted concentrations of e' and $h^{\cdot}$ is obvious. The measurements can be used to evaluate the equilibrium constants K_D, K_A, K_i, K_g, and K_S or K_S'. In the first place, the electrical measurements at various temperatures from room temperature down give values of E_D and E_A equal to about 0.01 eV. From these values, K_A and K_D can be evaluated at the temperatures 1000, 1100, and 1200°K, taking the effective masses of electrons and holes in PbS to be 0.34 times the free electron mass and assuming that E_A and E_D do not change appreciably with temperature. We get

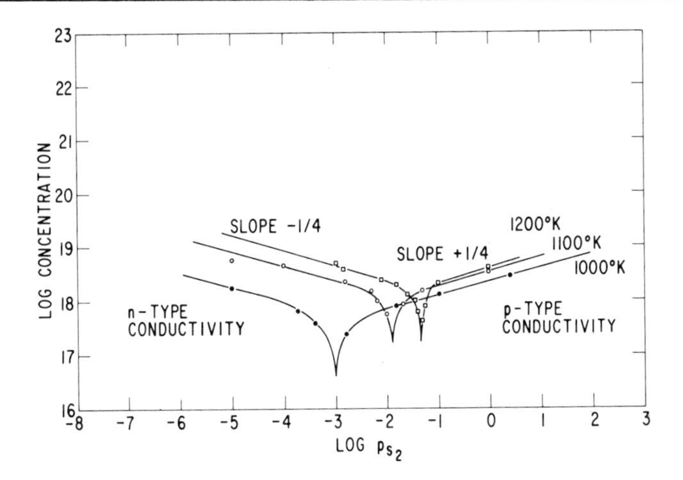

FIG. 14. Some experimental values of log n and log p in PbS measured at room temperature after quenching from 1000, 1100, and 1200°K. The samples were equilibrated at these temperatures under various pressures of sulfur [51].

$$K_A = K_D = 9.60 \times 10^{14}\, T^{\frac{3}{2}} \exp(-0.01/8.62 \times 10^{-5}\, T) \quad . \tag{119}$$

In Table 4 are listed the calculated values of K_A and K_D.

Referring to Fig. 13, we see that the n and p lines with slope = 1/4 (i.e., regions I and III) would intercept at the midpoint of region II (point A). The intercept would occur at a concentration value of $(K_i K_S')^{\frac{1}{4}}$

TABLE 4

Calculated Values of K_A (= K_D) for PbS[a] [51]

Temperature (°K)	$N_V = N_C$ (cm^3)	$K_A = K_D$ (cm^{-3})
1000	0.30×10^{20}	0.27×10^{20}
1100	0.35×10^{20}	0.31×10^{20}
1200	0.40×10^{20}	0.36×10^{20}

[a] $E_D = E_A$ is taken as 0.01 eV, and the electron and hole masses are taken equal to 0.34 times the free electron mass.

and a sulfur pressure value of K_g^2/K_S. These are deduced very simply by setting n in region I equal to p in region III. Furthermore, one can get an estimate of $K_i^{\frac{1}{2}}$ from the concentration values at the point where n or p begin to change slope from the value of 1/4 (point B). In this way, all the equilibrium constants may be evaluated for the different temperatures. The appropriate data and the calculated values of the constants are tabulated in Table 5. From the slopes of plots of log K vs 1/T, the values of the enthalpy changes for the reactions described by Eqs. (93), (95), and (105) can be determined. Thus, for K'_S we find

$$K'_S = 10^{43.4} \exp(-1.75/kT) \quad , \tag{120}$$

with an enthalpy change of 1.75 eV. K'_S is given by [see Eq. (107)]

$$K'_S = [L]^2 \exp(-\Delta\bar{G}^0/kT) = [L]^2 \exp[-(\Delta\bar{H} - T\,\Delta\bar{S}_\nu)/kT] \quad . \tag{121}$$

[L] for PbS is 1.9×10^{22}, so that $[L]^2 = 3.6 \times 10^{44} = 10^{44.6}$. This is in good agreement with the preexponential factor given in Eq. (120) and indicates that $T\,\Delta\bar{S}_\nu \approx -0.2$ eV at 1000°K. Schottky disorder is prominent in PbS because of the rather low value of 1.75 eV for $\Delta\bar{H}$, the change in enthalpy upon creation of a pair of charged Schottky defects. At the melting point (1400°K) of PbS, a vacancy concentration of 1.9×10^{19} per cm^3 is expected, and Bloem has reported this to be approximately the correct value.

The control of the native defect and charge carrier concentrations in PbS by control of the temperature and sulfur pressure is typical of the behavior of all binary compounds. Further examples and many more details can be found in the book by Kröger.

F. Impurities and the Mass Action Law

Impurity defects in solids can be treated in a manner entirely analogous to that used for native defects. Two examples will be given.

In Section II.G the role of some trivalent impurities in CdF_2 was discussed. It was pointed out that these impurities were charge com-

TABLE 5

Calculated Values of K_i, K'_S, and K_g for PbS

Temperature (°K)	K_i (cm^{-6})	K'_S (cm^{-6})	K_g (cm^{-3})
1000	1.0×10^{36}	4.0×10^{34}	2.3×10^{14}
1100	3.2×10^{36}	3.2×10^{35}	3.6×10^{15}
1200	6.3×10^{36}	1.0×10^{36}	1.5×10^{16}

pensated by interstitial fluoride ions (F'_i). After heating in Cd vapor at 500°C the crystals became colored and electrically conducting due to the presence of conduction band electrons e'. The proposed mechanism for the conversion of rare earth doped CdF_2 from an insulator to a semiconductor was given in Eq. (26). At 500°C the impurities which behave as donors are entirely ionized and present as $M^{\cdot}_{Cd}$ defects (M = rare earth, Y, In). We can write down a series of mass action equations for the various existing equilibria when CdF_2 containing some fixed concentration of M is annealed at a temperature T in Cd vapor at a cadmium pressure P_{Cd}:

$$0 \rightleftarrows F'_i + V^{\cdot}_F \quad \text{(assuming Frenkel disorder)} \quad ,$$

$$[F'_i][V^{\cdot}_F] = K_F \quad , \tag{122}$$

$$0 \rightleftarrows e' + h^{\cdot} \quad ,$$

$$n \cdot p = K_i \quad , \tag{123}$$

$$F'_i + \tfrac{1}{2}\,\text{Cd(vapor)} \rightleftarrows \tfrac{1}{2}\,Cd^{x}_{Cd} + F^{x}_F + e' \quad ,$$

$$n/p^{\frac{1}{2}}_{Cd}[F'_i] = K_g \quad . \tag{124}$$

The charge neutrality relation is

$$[M^{\cdot}_{Cd}] + [V^{\cdot}_{F}] + p = n + [F'_{i}] \quad . \tag{125}$$

We note that in Eq. (124) the quantities $[Cd^{x}_{Cd}]$ and $[F^{x}_{F}]$ do not appear since at low concentrations of rare earth impurities these remain essentially constant. The quantity $[Cd^{x}_{Cd}] = [L_{Cd}] - [M^{\cdot}_{Cd}] \approx [L_{Cd}]$ and $[F^{x}_{F}] = [L_{F}] = 2[L_{Cd}]$ and $[L_{Cd}]$ is a constant [see footnote accompanying Eq. (104)]. The range of interest to us in the charge neutrality relation is given by

$$[M^{\cdot}_{Cd}] = n + [F'_{i}] \quad . \tag{126}$$

We want to evaluate n since, as we shall see, this quantity can be determined from experiment. Solving for n directly, Eqs. (124) and (126) yield

$$n/([M^{\cdot}_{Cd}] - n) = K_{g}\, p_{Cd}^{\frac{1}{2}} \quad . \tag{127}$$

From the equilibrium represented by Eq. (124) we see that the concentration of "excess" Cd introduced into the crystal during anneal is just equal to n/2. Representing the "excess" Cd concentration by $[Cd_{ex}]$, we get

$$R/(1 - R) = K_{g}\, p_{Cd}^{\frac{1}{2}} \quad , \tag{128}$$

where $R = 2[Cd_{ex}]/[M^{\cdot}_{Cd}]$. The importance of this result is that we can determine R experimentally and thus compare the results with that predicted by Eq. (128). By reheating the colored, conducting crystals in a vacuum, the reaction represented by Eq. (124) is reversed and the "excess" Cd deposited on the walls of the vacuum system. This can be recovered and the Cd determined by microanalytical methods.

In Fig. 15 the experimental values of $\log[R/(1 - R)]$ are plotted against $\log p_{Cd}$ for a series of runs at 500°C using CdF_2 containing $[Y^{\cdot}_{Cd}] = 2.7 \times 10^{19}\ cm^{-3}$ [23]. It can be seen that the points fall on a straight line of slope equal to $\frac{1}{2}$, as required by Eq. (128). From these results the calculated value of K_{g} at 500°C is 4.8.

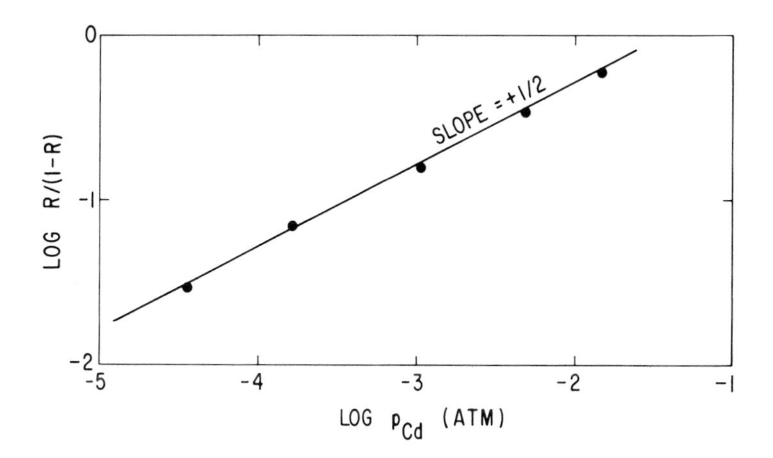

FIG. 15. Values of log[R/(1 - R)] (see text) vs log p_{Cd} for CdF_2 containing 2.7×10^{19} cm^{-3} Y impurities and annealed at 500°C under various cadmium pressures [23].

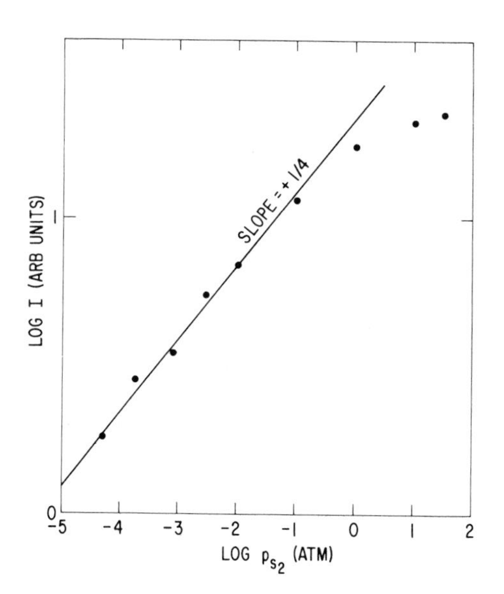

FIG. 16. Values of log I (I is the intensity of the infrared emission in ZnS doped with copper) vs p_{S_2} for ZnS containing 1.3×10^{18} cm^{-3} Cu impurities and annealed at 940°C under various sulfur pressures [53].

As a second example, we shall take ZnS doped with a small amount of Cu. When illuminated by infrared radiation in the range 0.7 to 1.0 μ, at liquid nitrogen temperatures, these materials emit fluorescent radiation in bands peaking at 1.67 and 1.55 μ. After these solids are annealed under various pressures of sulfur and then quenched rapidly to room temperature, marked variations in the intensity I of the infrared fluorescence are observed [53]. In Fig. 16 is shown a typical plot of log I vs log p_{S_2} for ZnS containing a total copper concentration $[Cu_T]$ of 1.3×10^{18} cm^{-3} and annealed at 940°C. It can be seen that at low p_{S_2} (< 1 atm), I varies as $p_{S_2}^{\frac{1}{4}}$, whereas at higher p_{S_2}, I tends to approach a constant value. A model had been proposed in which the fluorescence was due to internal transitions between states associated with the presence of Cu^{2+} ions ($3d^9$ configuration) in ZnS. In the effective charge symbolism Cu^{2+} ions at Zn sites in zinc sulfide are written as Cu_{Zn}^{x}. The problem then is to determine how $[Cu_{Zn}^{x}]$ varies with S_2 pressure. If we take as the important defects Cu_{Zn}^{x}, Cu_{Zn}', e', $h^{\cdot}$, V_S^{x}, $V_S^{\cdot}$, $V_S^{\cdot\cdot}$, V_{Zn}^{x}, V_{Zn}', V_{Zn}'' we can write the appropriate equilibria:

$$Cu_{Zn}^{x} \rightleftarrows Cu_{Zn}' + h^{\cdot} , \quad [Cu_{Zn}']p/[Cu_{Zn}^{x}] = K_c , \tag{129}$$

$$0 \rightleftarrows e' + h^{\cdot} , \quad np = K_i , \tag{130}$$

$$S_S^{x} \rightleftarrows V_S^{x} + \tfrac{1}{2}S_2(g) , \quad [V_S^{x}]p_{S_2}^{\frac{1}{2}} = K_g , \tag{131}$$

$$V_S^{x} \rightleftarrows V_S^{\cdot} + e' , \quad n[V_S^{\cdot}]/[V_S^{x}] = K_D , \tag{132}$$

$$V_S^{\cdot} \rightleftarrows V_S^{\cdot\cdot} + e' , \quad [V_S^{\cdot\cdot}]n/[V_S^{\cdot}] = K_D' , \tag{133}$$

$$0 \rightleftarrows V_{Zn}^{x} + V_S^{x} \text{ (Schottky defects)} , \quad [V_{Zn}^{x}][V_S^{x}] = K_S , \tag{134}$$

$$V_{Zn}^{x} \rightleftarrows V_{Zn}' + h' , \quad [V_{Zn}']h^{\cdot}/[V_{Zn}^{x}] = K_A , \tag{135}$$

$$V_{Zn}' \rightleftarrows V_{Zn}'' + h^{\cdot} , \quad [V_{Zn}'']h^{\cdot}/[V_{Zn}'] = K_A' . \tag{136}$$

The charge neutrality equation is given by

$$[Cu'_{Zn}] + n + [V'_{Zn}] + 2[V''_{Zn}] = p + [V^{\bullet}_S] + 2[V^{\bullet\bullet}_S] \quad . \tag{137}$$

At the relatively high Cu concentrations and high temperatures, we shall take, as the region to investigate, the one in which

$$[Cu'_{Zn}] >> n + [V'_{Zn}] + 2[V''_{Zn}] \text{ and } 2[V^{\bullet\bullet}_S] >> p + [V^{\bullet}_S] \quad . \tag{138}$$

Thus, the approximate charge neutrality equation becomes

$$[Cu'_{Zn}] = 2[V^{\bullet\bullet}_S] \quad . \tag{139}$$

If we combine Eqs. (129) to (133), we can obtain

$$\frac{p_{S_2}^{\frac{1}{2}} [V^{\bullet\bullet}_S][Cu'_{Zn}]^2}{[Cu^x_{Zn}]^2} = \frac{K_g K_D K_c^2}{K_i^2} = K \quad . \tag{140}$$

This is the equilibrium expression for the chemical reaction appropriate to this region under investigation:

$$2Cu^x_{Zn} = 2\,Cu'_{Zn} + V^{\bullet\bullet}_S + \tfrac{1}{2} S_2(\text{gas}) \quad . \tag{141}$$

After inserting Eq. (139) in (140), we find

$$p_{S_2}^{\frac{1}{2}} [Cu'_{Zn}]^3/[Cu^x_{Zn}]^2 = 2K \quad . \tag{142}$$

In addition to this final equation, we must have a mass balance condition on the copper impurity:

$$[Cu'_{Zn}] + [Cu^x_{Zn}] = [Cu_{total}] = \text{constant} \quad . \tag{143}$$

At low pressure of S_2, the reaction in Eq. (141) is displaced to the right, so that $[Cu'_{Zn}] >> [Cu^x_{Zn}]$ and we take in Eq. (143) that

$$[Cu'_{Zn}] = \text{constant} = [Cu_{total}] \quad . \tag{144}$$

From Eq. (142) we therefore see that, at low p_{S_2},

$$[Cu^x_{Zn}] \propto p_{S_2}^{\frac{1}{4}} \quad . \tag{145}$$

At high p_{S_2} values, the reaction is displaced to the left and $[Cu^x_{Zn}] >> [Cu'_{Zn}]$. Therefore, we can write

$$[Cu^x_{Zn}] = \text{constant} = [Cu_{total}] \quad . \tag{146}$$

Since the concentration of charge carriers is small in these regions of p_{S_2}, there will be no change in the concentration of defects on rapid quenching and we expect I to be proportional to $[Cu^x_{Zn}]$ as given by Eqs. (145) and (146). The results shown in Fig. 16 can therefore be explained by this model.

Many other informative examples of defect equilibria involving impurities and native defects can be found in the books mentioned in the Introduction. The book by Kröger, in particular, contains a great deal of detailed information on this subject.

G. Stability Limits of Nonstoichiometric Compounds

The group of nonstoichiometric compounds of interest to us in the previous sections have, as already explained, narrow limits of stability. Using CdTe as an example, we see from Fig. 9 that on either side of the solid phase region is a two-phase region. In these, the solid CdTe can coexist in equilibrium with a liquid phase of specified composition. In Fig. 9 the liquidus giving the composition of the liquid phase is not shown, due to the vastly expanded composition scale. Solid CdTe can be equilibrated at a temperature T (in a system like that shown in Fig. 10) with a liquid close to the proper liquidus composition at the same temperature T. Contact between the two condensed phases is made via the gas phase. According to the Phase Rule, such a three-phase, two-component system has only one degree of freedom, the temperature. After equilibrium is established, in which some Te or Cd may have been transferred between the solid and liquid phases, the composition of the solid will correspond to the stability limits of CdTe. There are obviously two stability limits. One composition limit represents CdTe in equilibrium with a liquid rich in Cd and the other with a liquid rich in Te. The solid will then contain those concentrations of native defects corresponding to the limits of stability of solid CdTe at the temperature T.

V. COMPOUNDS WITH LARGE DEVIATIONS FROM STOICHIOMETRY

As described in Section III. A, many compounds of the rare earths and of the transition elements can exist as homogeneous phases over rather large composition ranges. The large range of composition results from the presence of native defects in the solid. Since the composition range is large, the concentration of defects is high, and this is true, in many cases, even for the stoichiometric compositions. Thus titanium monoxide with the composition TiO, quenched from a high temperature, has the rock-salt structure with one Ti and one O missing in every seven. This same phase can exist over a composition range TiO_x $(0.7 < x < 1.2)$ with either an excess of Ti or of O vacancies in the structure [54].

From a chemical viewpoint one frequently speaks of the presence of metal atoms in more than one oxidation state. For example, a composition such as $Fe_{0.9}S$ has been written as $Fe^{2+}_{0.7}Fe^{3+}_{0.2}S^{2-}$. Electrical conduction, as described previously (Section II. G), is then considered to arise from an electron transfer, under the action of an applied field, from an Fe^{2+} ion to a neighboring Fe^{3+} ion, thus converting the latter to Fe^{2+}. The conduction process can then continue. This ionic picture may be too simple, however. In the NiAs structure of this solid, there are short metal-to-metal distances along the trigonal axis. It is therefore possible that the electrical conduction arises from electron motion in a partially filled 3d band arising from the overlap of the 3d orbitals of the metal atoms. Since these types of problems are discussed thoroughly in Chapter 3 , we need not concern ourselves with them here. For our present purposes, we need only consider that $Fe_{1-\delta}S$ has vacancies in the iron sublattice. The presence of iron vacancies has been shown by experimental studies involving the measurements of the bulk density D and the unit cell volume V_0. The mass per unit cell is given by the product DV_0, and it was found that this corresponded, on th average, to δ iron vacancies per unit cell. If Schottky disorder is predominant in the $Fe_{1-\delta}S$ phase, there must be some finite, although perhaps small, concentration of sulfur vacancies as well.

At the high defect concentrations present in the compounds under discussion, we can no longer make the assumption, as we have in the previous sections, that the defects do not interact with each other. This noninteraction at low defect concentrations is what enabled us to write down quantities such as $\Delta\bar{H}$ and $\Delta\bar{S}_\nu$ in Eq. (74) or (94), which were independent of composition. At high defect concentrations due to interaction between defects, the change in enthalpy or vibrational entropy upon the introduction of a single defect depends on the concentration of defects already present. A thermodynamic treatment of such systems has been given by Anderson [55]. In his treatment, Anderson used the approximation that the total interaction energy between defects could be represented as a sum of the contributions of those pairs of like defects that would occur as nearest neighbors in a completely random distribution. This assumption of a random distribution cannot, of course be completely accurate since the number of nearest neighbor defects in any given lattice will depend not only on the concentration of such defects (this alone yields a random distribution) but also on the value of E_i/kT where E_i is the total interaction energy. In fact, when the total interaction energy is much larger than the thermal energy, it can lead to an ordering of defects into superlattice structures and to the appearance of new phases of near stoichiometric compositions having ordered arrays of defects. Some of the structural consequences of this are discussed below. We will not describe Anderson's treatment becaus the many approximations may well limit its usefulness (see also Ref. [56]).

Compounds with large stoichiometric deviations can frequently be quenched from high temperatures (where $kT > E_i$) to yield disordered structures containing defects randomly distributed over the lattice sites. By annealing at lower temperatures (where $E_i > kT$), ordered phases can appear due to an ordering of the defects, such as vacancies, on the lattice sites. An excellent discussion of many such studies of order-disorder phenomena and the appearance of ordered phases can be found in Advances in Chemistry Series 39, Nonstoichiometric Compounds, American Chemical Society, Washington, D. C., 1963. We will discuss one interesting system which will illustrate some of the ideas concerning defect interactions and ordering.

There are a number of transition metal chalcogenides with compositions near MX which crystallize in the nickel arsenide structure. Among these are FeS, VSe, CoTe, NaTe, CrS, CrSe, and CrTe. There are also a number of compounds near the composition MX_2 which crystallize in the CdI_2 structure. Among these are VSe_2, $CoTe_2$, and $NiTe_2$. Both the NiAs and CdI_2 structures are based on a hexagonal unit cell. In the NiAs structure there are metal atoms at (0, 0, 0), (1, 0, 0), (0, 1, 0), and (1, 1, 0) and similar layers at $z = 1/2$ and 1 (the hexagonal c axis corresponds to the z direction). The two nonmetal atoms are located at (2/3, 2/3, 1/4) and (1/3, 1/3, 3/4). The CdI_2 structure is derived from the NiAs structure by the removal of the layer of metal atoms at $z = 1/2$. In the nonstoichiometric compounds of composition (metal deficiency) $M_{1-\delta}X$, crystallizing in the NiAs structure, x-ray and density studies have shown that vacancies are present in the metal sublattice. In the nonstoichiometric compounds of composition $M_{1-\delta}X_2$ (metal excess), crystallizing in the CdI_2 structure, interstitial metal atoms are present at $z = 1/2$. It is possible to pass continuously from the NiAs to the CdI_2 structure by forming, first, metal vacancies in the MX compound randomly distributed over all the metal sites. As the concentration of metal vacancies becomes larger, the increasing interaction energy can lead to an ordering of these vacancies. They may tend to segregate in every second metal layer, leading, at the limit, to the compound $M_{\frac{1}{2}}X$ or MX_2 with the CdI_2 structure. By this process one can pass continuously from one phase with the NiAs structure to another phase with the CdI_2 structure without a first-order phase transition. The composition of all the solids may be represented by $M_{1-\delta}X$ with $\frac{1}{2} \geq \delta \geq 0$. Behavior of this type has indeed been found in the systems CoTe-$CoTe_2$ [57] and NiTe-$NiTe_2$ [58] quenched from high temperatures.

An x-ray study of the Cr-S system [59] indicates the behavior of a system annealed at lower temperatures, thus allowing the defects to order. This ordering leads to low symmetry structures with small homogeneity ranges, related to the NiAs-CdI_2 structures. The compounds found were CrS, Cr_7S_8, Cr_5S_6, Cr_3S_4, and Cr_2S_3, and at least three of these (Cr_7S_8, Cr_3S_4, and Cr_2S_3) exist over a range of composition as homogeneous phases. All of the compounds (except CrS) have structures related to those intermediate between those of NiAs and CdI_2. Cr_7S_8 (i.e., $Cr_{0.875}S$) and compositions lying within the single-phase region around Cr_7S_8 have a random distribution of metal vacancies in every second layer of the NiAs structure. At high temperatures, the range of homogeneity about the Cr_7S_8 composition is much broader, and the defects apparently become completely disordered over all the metal sites of the NiAs structure [60]. The compounds having the compositions Cr_5S_6, Cr_3S_4, and Cr_2S_3 can be considered as having larger metal vacancy concentrations than Cr_7S_8. In these, the vacancies are confined to every second layer of the NiAs structure but are ordered within these layers. This leads to structures that have lower symmetries and larger unit cells than the NiAs or CdI_2 structures but are clearly related to them. Every second metal layer is incompletely filled (i.e., it has metal vacancies) when referred to the NiAs structure or it has extra interstitial metal atoms when referred to the CdI_2 structure. Nonstoichiometry in these lower symmetry phases leading to compositions centered about Cr_2S_3 and Cr_3S_4 arises from adding or removing metal atoms randomly from the ordered array in every second metal layer.

REFERENCES

1. C. Kittel, Introduction to Solid State Physics, Wiley, New York, 1956, pp. 103-156.
2. F. Seitz, J. Chem. Phys., 6, 150 (1938).
3. H. Gründig, Z. Physik, 158, 577 (1960).
4. F.A. Kröger, The Chemistry of Imperfect Crystals, North-Holland Publ., Amsterdam, 1964, pp. 192-210.
5. D.G. Thomas and J.J. Lander, J. Chem. Phys., 25, 577 (1960).
6. J.J. Lander, J. Phys. Chem. Solids, 15, 324 (1960).
7. J.S. Prener and W.W. Piper, "Electronic Properties of Localized Defects in Apatites," 1969, to be published.
8. A.F. Wells, Structural Inorganic Chemistry, Clarendon Press, Oxford, 1962, pp. 29-35.
9. H. Welker, Z. Naturforsch., 7a, 744 (1952).
10. E.M. Rhoderick, J. Phys. Chem. Solids, 8, 498 (1959).
11. C. Kolm, S.A. Kalin, and B.A. Auerbach, Phys. Rev., 108, 965 (1957).
12. D. Curie and J.S. Prener, in Physics and Chemistry of II-VI Compounds (M. Aven and J.S. Prener, eds.), Chap. 9, North-Holland Publ., Amsterdam, 1967, pp. 445-452.
13. R.W. Ure, Jr., J. Chem. Phys., 26, 1363 (1957).
14. J.D. Kingsley and J.S. Prener, Phys. Rev., 126, 458 (1962).
15. D. de Nobel, Philips Res. Repts., 14, 361, 430 (1959).
16. C.J. Delbec, W. Hayes, M.C.M. O'Brien, and P.H. Yuster, Proc. Roy. Soc. (London), A271, 243 (1963).
17. J. Bloem and F.A. Kröger, Philips Res. Repts., 12, 303 (1957).
18. J. Bloem, Philips Res. Repts., 13, 167 (1958).
19. A.N. Murin, S.N. Banasevich, and Yu. S. Grushko, Soviet Phys.-Solid State, 3, 1762 (1962).
20. B.G. Lure'e, A.N. Murin, and R.F. Brigevich, Soviet Phys.-Solid State, 4, 1432 (1963).

21. J. Teltow, Ann. Physik, 5, 63, 71 (1941).
22. H. Layer and L. Slifkin, J. Phys. Chem., 66, 2396 (1962).
23. J.S. Prener and J.D. Kingsley, J. Chem. Phys., 38, 667 (1963).
24. W.W. Piper, L.C. Kravitz, and R.K. Swank, Phys. Rev., 138, A1802 (1965).
25. J.S. Prener and W.W. Piper, J. Phys. Chem. Solids, 30, 1465 (1969).
26. A.B. Lidiard, Phys. Rev., 94, 29 (1954).
27. H. Reiss, C.S. Fuller, and F.J. Morin, Bell Syst. Tech. J., 35, 535 (1956).
28. J.S. Prener, J. Chem. Phys., 25, 1294 (1956).
29. F.A. Kröger, The Chemistry of Imperfect Crystals, North-Holland Publ., Amsterdam, 1964, pp. 257-312.
30. A.B. Lidiard, Phys. Rev., 112, 54 (1959).
31. H.A. Bethe, NDRC Rept. 43-12 (Publications Board, U.S. Department of Commerce, Dec. 1942).
32. C.A. Coulson, Valence, Oxford Univ. Press, London and New York, 1952, p. 263.
33. M. Rubenstein and E. Banks, J. Electrochem. Soc., 106, 404 (1959).
34. J.D. Kingsley and J.S. Prener, Phys. Rev. Lett., 8, 315 (1962).
35. P.F. Weller, Inorg. Chem., 4, 1545 (1965); 5, 736, 739 (1966).
36. J. Lambe and C. Kikuchi, Phys. Rev., 118, 71 (1960).
37. F.A. Kröger, Some Aspects of the Luminescence of Solids, Elsevier, Amsterdam, 1948, pp. 57-106.
38. F. Moser, J. Appl. Phys., 33, 343 (1962).
39. J.S. Prener, J. Chem. Phys., 21, 160 (1953).
40. I. Broser, H. Maier, and H.J. Schulz, Phys. Rev., 140, A2135 (1965).
41. D. Curie and J.S. Prener, in Physics and Chemistry of II-VI Compounds (M. Aven and J.S. Prener, eds.), Chap. 9, North-Holland Publ., Amsterdam, 1967, pp. 455-464.

42. D.G. Thomas, M. Gershenzon, and F.A. Trumbore, Phys. Rev., 133, A269 (1964).

43. R. Schenck and T. Dingmann, Z. Anorg. Chem., 166, 133 (1927).

44. W. Biltz and R. Juza, Z. Anorg. Allgem. Chem., 190, 161 (1930).

45. L.S. Darken and R.W. Gurry, J. Am. Chem. Soc., 68, 799 (1946).

46. C. Wagner and W. Schottky, Z. Physik. Chem., B11, 163 (1931).

47. J. Frenkel, Z. Physik, 35, 652 (1926).

48. R. Fowler and R.A. Guggenheim, Statistical Thermodynamics, Cambridge Univ. Press, London and New York, 1949, pp. 88-89.

49. See any textbook on chemical thermodynamics. For example, F.H. MacDougall, Thermodynamics and Chemistry, Wiley, New York, 1939, pp. 136-142.

50. F.H. MacDougall, Thermodynamics and Chemistry, Wiley, New York, 1939, pp. 81-86.

51. J. Bloem, Philips Res. Repts., 11, 273 (1956).

52. F.A. Kröger and H.J. Vink, in Solid State Physics (F. Seitz and D. Turnbull, eds.), Vol. 3, Academic, New York, 1956, pp. 335-342.

53. E.F. Apple and J.S. Prener, J. Phys. Chem. Solids, 13, 81 (1960).

54. P. Ehrlich, Z. Elektrochem., 45, 362 (1939).

55. J.S. Anderson, Proc. Roy. Soc. (London), A185, 69 (1946).

56. J.S. Prener, in Nonstoichiometric Compounds, Advances in Chemistry Series 39, American Chemical Society, Washington, D.C., 1963, paper 15, pp. 174-178.

57. S. Tengner, Z. Anorg. Allgem. Chem., 239, 127 (1938).

58. W. Klemm and N. Fratini, Z. Anorg. Allgem. Chem., 251, 222 (1943).

59. F. Jellinek, Acta Cryst., 10, 620 (1957).

60. H. Haraldsen and F. Mehmed, Z. Anorg. Allgem. Chem., 239, 369 (1938).

Chapter 9

ATOM MOVEMENTS: DIFFUSION*

R. F. Brebrick†

Lincoln Laboratory
Massachusetts Institute of Technology
Lexington, Massachusetts

*This work was sponsored by the Department of the Air Force.

†Present address: Marquette University, Milwaukee, Wisconsin.

I. INTRODUCTION

Diffusion is the process of intermixing on an atomic or molecular scale. It arises because of gradients in the chemical potentials of the system components and continues until these are eliminated. In the absence of external forces, such as an applied electrostatic field or an applied centrifugal field, the final result of the diffusion process is the attainment of a uniform concentration in each phase present.

As a practical example consider the T-x projection of the Pb-Se phase diagram shown in Fig. 1. The single compound formed, PbSe, is a semiconductor whose homogeneity range includes the stoichiometric, 1 to 1, composition but is extremely narrow. The range is sufficiently wide, however, to entail a significant change in electrical properties as it is traversed. Lead selenide can be made Pb-rich and n-type up to about 10^{19} electrons/cm^3 or Se-rich and p-type up to about 10^{19} holes/cm^3, depending upon the temperature. Suppose an initially uniform, Se-rich, 1×10^{18} p-type, single crystal of PbSe is placed into an evacuated, sealed, silica ampoule along with a powdered mixture whose composition is 40 at. % Se and heated to 600 °C as indicated in Fig. 2. According to the phase diagram in Fig. 1, the 40 at. % Se mixture consists of a 4 at. % Se liquid phase and PbSe that is as Pb-rich as possible at 600 °C. The crystal-mixture system will change to attain equilibrium, part of this change involving a transport of material between the mixture and the single crystal through the vapor phase and part of it involving solid state diffusion within the single crystal. If the relative weights of the mixture and single crystal are such that the overall composition lies in the Pb-rich three-phase field, then at equilibrium the PbSe single crystal will be as Pb-rich as possible at 600 °C. The partial pressures of the various selenium species and of Pb(g) will be quickly established by the mixture and the surface of the single crystal will become Pb-saturated. There is then a concentration gradient within the crystal which tends to smooth itself out by the movement of Se from within the

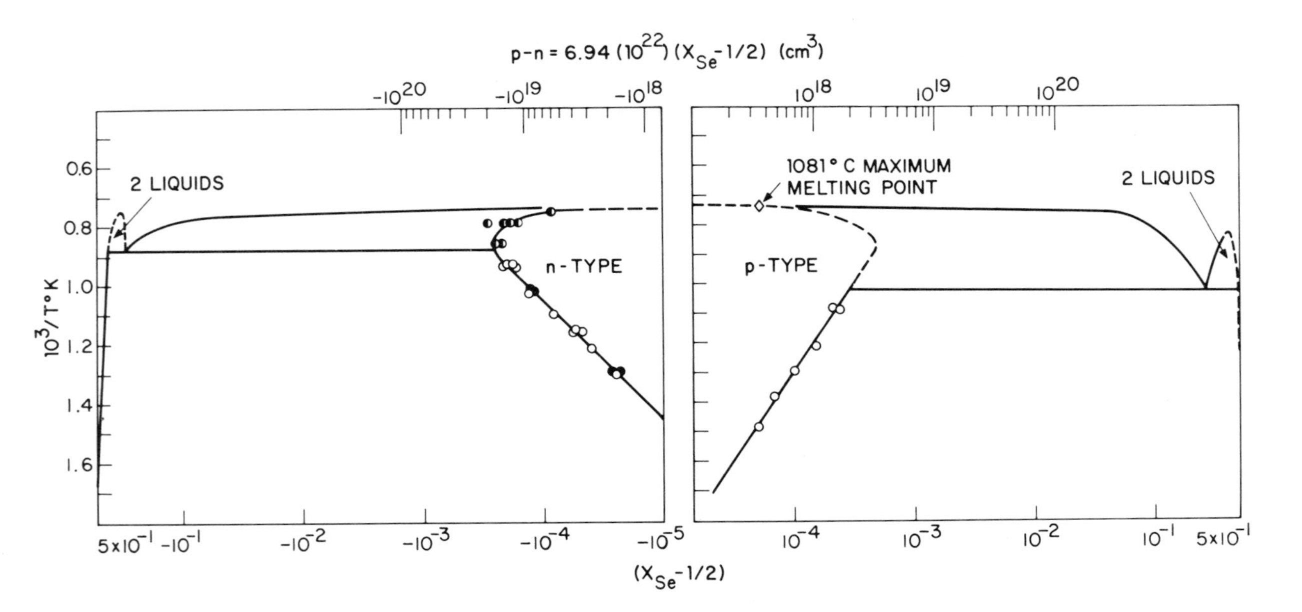

FIG. 1. Temperature-composition projection for the Pb-Se system. To show the homogeneity range for PbSe(c) the composition variable plotted is log ($\frac{1}{2}$ - x_{Se}) when the atom fraction of Se, x_{Se}, is less than $\frac{1}{2}$ and log (x_{Se} - $\frac{1}{2}$) when it is greater. ○ Ref. [26]; ● R.F. Brebrick and E. Gubner, J. Chem. Phys., 36, 170 (1962); ◑ R.F. Brebrick, unpublished; ◊ A.E. Goldberg and G.R. Mitchell, J. Chem. Phys., 22, 220 (1954).

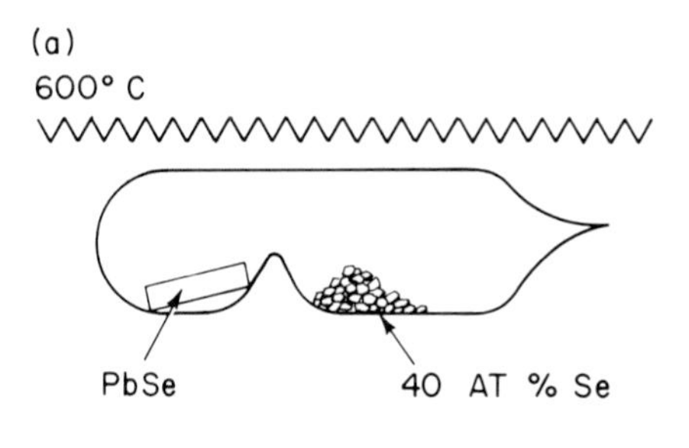

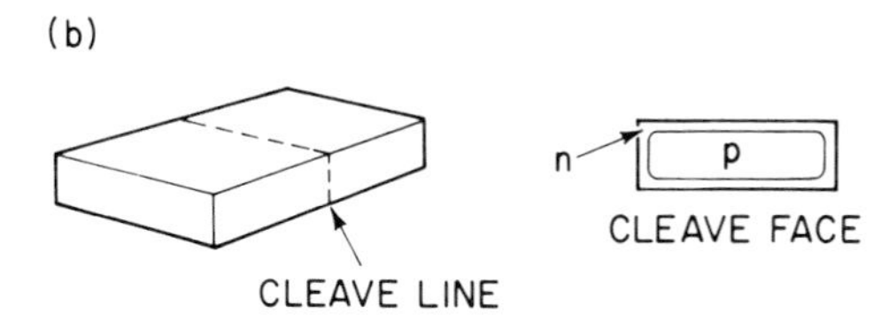

FIG. 2. (a) Experimental arrangement for changing the composition of a PbSe crystal to that corresponding to the Pb-rich homogeneity limit. At 600 °C the powder with an overall composition of 40 at. % Se is partially molten. The critical point is that the average composition of the crystal and powder taken together fall in the Pb-rich three-phase field. (b) Schematic showing the cleave line on an initially p-type PbSe crystal and the contour of the p-n junction on the resulting cleave face after a short anneal under the conditions shown in Fig. 2(a).

crystal to the surface and the movement of Pb from the vapor phase to the crystal surface and into its interior. Since the mixture maintains the crystal surface in a Pb-saturated condition, the crystal will eventually become entirely Pb-saturated.

If the process is halted after 60 min by cooling the tube and contents, a p-n junction dividing the still Se-rich, p-type, interior from the Pb-rich, n-type exterior will be found 0.1 mm in from the crystal surface. In this way, p-n junctions are conveniently made in PbSe, PbS, and PbTe crystals. These are useful for such devices as injection lasers. If the diffusion process is allowed to continue to about 400 hr, a single crystal 2 mm thick will be entirely converted to Pb-rich, 4×10^{18} n-type. In this way, by varying the temperature and also by changing the composition of the mixture to one in the Se-rich three-phase field, uniform PbSe crystals can be made with carrier concentrations over a wide range. It should be emphasized that, under the isothermal conditions imagined, the system will cease to change when the PbSe single crystals become Pb-saturated, for then the chemical potentials of Pb and Se will be the same in the single crystal as they are in the small PbSe crystals and Pb-rich melt of the mixture. The single crystal need not become partially molten as the mixture is.

II. THE ESSENTIAL ROLE OF POINT DEFECTS FOR DIFFUSION IN SOLIDS

In solids, diffusion may occur predominantly along the surface, along grain boundaries, or through the bulk. We shall be primarily concerned with bulk or volume diffusion. Volume diffusion can readily be distinguished from surface diffusion if the diffusing sample can be sectioned, concentration gradients within the bulk indicating the presence of volume diffusion. Volume diffusion becomes increasingly important relative to grain boundary diffusion as the grain size and/or the temperature is increased.

In gaseous diffusion, atoms or molecules travel in straight lines at high speeds between collisions. The collisions tend to randomize the direction of the atom motion and occur after the atoms have traveled an average distance of about 10^{-6} cm (at room temperature and atmospheric pressure). In crystalline solids the atoms, molecules, or ions are more densely packed by a factor of about 10^3. They are trapped in small regions by the potential field of the crystal, vibrating about equilibrium positions with root mean square amplitudes that depend upon the temperature and specific crystal, but which are of the order of 10^{-9} cm. The vibrating atoms exchange energy and occasionally one may obtain a much higher energy than the average, allowing it to move out of its site into an adjacent, unoccupied site. Here it is again trapped in a potential energy well until a subsequent jump. Thus the atom jumps are activated processes and have been analyzed using absolute reaction rate theory. The probability of an atom jump per unit time, w, assuming an available site for the atom to jump into, depends exponentially upon the free energy of activation, g_m:

$$w = \nu \exp(-g_m/kT) \quad , \tag{1}$$

where ν is the vibration frequency in the direction of the jump. Activation energies are typically in the range 30-80 kcal/mole although for some materials with open structures they are as low as 2-4 kcal/mole.

In particular substances interstitial sites may be involved in diffusion. These are sites not occupied in the perfect crystal, but occupied to some, generally small, extent in the real crystal because of thermal disorder or deviations from stoichiometry. The interstitial atom may move directly by jumping into a neighboring interstitial site. It may also move indirectly by knocking an adjacent atom from a normal lattice site into an interstitial site and occupying the vacant site itself. The direct interstitial, indirect collinear interstitial, and indirect noncollinear interstitial mechanisms are illustrated in Fig. 3.

In other substances the interstitial sites may be occupied, but because the atom is too large for the interstitial site or because of more general repulsive forces, the concentration of occupied interstitial sites is negligible even when compared to that, generally small, concentration of normal sites that are unoccupied. These unoccupied normal sites or vacancies are then the predominant atomic point defects. The vacancy may move by having a neighboring atom jump into it. It may also be involved in more complicated arrangements such as vacancy-vacancy pairs or vacancy-impurity atom complexes. In these latter arrangements a pair of point defects are bound together strongly enough that they tend to remain nearest neighbors while diffusing. The vacancy mechanism is illustrated in Fig. 4.

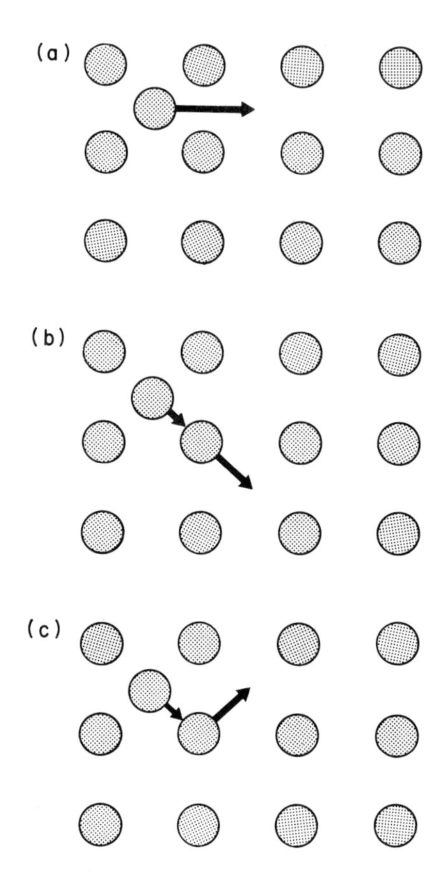

FIG. 3. (a) Direct interstitial mechanism. (b) Indirect collinear interstitial or collinear interstitialcy mechanism. (c) Noncollinear interstitialcy mechanism.

Other mechanisms have also been considered such as the direct place exchange in which two atoms move by squeezing by their neighbors to interchange places. More complicated ring mechanisms involving three, four, or more atoms have also been considered. Currently, the tendency is to consider the diffusion mechanism in most crystalline solids to be one of those associated with vacancies or interstitials.

If, as stated above, volume diffusion in crystalline solids occurs through point defects, then diffusion studies can be used to learn some-

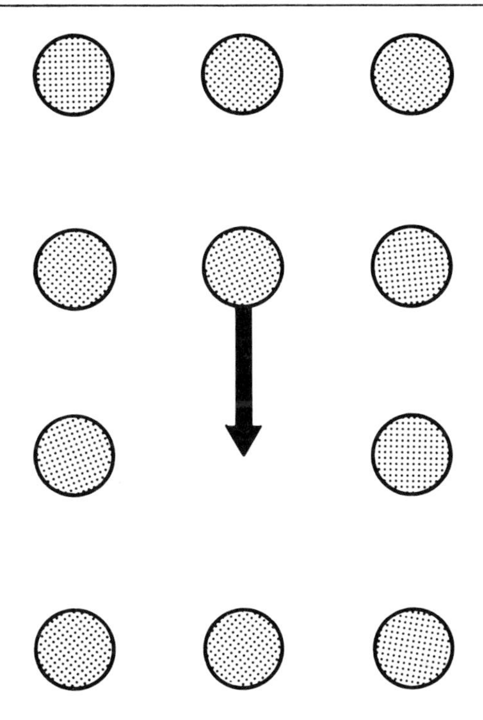

FIG. 4. Vacancy mechanism.

thing about the nature of atomic point defects and how they move, in addition to obtaining the macroscopic rate of matter transport. As discussed in Chapter 8, equilibrium studies can also be used to characterize point defects in a manner generally complementary to diffusion studies and conversely point defects are important in determining the equilibrium partial molar thermodynamic quantities, at least for crystalline compounds.

III. A SIMPLE KINETIC TREATMENT OF DIFFUSION

A good deal of the nature of diffusion can be understood in terms of a simplified picture based upon the assumed independent, random, nature of atomic motion. Suppose a concentration gradient of some

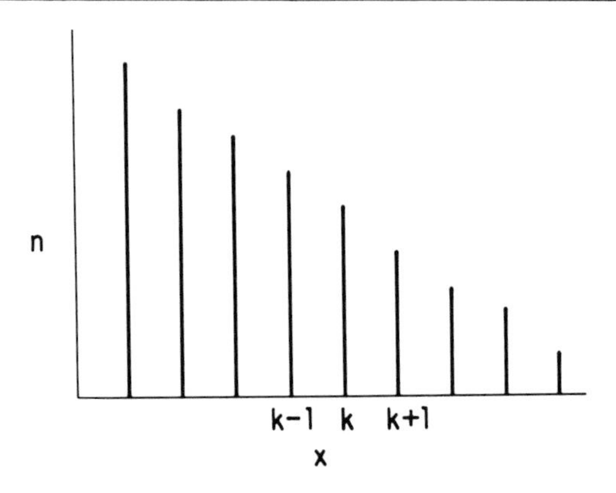

FIG. 5. Number of atoms per cm^2 as a function of distance in a crystal. The vertical lines mark the positions of the atomic planes perpendicular to the concentration gradient.

species is set up in a phase as shown in Fig. 5. Let $k-1$, k, and $k+1$ be consecutive atomic planes with, respectively, n_{k-1}, n_k, and n_{k+1} atoms/unit area. Assume (1) every atom has a probability ν, per unit time, of jumping to the right or left regardless of the concentration or the concentration gradient, (2) the atomic jumps are independent. Then the number of atoms jumping per unit time and area from plane k to plane $k+1$ is given by $\frac{1}{2}\nu n_k$, while the number jumping from $k+1$ to k is given by $\frac{1}{2}\nu n_{k+1}$. The net flux or number of atoms jumping from plane k to plane $k+1$ per unit time and area is then

$$J_{k,k+1} = \frac{\nu}{2}(n_k - n_{k+1}) \tag{2}$$

and the net flux from plane $k-1$ is

$$J_{k-1,k} = \frac{\nu}{2}(n_{k-1} - n_k) \quad . \tag{3}$$

We see the net flow of atoms is in the direction of decreasing concentration and that this flux must act in such a way as to eventually smooth out the concentration gradient. At this point the constant atomic interchange continues but the net flow ceases.

We define a concentration per unit volume as

$$c = n/a_0 \quad , \tag{4}$$

where a_0 is the interplanar spacing. We then expand c_{k+1} in a Taylor series about c_k:

$$c_{k+1} = c_k + \left(\frac{\partial c_k}{\partial x} \right) a_0 + \cdots \quad . \tag{5}$$

Neglecting the higher order terms that are unexpressed in Eq. (5) and substituting into Eq. (2) we obtain

$$J_{k,k+1} = - \frac{\nu a_0^2}{2} \frac{\partial c_k}{\partial x} \quad . \tag{6}$$

Since Eq. (6) holds for any planes k and $k+1$ we can drop the subscripts identifying the planes and rewrite Eq. (6) as

$$J = - \frac{\nu a_0^2}{2} \frac{\partial c}{\partial x} \quad . \tag{7}$$

Equation (7) is Fick's first law for the case of planar diffusion, with the diffusion coefficient given by $\nu a_0^2/2$.

The above treatment is basically more correct for gaseous diffusion than for that in crystalline solids. For gaseous diffusion one would speak of elements of volume rather than atomic planes, mean free paths rather than interplanar spacing, and average velocity rather than jump frequency. These are minor matters, however, and can easily be taken care of. The point is that the basic assumption (1) is generally more appropriate for gases than for crystalline solids. We can do somewhat better in a still simple kinetic treatment. Let us assume the vacancy mechanism of diffusion is predominant and amend assumption (1) to (1'): The probability per unit area and time that an atom in plane k jump to plane $k+1$ is equal to a jump frequency $\nu/2$ times the probability that there be a vacancy in plane $k+1$. We maintain assumption (2), that the atomic jumps are independent. The net flux from plane k to plane $k+1$ per unit area and time is then given by

$$J_{k,k+1} = \frac{\nu z}{2N}(n_k v_{k+1} - n_{k-1} v_k) \quad , \tag{8}$$

where v_{k+1} is the number of vacancies per unit area in plane $k+1$ and N is the total number of sites per unit area in each plane. The vacancy distribution has been assumed to be random so that v_k/N is the probability that a particular site on plane k be a vacant site. The constant z is the number of sites in plane $k+1$ that a particular atom in plane k could reach in one jump. It is therefore a modified coordination number. The average number of vacant sites in plane $k+1$ that can be reached in one jump by an atom in plane k is then zv_{k+1}/N. Changing to concentrations per unit volume as in Eq. (4), expanding c_{k+1} and v_{k+1}/a_0 in Taylor's series about c_k and v_k/a_0, respectively, and substituting into Eq. (8), we have

$$J = \frac{\nu z a_0^2}{2S}\left(c\,\frac{\partial c_v}{\partial x} - c_v\,\frac{\partial c}{\partial x}\right) \quad . \tag{9}$$

The subscript indices identifying the atom planes have been dropped since the equation is equally valid for every atomic plane.

Equation (9) gives the atom flux from any plane to an adjacent plane in terms of the values of the concentrations and gradients at the plane in question. The concentration of vacancies is c_v, that of sites is S. We have already assumed the concentration of sites, S, is a constant. (This is not necessarily accurate in every case, but on the other hand it is not a bad approximation in solids. At any rate it enables us to obtain some basically correct results from a simple treatment.) Thus we have

$$c + c_v = S \quad , \tag{10}$$

$$\frac{\partial c}{\partial x} = -\frac{\partial c_v}{\partial x} \quad . \tag{11}$$

In a metallic alloy or in a slightly impure element it is more appropriate to consider c_v a constant at constant temperature. The assumption entailed in Eq. (10) is appropriate for either component of an ordered binary compound. Our treatment is therefore restricted to this case from this point. Substituting Eqs. (10) and (11) into (9) gives

$$J = -\frac{\nu z a_0^2}{2} \frac{\partial c}{\partial x} \quad . \tag{12}$$

We have the same result as obtained from our earlier treatment and expressed by Eq. (7) except for the inclusion of the modified coordination number z. The derivation has been different, however, the probability of an atom jump depending upon the probability of a vacancy being present within the coordination shell of the atom. Again the proportionality factor between the concentration gradient and the flux is called a diffusion coefficient:

$$D_v = \frac{\nu z a_0^2}{2} \quad . \tag{13}$$

We can call this diffusion coefficient the vacancy diffusion coefficient.

Let us consider the movement of a tracer atom under assumptions (1') and (2) and with the condition that there is no gradient in the overall chemical compositions, i.e., the concentration of vacancies is a constant independent of position, and the sum of the concentrations of tracer and normal atoms is therefore also a constant. Since tracers can and are also used when a composition gradient is present, we shall call this special case, tracer self-diffusion. Then Eq. (9) becomes

$$J^* = \frac{\nu z a_0^2}{2S} \left(c^* \frac{\partial c_v}{\partial x} - c_v \frac{\partial c^*}{\partial x} \right) \quad , \tag{14}$$

where c^* is the concentration of tracer atoms. Since c_v is a constant, Eq. (14) reduces to

$$J^* = -\frac{\nu z a_0^2}{2} \frac{c_v}{S} \frac{\partial c^*}{\partial x} \quad , \tag{15}$$

$$D^* = \frac{\nu z a_0^2}{2} \frac{c_v}{S} = \frac{c_v}{S} D_v \quad . \tag{16}$$

Equation (16) states that, for self-diffusion measured by tracers, the diffusion coefficient is proportional to the fraction of vacancies and is smaller than the vacancy diffusion coefficient by the same fraction. The fraction of vacancies is of the order of 10^{-3} or less except for the special case of defect structures (those in which only a fraction of sites that are crystallographically equivalent, or almost so, are occupied by atoms, e.g., AgI above 147°C). Therefore concentration changes are accomplished much more rapidly in a composition gradient than in tracer diffusion in the absence of an overall composition gradient. A more complete analysis in Section VII yields the same result. Moreover in a binary compound, MN, c_v depends exponentially upon temperature for a fixed partial pressure of one component, so that in view of Eq.(1), giving the exponential temperature dependence of the jump frequency w, we can write

$$D^*(p_M \text{ fixed}) = A \exp[-(g_m + g_v)/kT] \quad , \tag{17}$$

where g_v is excess free energy for the creation of a vacancy and g_m is the free energy of motion mentioned earlier.

Finally c_v can be a very strong function of composition for a binary compound MN. If the square root of the Schottky constant, $k_s^{\frac{1}{2}}$, giving the concentration of M-vacancies and the equal concentration of N-vacancies present in the stoichiometric compound is not too large and the width of the homogeneity range is not too small, then c_v can vary greatly with change of composition and D^* varies in the same way. Diffusion by the interstitial mechanisms could be discussed similarly and with analogous results.

This treatment is a primitive one, and it is necessary to emphasize its limitations. It does not cover diffusion in a binary solid solution in which the two types of atoms occupy and move on the same sites. Here two types of diffusing species would have to be considered and it would be more correct to consider the vacancy concentration as a constant independent of composition. Finally, and more important, the treatment is not quite correct in the analysis of tracer diffusion. The critical point is the assumed independence of the tracer atom jumps as expressed in assumption (2) and is associated with the subject of correlation coefficients. However for a compound in which the two types of atoms occupy sites in separate sublattices, and do not move into sites in the other sublattice, the tracer diffusion coefficient given by Eq. (16) is only in error by the omission of a factor, the correlation factor, whose numerical value depends only upon the diffusion mechanism and the crystal structure.

IV. CORRELATION COEFFICIENTS

A. Significance and Determination in Ionic Crystals

The assumed independence of atomic jumps means the number of atoms jumping from one atom plane to an adjacent one is given by the product of the probability of a jump times the number of atoms per unit area on the plane. This is a good assumption in the case of diffusion in a composition gradient if the vacancy concentrations are small and if there is only one species of atom moving through a lattice or sublattice. The nearest neighbor coordination shell for each vacancy will then almost always contain no vacancies and the coordinating atoms will all be the same, so that the vacancy jumps with equal probability in any direction. As a result the jumps of the atoms are also random. The situation is different in the case of tracer diffusion, however. The jumps of the vacancy itself are random, but those of the tracer atom are not. Suppose a tracer atom has just made a jump, as indicated in Fig. 6. Neglecting the small difference in the jump frequencies between the tracer and normal atom attributable to their difference in mass, the vacancy will jump with equal probability to each of the surrounding sites. The tracer atom itself, whose diffusion is being followed, will most likely jump back into the vacancy (the probability of another vacancy being near the tracer being small), the net result of two jumps being no displacement of the tracer at all. Eventually, a normal atom will jump into the

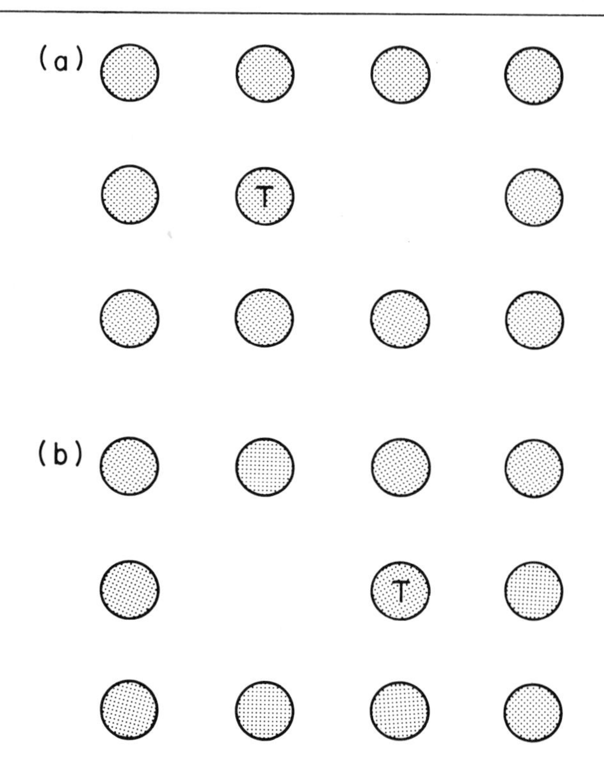

FIG. 6. Illustration of the correlation between successive jumps of a tracer atom moving via the vacancy mechanism.

vacancy, moving it away from the tracer. As a result the tracer diffusion constant is a fraction f of that calculated assuming independent, random jumps, and Eq. (16) must be written as

$$D^* = f \frac{\nu z a_0^2}{2} \frac{c_v}{S} \quad . \tag{18}$$

The factor f is called the correlation factor. Its value depends only upon the geometry of the crystal structure and the diffusion mechanism for (1) tracer self-diffusion in a pure element or for (2) tracer self-diffusion in an ordered compound MN in which the M-atoms do not move into N-sites and N-atoms do not move into M-sites. Values of the correlation factors have been calculated for a number of crystal structures, and some are given in Table 1 under the assumption that only nearest

TABLE 1

Correlation Factor for Tracer Self-Diffusion

A. Vacancy Mechanism

Structure	Coordination number	f, correlation factor
Diamond	4	1/2
Simple cubic	6	0.6531
Body centered cubic (bcc)	8	0.7272
Face centered cubic (fcc)	12	0.7815
Hexagonal close packed (hcp)	12	$f_x = f_y = 0.7812$ $f_z = 0.7815$

B. Interstitial Mechanisms in a Rock Salt Structure with Lattice Parameter a_0

Mechanism	Jump distance	Distance of tracer displacement/distance of charge displacement in a unit jump	f, correlation factor
Direct interstitial	$a_0/2$	1	1
Collinear interstitialcy	$\sqrt{3}\, a_0/2$	1/2	2/3
Noncollinear interstitialcy	$\sqrt{2}\, a_0/2$	3/4	0.9697

neighbor jumps are allowed. As indicated above, the correlation factor of 0.7815 for the vacancy mechanism in a fcc structure applies not only to self-diffusion in a fcc structure of a pure metal but also to diffusion of Na tracers in NaCl via vacancies in the Na ion sublattice (since this sublattice is fcc). It is seen that the larger the coordination number, which gives the number of atoms that can reach a vacancy in one jump, the closer the correlation factor is to unity. As we shall see shortly, the correlation factor for vacancy diffusion in a disordered solid solution, AB, in which the A and B atoms reside on and jump between sites in the same lattice is more complicated. It depends upon the relative jump frequencies of A and B as well as the geometry of the crystal structure.

The correlation factor for the direct interstitial mechanism is unity when the occupancy of an interstitial site constitutes a point defect and the fraction of unoccupied interstitial sites is close to unity. In this case the interstitial sites immediately surrounding an interstitial atom are almost always all unoccupied. Thus the direction the interstitial atom jumps is independent of the direction of the previous jump.

The concept of a correlation factor for diffusion in crystalline solids was pointed out theoretically by Bardeen and Herring [1] and was then applied to ionic crystals [2]. By ionic crystals we mean those which conduct an electrical current almost completely by ion flow rather than by electronic carriers (e. g., alkali and silver halides). It is easiest, with this type of solid, for us to determine the partial conductivity due to one type of ion. The partial conductivity and tracer self-diffusion coefficient for a particular ion then give two measures of the movement of this ion in the compound. When a single microscopic mechanism of diffusion is operative,

$$D_A^* = f_A kT\sigma_A / c_A q_A^2 \quad , \tag{19}$$

where D_A^* is the tracer self-diffusion coefficient of ion A with effective charge q_A, σ_A is that part of the electrical conductivity attributable to A, c_A is the total concentration of A ions, k is the Boltzmann constant, and T is the absolute temperature. With the correlation factor f_A set equal to unity, Eq. (19) is the so-called normal Einstein relation. In order to characterize the diffusion mechanism in ionic compounds, one essentially solves for f_A from Eq. (19) using experimental values for

D_A^* and σ_A. One way of viewing Eq. (19) is to realize that the partial conductivity of A ions is determined by the random or uncorrelated motion of the defect by which the A ions diffuse. The tracer self-diffusion coefficient only measures that fraction of the defect jumps that involve a tracer ion, and the tracer ion jumps are correlated to some extent. As a result the tracer self-diffusion coefficient is smaller than one would calculate from the partial conductivity using Eq. (19) with $f_A = 1$. For the interstitialcy mechanisms another effect also enters, as can be seen by reference to Figs. 3(b), (c). If the diffusing species are ions, the net charge displacement as a result of a unit jump is larger than that of either ion involved. This also increases the conductivity over the tracer diffusion coefficient, so that the correlation factor f_A in Eq. (19) is the product of a displacement factor and a proper correlation factor for the interstitialcy mechanisms. Both factors are listed in Part B of Table 1 for the rock salt structure. The displacement factors for the vacancy and interstitial mechanisms, involving only one ion per unit jump, are unity. Thus if the cation M in the compound MN with a rock salt structure diffuses by a vacancy mechanism on the fcc M-sublattice, an experimental value of f_M of 0.7812 should be obtained from Eq. (19). If M diffuses by a collinear interstitialcy mechanism, a value of $(1/2)(2/3) = 1/3$ is obtained. Although in actual practice the situation is often complicated by the simultaneous occurrence of more than one diffusion mechanism [3-5], the above illustrates how diffusion mechanisms can be determined in pure ionic compounds. The subject is discussed again in Section VIII.

B. Correlation Coefficients from the Isotope Effect

Since in metals an electrical current is carried entirely by electronic carriers the procedure described above to measure the correlation coefficient and determine the diffusion mechanism in inapplicable. Instead one measures and compares the diffusion coefficients for two isotopes. (The isotope effect of course is general and can be used with all solids.) Because of their different masses, the isotopes vibrate with different frequencies. In particular, the vibration frequencies are different for that particular mode which, when excited, leads to a passage of the atom over the energy barrier associated with a jump to a new position. Consequently the jump frequencies differ for the isotopes. The jump frequencies w_i are related to the isotope masses by [6, 7]

$$w_\alpha/w_\beta - 1 = \Delta K[(m_\beta/m_\alpha)^{\frac{1}{2}} - 1] \quad . \tag{20}$$

Here ΔK is a positive number less than or equal to one. Of the kinetic energy associated with the vibrational mode leading to a jump, it represents the fraction residing in the migrating atom. Thus if the migrating atom is loosely coupled to the rest of the structure in its jump, ΔK is near unity. This is the case for self-diffusion in a number of fcc metals. On the other hand, if the migrating atom is tightly coupled to the rest of the structure, ΔK is near zero. A great many atoms are then involved when one atom jumps, and the effective mass of this group is so large that changing the mass of the migrating atom does not significantly change this effective mass or the jump frequency.

For both isotopes the diffusion coefficient is of the form

$$D_i = a_0^2 k f_i w_i \quad , \tag{21}$$

where a_0^2 is the jump distance, f_i and w_i are the correlation factor and jump frequency for the ith isotope, and k contains numerical constants and the concentration of the defect involved in the diffusion mechanism and is the same for both isotopes. Using Eq. (21) we can write

$$\frac{D_\alpha}{D_\beta} - 1 = \frac{w_\alpha f_\alpha - w_\beta f_\beta}{w_\beta f_\beta} = \frac{\Delta w}{w_\beta}\left(1 + \frac{w_\beta}{f_\beta}\frac{\Delta f}{\Delta w} + \frac{\Delta f}{f_\beta}\right) \quad , \tag{22}$$

where

$$\Delta f = f_\alpha - f_\beta \quad , \qquad \Delta w = w_\alpha - w_\beta \quad .$$

The last term in the parenthesis of Eq. (22), $\Delta f/f_\beta$, is small compared to the other terms and is dropped. The second term is small enough to be well approximated by a derivative, so that Eq. (22) becomes

$$\frac{D_\alpha}{D_\beta} - 1 = \frac{\Delta w}{w_\beta}\left(1 + \frac{d \ln f_\alpha}{d \ln w_\alpha}\right) \quad . \tag{23}$$

Now for the vacancy mechanism in cubic structures, kinetic treatments show [8] the correlation factor for either self-diffusion or for diffusion of a dilute substitutional impurity is of the form

$$f_i = \frac{u}{2w_i + u} \; , \tag{24}$$

when u is a sum over the various jump frequencies of the solvent atoms and is generally independent of the isotopic masses. When u is much larger than w_i, f_i is equal to one. This is physically reasonable since the jump frequencies of the solvent atom are then much greater than that of the isotope. As a consequence, after an isotopic atom has jumped into a vacancy, it is most likely that solvent atoms will jump into that vacancy and move it away from the isotopic atom. The probability of the isotopic atom returning to the same vacancy, an event that reduces the value of the correlation coefficient, is thereby diminished. Assuming Eq. (24) holds in general, it and Eq. (20) can be substituted into Eq. (23) to give

$$(D_\alpha/D_\beta - 1)/([m_\beta/m_\alpha]^{\frac{1}{2}} - 1) = f_\alpha \, \Delta K \quad . \tag{25}$$

[The left-hand member of Eq. (25) is called the strength of the isotope effect.] The value of $f_\alpha \, \Delta K$ obtained from Eq. (25) provides a lower limit for the value of f for the isotope being studied. For self-diffusion, the value of f depends only upon the diffusion mechanism and crystal structure. Thus all mechanisms with a value of f smaller than the measured value of $f \, \Delta K$ are excluded. If the diffusion mechanism is known, ΔK can be determined. In this way it has been shown that self-diffusion in many fcc metals occurs by a vacancy mechanism with ΔK close to unity.

For diffusion of a dilute substitutional impurity, the value of $f_\alpha \Delta K$ from Eq. (25) allows a further characterization of the possible diffusion mechanism, if not a unique choice. In this case the value of f_i depends upon the relative jump frequencies as well as the crystal structure and diffusion mechanism. Thus for vacancy diffusion in a fcc structure or sublattice, an approximate kinetic treatment [9] gives the correlation coefficient of an impurity isotope as

$$f_\alpha = (k_2 + 3.5k_1)/(w_\alpha + k_2 + 3.5k_1) \quad . \tag{26}$$

A more exact treatment [10] results in the replacement of the constant 3.5 in Eq. (26) by 2.58. This equation is of the form of Eq. (24) since w_α is the jump frequency for the impurity isotope and k_2 and k_1 are the jump frequencies for the solvent atom. The jump frequencies are defined in Fig. 7, which shows part of a fcc structure. A vacancy is at the center of the cube and is shown surrounded by its 12 nearest neighbors, one of which is an isotope, α, of the impurity atom. Four of the nearest neighbors are simultaneously nearest neighbors of the impurity and are assumed to have a jump frequency k_2, which is different from the jump frequency k_1 of the other 7 solvent atoms. As an illustration we cite the experimental determination of the diffusion coefficients for ^{109}Cd and ^{115m}Cd in both pure Ag and pure Cu [11]. In both cases, $f\Delta K$ determined from Eq. (25) is zero within experimental error. If, consistent with the observations on self-diffusion in Ag, one assumes $\Delta K \approx 1$ and that the vacancy mechanism predominates, then the correlation factor is near zero, and using Eq. (26), $w_\alpha >> k_2 + 3.5k_1$, i.e., the jump frequencies for the Cd isotopes are much larger than those for either Cu in essentially pure Cu or those for Ag in essentially pure Ag. The diffusion coefficient for the Cd isotope using Eq. (21) is then

$$D_{Cd} = a_0^2 k(k_2 + 3.5k_1) \quad . \tag{27}$$

Thus the diffusion coefficient for a dilute solution of Cd in either Ag or Cu does not depend upon the jump frequency nor isotopic mass of the

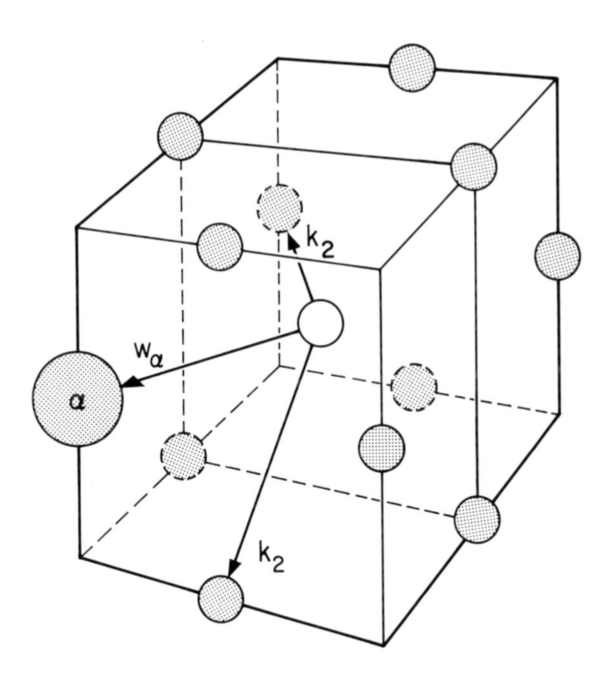

FIG. 7. Various jump frequencies for the vacancy mechanism in a dilute binary system with a fcc structure. A unit cell with a vacancy in the central position and an impurity atom in an edge position is shown. The jump frequency for the impurity atom is w_α.

Cd. The slow step in the diffusion process is the solvent-vacancy exchange.

V. PRACTICAL DIFFUSION COEFFICIENTS

A. Introduction

In this and the following section we give a phenomenological description of planar or one-dimensional diffusion in an isothermal, binary system. Practical diffusion coefficients in this case are defined by the ratio of the diffusion flow of a given species to the concentration gradient of that species and are closely related to experiment. This is in contrast to fundamental diffusion coefficients, which are defined in terms of chemical potential gradients. A strictly logical order would place the present section shortly after the beginning of a description of diffusion based upon irreversible thermodynamics, as given in Section VII following, since the latter provides the justification for the starting equation, Eq. (29), below. We have chosen to proceed otherwise in view of the length of the discussion of practical diffusion coefficients and in view of its intimacy with experiment.

The object of the present section is to obtain an equation giving the temporal and spatial variation of the concentrations in a binary system which, unlike Fick's second law, is valid even when the partial atomic volumes depend upon composition and to put this equation in a form that allows the determination of the chemical diffusion coefficient as a function of composition from experimental data. In the process we are led naturally to a consideration of intrinsic diffusion coefficients and the Kirkendahl effect.

B. Definition of Practical Diffusion Coefficients

For a binary system A-B the diffusion flows J_i are given by

$$J_i^R = c_i(u_i - u_{RL}) \quad , \tag{28}$$

$$J_i^R = -D_i^R \frac{\partial c_i}{\partial z} \quad , \tag{29}$$

where the subscript i stands for either A or B. The diffusion flow of i measured relative to an as yet unspecified flow reference frame R is J_i^R; u_i and u_{RL} are, respectively, the velocity of species i and that of the flow reference frame, both measured relative to a fixed laboratory coordinate system L; c_i is the concentration of species i; z is the spatial coordinate in the laboratory coordinate system; and D_i^R is the practical diffusion coefficient for species i when diffusion is defined as flow relative to the flow reference frame R. Equation (28) follows by definition from a continuum description of diffusion and allows for the possibility that the flow reference plane and the fixed laboratory frame may not coincide; i.e., not only can c_i and u_i vary with time and position but u_{RL} is generally not zero and can also vary. Where diffusion is not occurring, u_i, u_{RL}, and dc_i/dz are all zero. Equation (29) follows from a consideration of irreversible thermodynamics, Eq. (68), and applies to isothermal diffusion in a binary system in the absence of externally applied forces. In addition to the specifications given we assume the diffusion cross section perpendicular to the z axis is constant. The units of length and time are taken as the centimeter and second, respectively. The units of amount (grams, number of atoms, number of gram-atomic or gram-molecular weights) in which the concentrations and flows are to be expressed will be taken throughout as the number of atoms. The diffusion flows are then given as the number of A or B atoms crossing the reference plane R per cm^2 and sec. The concentrations are given as the number of atoms/cm^3, and the diffusion coefficients have the dimensions cm^2/sec. We emphasize that the values for the diffusion flows and diffusion coefficients depend upon the choice of the flow reference frame R, as indicated explicitly by the superscript attached to these quantities in Eqs. (28) and (29). As might be implied from our discussion thus far, the definition of the flow reference frame and of what is to be called diffusion is somewhat arbitrary. We shall present arguments for a specific choice later.

The flows relative to any two flow reference frames R and S are related by

$$J_i^S = J_i^R + c_i(u_{RL} - u_{SL}) \quad , \tag{30}$$

where u_{RL} and u_{SL} are, respectively, the local velocities of the R reference frame and S reference frame, both measured relative to the fixed laboratory coordinate system. For a flow reference frame L, fixed in the laboratory coordinate frame, we can apply the equation of continuity to Eq. (29), expressing the conservation of the number of A atoms and of B atoms in the assumed absence of chemical reactions, to obtain

$$\frac{\partial c_i}{\partial t} = -\frac{\partial}{\partial z} J_i^L = \frac{\partial}{\partial z}\left(D_i^L \frac{\partial c_i}{\partial z}\right) . \tag{31}$$

However, for a binary system there are usually two such equations, one for A atoms and one for B atoms, and more importantly, D_A^L and D_B^L are in general different. It is advantageous to arrange things so that one independent equation serves to describe the concentration redistribution that occurs during diffusion. Using Eq. (30) we can replace the diffusion flow J_i^L in Eq. (31) by one relative to an arbitrary reference frame R:

$$\frac{\partial c_i}{\partial t} = \frac{\partial}{\partial z}\left(D_i^R \frac{\partial c_i}{\partial z} - c_i u_{RL}\right) . \tag{32}$$

We are interested in eliminating u_{RL} from this equation.

C. The Mutual or Chemical Diffusion Coefficient

Some auxiliary relations involving the partial atomic volumes, $\bar{v}_A$ and $\bar{v}_B$, are necessary to proceed. The volume of a homogeneous element of the system is equal to the number of A atoms times the partial atomic volume of A plus the number of B atoms times the partial atomic volume of B. It immediately follows that

$$c_A \bar{v}_A + c_B \bar{v}_B = 1 . \tag{33}$$

The differential of Eq. (33) reduces to

$$\bar{v}_A \, dc_A + \bar{v}_B \, dc_B = 0 \tag{34}$$

since at constant temperature and pressure the partial atomic volumes, like other partial atomic or partial molar quantities of thermodynamics, satisfy the relation

$$c_A \, d\bar{v}_A + c_B \, d\bar{v}_B = 0 . \tag{35}$$

When Eqs. (33)-(35) are used, we are assuming that local thermodynamic equilibrium prevails during the diffusion process. That is, for macroscopically small elements of the nonequilibrium, diffusing system, the thermodynamic state variables are assumed to be defined, and the equilibrium relations among these variables and the thermodynamic functions are assumed to hold. (This assumption, although occasionally partially relaxed, is basic to the whole of irreversible thermodynamics also.) Equation (34) will be used often in what follows to express a partial derivative of c_B in terms of that in c_A.

Multiplying Eq. (32) with $i = A$ by $\bar{v}_A$, Eq. (32) with $i = B$ by $\bar{v}_B$, adding, and noting that the left-hand member is zero by Eq. (34), we have

$$0 = \bar{v}_A \frac{\partial c_A}{\partial t} + \bar{v}_B \frac{\partial c_B}{\partial t}$$

$$= \bar{v}_A \frac{\partial}{\partial z}\left(D_A^R \frac{\partial c_A}{\partial z}\right) + \bar{v}_B \frac{\partial}{\partial z}\left(D_B^R \frac{\partial c_B}{\partial z}\right) - \bar{v}_A u_{RL} \frac{\partial c_A}{\partial z}$$

$$- \bar{v}_B u_{RL} \frac{\partial c_B}{\partial z} - (\bar{v}_A c_A + \bar{v}_B c_B) \frac{\partial u_{RL}}{\partial z} \quad . \qquad (36)$$

By Eq. (33), the coefficient of $\partial u_{RL}/\partial z$ in the 5th or last term on the right-hand side of Eq. (36) is unity, while the sum of the 3rd and 4th terms is zero by Eq. (34). Using these facts and solving for $\partial u_{RL}/\partial z$ gives

$$\frac{\partial u_{RL}}{\partial z} = -\bar{v}_B \frac{\partial}{\partial z}\left(\frac{\bar{v}_A D_B^R}{\bar{v}_B} \frac{\partial c_A}{\partial z}\right) + \bar{v}_A \frac{\partial}{\partial z}\left(D_A^R \frac{\partial c_A}{\partial z}\right) \quad , \qquad (37)$$

where a spatial derivative of c_B has been replaced by one of c_A in the first parenthesis, again using Eq. (34). Integrating Eq. (37) by parts with respect to dz from $-\infty$ (physically, from a point where diffusion is not occurring and u_{RL} and $\partial c_A/\partial z$ are zero) to z gives the local velocity of the arbitrary flow reference frame R as

$$u_{RL} = \bar{v}_A (D_A^R - D_B^R) \frac{\partial c_A}{\partial z} + \int_{-\infty}^{z} \left(\frac{\mathcal{D}}{c_A \bar{v}_B}\right)\left(\frac{\partial \bar{v}_B}{\partial c_A}\right)\left(\frac{\partial c_A}{\partial z}\right)^2 dz' \quad , \qquad (38)$$

where

$$\mathcal{D} = c_B \bar{v}_B D_A^R + c_A \bar{v}_A D_B^R \quad . \qquad (39)$$

Therefore in the diffusion zone where $\partial c_A/\partial z$ is not zero, there are two contributions to u_{RL}. One is represented by the integral of Eq. (38) and arises from a dependence of the partial atomic volumes upon composition. The second arises from an inequality of the diffusion coefficients D_B^R and D_A^R and is represented by the first term on the right-hand side of Eq. (38).

Substituting Eq. (38) into Eq. (32) with $i = A$ and using Eq. (33) gives

$$\frac{\partial c_A}{\partial t} = \frac{\partial}{\partial z}\left(\mathscr{D}\frac{\partial c_A}{\partial z}\right) - \frac{\partial}{\partial z}\left\{c_A \int_{-\infty}^{z}\left(\frac{\mathscr{D}}{c_A\bar{v}_B}\right)\left(\frac{\partial \bar{v}_B}{\partial c_A}\right)\left(\frac{\partial c_A}{\partial z}\right)^2 dz'\right\} . \qquad (40)$$

If the A and B subscripts are interchanged everywhere in Eq. (40), except under the integral sign, another equally valid equation giving $\partial c_B/\partial t$ results. These are not independent, of course, being related by Eq. (34). In view of these considerations, Eq. (40) is one of the results we have been seeking. The macroscopic change with time and position of the concentrations in an isothermal, binary system is given in terms of a single or mutual diffusion coefficient $\mathscr{D}$, which we shall call the chemical diffusion coefficient (so-called because it is connected with the process that occurs when concentration gradients in A and B are present in contrast to the tracer self-diffusion coefficients which describe the process when there is no gradient in the overall chemical concentration of A or B, only in those of various isotopes). We note that insofar as Eq. (29), Fick's first law, is valid for an arbitrary flow reference frame, the value of $\mathscr{D}$ is unaffected by what choice is made for this frame and is independent of any knowledge concerning the local velocity of this frame. However, Eq. (40) is more general than Fick's second law, which is commonly assumed to be a direct consequence of Fick's first law and reduces to it only when the 2nd term on the right-hand side of Eq. (40) is zero. This occurs when the partial atomic volumes are independent of composition, as is the case for ideal gases and indeed is approximately true for many solid systems. This is also approximately the case when two initially homogeneous parts of a system with only slightly different compositions are brought together to form a diffusion couple, for then $(\partial c_A/\partial z)^2$ is small. Such a procedure has been adopted in precision measurements of $\mathscr{D}$ in aqueous solutions [12]. We will use Eq. (40) in its full form to arrive at an equation allowing the determination of $\mathscr{D}$ from experimental data. First, however, we digress to make a specific choice, based on physical considerations, for the hitherto arbitrary flow reference frame.

D. Intrinsic Diffusion Coefficients and the Kirkendahl Effect

Even in a gaseous system in which the partial atomic volumes are constant at constant total pressure and temperature, the description of diffusion in terms of $\mathscr{D}$ still contains contributions from local bulk flow as well as from what might be termed diffusion. The lighter, more rapid molecules tend to build up pressure in the regions into which they are diffusing. This pressure build-up is relieved by the bulk flow of both types of molecules together. The description of the entire process in terms of $\mathscr{D}$ lumps both the bulk flow and diffusion together. The existence of bulk flow during diffusion was originally demonstrated experimentally in metallic [13] and in solvent-polymer [14] systems. It is now known to be a common phenomenon and is called the Kirkendahl effect. Smigelskas and Kirkendahl [13] placed fine molybdenum wires on the surface of a block of 70 at. % Cu, 30 at. % Zn and then plated the block with a 0.25-cm thick layer of Cu. After heating to allow diffusion to occur, the block was cooled, sectioned, and examined. The chemically inert Mo wires on opposite sides of the block were found to be displaced relative to one another, the displacement varying as the square root of the diffusion time, and showing the wires to be 0.056 cm closer after 56 days. These findings were originally viewed skeptically and prompted considerable experimental work which confirmed the effect. In the later work, two initially homogeneous bars of constant cross section were welded to either end of a third homogeneous bar of a different composition, fine wires of some chemically inert material being first placed at the ends of the middle bar [15]. The common opinion is that the chemically inert markers move with the local crystal lattice in which they are embedded. Thus, diffusion defined relative to inert markers as a flow reference frame is taken to be the same as diffusion defined relative to the atomic planes of a crystal and does not therefore include as diffusion any motion in which the plane participates as a whole (bulk flow). Such diffusion is called intrinsic diffusion, and a flow reference frame consisting of the atomic planes or, from a phenomenological point, defined by inert markers, will be designated by I. The equations we have developed in the last subsection for an arbitrary flow reference frame R can be taken over unchanged except for the replacement of the R superscript by I. We see immediately from Eq. (38) that the local velocity of the atomic planes will in general be nonzero, and there will be a Kirkendahl effect where diffusion is occurring since the intrinsic diffusion coefficients, D_A^I and D_B^I, will in general not be equal.

E. Evaluation of the Chemical Diffusion Coefficient

Currently the emphasis is not in solving Eq. (40) to obtain $c_i(z, t)$ for an assumed composition dependence of $\mathscr{D}$, but rather in solving for $\mathscr{D}$ from Eq. (40) so that it can be determined from experimental concentration-penetration data. There are approximate methods for so doing, but Baluffi [16] has shown how to obtain an exact solution.

We define auxiliary functions θ_i, $i = A, B$:

$$\theta_i = D_i^I - u_{IL} c_i \left(\frac{\partial z}{\partial c_i}\right) \tag{41}$$

so that the flow relative to the laboratory coordinate system can be written as

$$J_i^L = -\theta_i \frac{\partial c_i}{\partial z} \tag{42}$$

and Eq. (31) can be written as

$$\frac{\partial c_i}{\partial t} = \frac{\partial}{\partial z}\left(\theta_i \frac{\partial c_i}{\partial z}\right) . \qquad (43)$$

In view of Eq. (41), the chemical diffusion coefficient given by Eq. (39) can be written in terms of θ_A and θ_B as

$$\mathscr{D} = c_B \bar{v}_B \theta_A + c_A \bar{v}_A \theta_B . \qquad (44)$$

We now apply the Boltzmann transformation $\lambda = z/t^{\frac{1}{2}}$ and assume a solution $c_i(\lambda)$. Equation (43) then becomes an ordinary partial differential equation with λ the independent variable. If the boundary conditions can be written solely in terms of λ, with neither z nor t appearing separately, we may proceed. Two such boundary conditions and their expression in terms of λ are:

(a) diffusion between two infinitely long bars each of uniform but different initial composition

$$c = c_1, \quad z < 0, \; t = 0; \qquad c = c_1, \quad \lambda = -\infty$$

$$c = c_2, \quad z > 0, \; t = 0; \qquad c = c_2, \quad \lambda = +\infty;$$

(b) diffusion into a semiinfinite bar of uniform initial composition and with a fixed surface composition

$$c = c_s, \quad z = 0; \qquad c = c_s, \quad \lambda = 0$$

$$c = c_1, \quad z > 0, \; t = 0; \qquad c = c_1, \quad \lambda = +\infty .$$

(As usual, these boundary conditions apply with finite samples in which one-dimensional or planar diffusion is occurring if the diffusion is not allowed to proceed throughout the entire sample.) Equation (43) then becomes

$$-\tfrac{1}{2}\lambda \frac{dc_i}{d\lambda} = \frac{d}{d\lambda}\left(\theta \frac{dc_i}{d\lambda}\right) .$$

Upon integration and solving for θ_i one obtains

$$\theta_i = -\frac{1}{2}\left(\frac{d\lambda}{dc_i}\right) \int_{c_i(-\infty)}^{c_i(z)} \lambda \, dc_i . \qquad (45)$$

If we confine ourselves to data for a single fixed value of t we can rewrite Eq. (45) as

$$\theta_i = -\frac{1}{2t}\left(\frac{dz}{dc_i}\right)\int z\,dc_i \quad . \tag{46}$$

Substituting for this value of θ_i into Eq. (44) for $\mathscr{D}$, using Eqs. (33) and (34), and rearranging, we have

$$\mathscr{D} = -\frac{1}{2t}\left(\frac{\partial z}{\partial c_A}\right)\left\{[1 - c_A(\bar{v}_A - \bar{v}_B)]\int_{c_A(-\infty)}^{c_A(z)} z\,dc'_A + \bar{v}_B c_A \int_{c_A(-\infty)}^{c_A(z)} \left(\frac{\bar{v}_A - \bar{v}_B}{\bar{v}_B}\right) z\,dc'_A\right\} \quad . \tag{47}$$

We see that no knowledge of the movement of the inert markers is required to use this equation. The first integral can be obtained graphically by plotting the concentration-penetration data. The second integral is usually small, but in any event can be obtained by scaling down the z coordinate of the penetration curve by the factor $(\bar{v}_A - \bar{v}_B)/\bar{v}_B$. Thus by finding the area under the curve giving z as a function of c_A, from a point where no diffusion has occurred up to a particular value of c_A, say c''_A, evaluating the slope of the curve, $\partial c_A/\partial z$, at c''_A, and performing the related steps to obtain the second integral, Eq. (47) can be used to obtain the value of $\mathscr{D}$ at c''_A. This of course can be repeated over the range in c_A covered by the diffusion process in the particular experiment analyzed. There are alternate methods of plotting the experimental c_A - z data that allow a more accurate integration [17]. We stress that the laboratory coordinate frame is used to measure z, so that that the origin is fixed relative to the ends of the sample which have undergone no diffusion. Diffusion coefficients have often been obtained in the past using the simplified form of Eq. (47) that results when $\bar{v}_A = \bar{v}_B$ = constant and a coordinate frame whose origin (the Matano interface) is given by

$$\int_{c_A(-\infty)}^{c_A(+\infty)} z\, dc'_A = 0 \quad . \tag{48}$$

This origin coincides with that in the laboratory coordinate frame appropriate for Eq. (29) only when $\bar{v}_A = \bar{v}_B$ = constant. (The effect of using a Matano interface to define the origin of the coordinate frame when the partial atomic volumes are not constant is to compensate to some extent for the error incurred in using the simplified version of Eq. (29) obtained by setting $\bar{v}_A = \bar{v}_B$ [17, pp. 238-240].

The equation for the chemical diffusion coefficient $\mathscr{D}$ used by Darken [18] in his original analysis of the Kirkendahl effect is obtained by setting $\bar{v}_A = \bar{v}_B$ = constant in Eq. (47). The origin of the coordinate frame was taken by him as the Matano interface defined by Eq. (48). As stated above, this is the same as the origin of the fixed laboratory frame if the assumption of equal and constant atomic volumes is valid.

F. Evaluation of Intrinsic Diffusion Coefficients

The intrinsic diffusion coefficients can be evaluated if marker movement data are also available. Substituting Eq. (46) for θ_i and Eq. (38) for u_{IL} into Eq. (41) we obtain, for a fixed diffusion time t, the result

$$D_A^I = \frac{1}{2t}\left(\frac{\partial z}{\partial c_A}\right)\left\{2tu_{IL}c_A - \int_{c_A(-\infty)}^{c_A(z)} z\, dc'_A\right\} \quad , \tag{49}$$

$$D_B^I = -\frac{1}{2t}\left(\frac{\partial z}{\partial c_A}\right)\frac{\bar{v}_B}{\bar{v}_A}\left\{2tu_{IL}c_B + \int_{c_A(-\infty)}^{c_A(z)} z\, dc'_A + \int_{c_A(-\infty)}^{c_A(z)}\left(\frac{\bar{v}_A - \bar{v}_B}{\bar{v}_B}\right) z\, dc'_A\right\} \quad . \tag{50}$$

The slopes and the integrals in Eqs. (49) and (50) are the same as those in Eq. (47) for $\mathscr{D}$, so D_A^I and D_B^I are readily determined if the marker velocity u_{IL} is determined.

The behavior of a marker placed initially at the origin is particularly simple. Making the Boltzmann transformation $\lambda = z/t^{\frac{1}{2}}$, the velocity of any marker as given by Eq. (38) can be written as

$$u_{IL} = \frac{1}{t^{\frac{1}{2}}}\left\{(D_A^I - D_B^I)\bar{v}_A \frac{dc_A}{d\lambda} + \int_{-\infty}^{\lambda} \frac{\mathscr{D}}{c_A\bar{v}_B}\left(\frac{d\bar{v}_B}{dc_A}\right)\left(\frac{dc_A}{d\lambda}\right)^2 d\lambda'\right\} \quad . \tag{51}$$

Now the quantities in the parentheses are all functions of λ; D_i^I and $\bar{v}_i$ are directly functions of c_A and c_B, but these in turn are functions of λ. Thus a particular solution of Eq. (51) is obtained for any marker that moves in such a way that the value of λ at the marker, say λ_m, is a constant. The concentrations at the position of this type of marker remain fixed since they are a function of λ, which is fixed. Thus if the displacement of the marker, z_m, is given by

$$z_m = \lambda_m t^{\frac{1}{2}} \quad , \tag{52}$$

where λ_m is a constant equal to twice the sum of the quantitites within the parentheses in Eq. (51), then Eq. (51) is satisfied. Since $z_m = 0$ when $t = 0$, Eq. (52) applies only to a marker initially at $z = 0$. Differentiating Eq. (52), we have

$$\frac{dz_m}{dt} = u_{IL} = \tfrac{1}{2}\lambda_m t^{-\frac{1}{2}} \quad . \tag{53}$$

Thus for a marker positioned initially at the interface of a diffusion couple, the marker velocity can be obtained from a measurement of the marker displacement using Eqs. (52) and (53). The values of D_A^I and D_B^I for the fixed composition prevailing at this marker can then be calculated from Eqs. (49) and (50), respectively.

G. Summary of Practical Diffusion Coefficients for Binary Systems

Assuming that local thermodynamic equilibrium is maintained during the diffusion process, we have obtained a phenomenological description of diffusion in a binary system. The movement of both components is given in terms of a single mutual diffusion coefficient, called the chemical diffusion coefficient, by Eq. (40). (When all the A's are changed to B's and conversely in Eq. (40), an equation for dc_B/dt is obtained.) This equation reduces to Fick's second law only when the partial atomic volumes are constant, the complete equation being valid in the more general case. This approach has been developed within the past fifteen years. Traditionally either Fick's second law has been accepted as an immediate consequence of the first law for nonreacting systems or diffusion has been defined as motion relative to a local center of volume, or mass, or number of atoms (molecules or ions) and the assumption made that this center is fixed relative to a laboratory coordinate frame for a particular experimental situation or nearly enough so that its velocity can be neglected. Both views lead to Fick's second law, but the latter is more explicit in its assumptions. The generality of the extended form of Fick's second law given by Eq. (40) is restricted by the assumption that local thermodynamic equilibrium is maintained throughout the diffusion system. This simple view is apparently valid for a wide variety of physical phenomena, but there are exceptions. One difficulty, that arises in diffusion phenomena and has been particularly noted in studies of the Kirkendahl effect, is the development of porosity within metallic diffusion specimens. This might be viewed as arising from a departure from local equilibrium, a temporary supersaturation of vacancies. Regardless of the cause, the presence of porosity within the diffusion system is not taken into account in the description presented here.

Having obtained an extension of Fick's second law in Eq. (40) that is valid for any thermodynamically consistent variation of the partial atomic volumes, we are able to express the chemical diffusion coefficient in terms of experimental observables in Eq. (47) which can be applied whenever the Boltzmann transformation can be made. This equation, too, is valid for any thermodynamically consistent variation of the partial atomic volumes with composition. It reduces to those used previously only when the partial atomic volumes are constant and equal. The widely used concept of a Matano interface is not necessary in the application of Eq. (47).

The intrinsic diffusion coefficients discussed here are related to the chemical diffusion coefficient by Eq. (39). However, the latter requires only a concentration-penetration curve and a knowledge of the partial atomic volumes for its evaluation. Information concerning the movement of inert markers is necessary to evaluate the intrinsic diffusion coefficients.

VI. THE PHENOMENOLOGICAL DESCRIPTION OF DIFFUSION BASED UPON IRREVERSIBLE THERMODYNAMICS

A. Introduction

Irreversible thermodynamics is the extension of thermodynamics to nonequilibrium systems [19, 20]. It is assumed for each small element of the system that the thermodynamic state variables such as temperature, pressure, and composition are well-defined and that all thermodynamic quantities are the same functions of these state variables as they are for an equilibrium system. This is called the assumption of local equilibrium and was used in the last section to relate the partial molar volumes and concentrations in a binary system. Using equations describing the conservation of energy and mass, the equation of motion for a continuous system, and thermodynamic laws, an equation for the local rate of internal entropy production (per unit time and volume) is obtained, which is of the form

$$T\sigma = \sum J_i X_i \quad , \tag{54}$$

where the J_i are flows of various kinds (diffusion of matter, heat flow, electron flow, rate of production of a particular chemical species by chemical reaction, etc.) and X_i is the generalized force conjugate to the flow J_i. The flows are generally assumed to be linear functions of all the forces of the same tensorial character:

$$J_i = -\sum_j \mathscr{L}_{ij} X_j \quad , \tag{55}$$

where the so-called phenomenological coefficients $\mathscr{L}_{ij}$ are functions of the thermodynamic state variables but not of the gradients composing the forces. [Equation (55) differs from the usual form presented in the presence of the minus sign. This amounts to a redefinition of the $\mathscr{L}_{ij}$ that is adopted here to avoid an excess of minus signs in the diffusion flow equations.] Thus the vector diffusion flow of species A depends not only upon the forces for all the diffusion flows but also upon the force for the vector heat flow. A simpler version of this type of cross effect is seen in diffusion in anisotropic materials when a composition gradient in one direction in general gives rise to a diffusion flow with components in all three directions. On the other hand, a diffusion flow does not depend upon the force conjugate to the chemical reaction rate, which is a scalar. Using Eq. (55), the rate of entropy production can be written as a quadratic form in the forces X_i that is positive definite. That is, the local entropy production is zero only when all the forces X_i are zero and the system is in equilibrium. The interaction of the various processes implied in Eq. (55) is an important part of the content of irreversible thermodynamics. A second part is in the explicit equations for the forces X_i, as will be discussed shortly. A third part is contained in the Onsager reciprocity relations.

The phenomenological coefficients $\mathscr{L}_{ij}$ relating the flows and forces in Eq. (55) form a symmetric matrix in the absence of an applied magnetic field and in the absence of Coriolis forces (such as encountered by matter in a centrifuge), i. e.,

$$\mathscr{L}_{ij} = \mathscr{L}_{ji} \quad . \tag{56a}$$

More generally,

$$\mathscr{L}_{ij}(B, \omega) = \mathscr{L}_{ji}(-B, -\omega) \quad , \tag{56b}$$

where B is the magnetic induction and ω the angular velocity of rotation.

For diffusion in solids one usually neglects viscosity terms in the entropy production given by Eq. (54) and assumes that the system is in mechanical equilibrium in the absence of external forces. For mechanical equilibrium, the diffusion flows, which are originally defined relative to the local center of mass in the development of the equation for the rate of internal entropy production, can be expressed relative to an arbitrary flow reference frame (Prigogine's theorem)[19, pp. 44, 239]. We take the diffusion flows as measured relative to inert markers, as in the discussion of intrinsic diffusion in the last section, and make the usual assumption that this is the same as measuring the diffusion flows relative to the local crystal lattice.

The force conjugate to the ith diffusion flow for diffusion in the absence of external forces is the negative of the gradient of the ith chemical potential:

$$X_i = -\nabla(\mu_i/T) \quad . \tag{57}$$

The force conjugate to the heat flow is

$$X_q = -\frac{1}{T}\nabla T \quad . \tag{58}$$

For isothermal diffusion, to which we shall limit our considerations, the diffusion flows are

$$J_i = \sum_i L_{ij} \nabla \mu_i \quad , \tag{59}$$

where $L_{ij} = \mathscr{L}_{ij}/T$.

B. The Thermodynamic Factor for Isothermal Diffusion in a Binary System

We shall utilize irreversible thermodynamics to extend our discussion of practical diffusion coefficients in an isothermal binary system. We have seen that the chemical diffusion coefficient $\mathscr{D}$, which can be determined from concentration-penetration curves, depends upon the intrinsic diffusion coefficients D_A^I and D_B^I, as shown in Eq. (39). Using irreversible thermodynamics we are able to obtain part of the composition dependence of these intrinsic diffusion coefficients, that contained in a so-called thermodynamic factor. Secondly, we are able to show the general form of the relation between the diffusion coefficient determined using tracers and the intrinsic diffusion coefficients.

We consider isothermal diffusion in a binary system A-B. Writing out Eq. (59) explicitly, the diffusion flows are

$$J_A = L_{AA} \nabla \mu_A + L_{AB} \nabla \mu_B \quad , \tag{60a}$$

$$J_B = L_{BA} \nabla \mu_A + L_{BB} \nabla \mu_B \quad . \tag{60b}$$

At constant temperature and pressure the chemical potentials are related by the Gibbs-Duhem relation

$$x_A \nabla \mu_A + x_B \nabla \mu_B = 0 \quad , \tag{61}$$

where x_A and x_B are the respective atom fractions, the sum of which is always unity. Using Eq. (61) we can rewrite Eq. (60) so that the flow of each species depends only upon the chemical potential gradient of that species, i.e.,

$$J_A^I = \left(L_{AA} - \frac{x_A}{x_B} L_{AB} \right) \nabla \mu_A \quad , \tag{62a}$$

$$J_B^I = \left(- \frac{x_B}{x_A} L_{BA} + L_{BB} \right) \nabla \mu_B \quad . \tag{62b}$$

Equations (62) tell us that, in the absence of external forces, a diffusion flow in an isothermal binary system is proportional to the gradient in the chemical potential of the species whose flow is being considered. The intrinsic diffusion coefficients have been defined in terms of concentration gradients as

$$J_A^I = -D_A^I \nabla c_A \quad , \tag{63a}$$

$$J_B^I = -D_B^I \nabla c_B \quad . \tag{63b}$$

We now wish to express the chemical potential gradients in the equations for the intrinsic diffusion flows given by Eqs. (62) in terms of concentration gradients. We can then compare the results with Eqs. (63) and see what factors compose the intrinsic diffusion coefficients. At the same time we justify Eqs. (63) and the equivalent equation (29), which was the starting point for Section V on practical diffusion coefficients. To make this transformation we develop some necessary simple relations.

The chemical potential can be written in terms of atom fraction and activity coefficient as

$$\mu_i = kT \ln \gamma_i x_i + \mu_i^0(T) \quad . \tag{64}$$

Since we maintain the number of atoms as the unit of amount in which the diffusion flow is expressed, the units of the chemical potential are taken as energy units/atom. Now the Gibbs-Duhem relation given by Eq. (61) can be written in terms of activity coefficients as

$$x_A \, d \ln \gamma_A + x_B \, d \ln \gamma_B = 0 \quad , \tag{65}$$

and an immediate consequence of Eq. (65) is that

$$\frac{d \ln \gamma_A}{d \ln x_A} = \frac{d \ln \gamma_B}{d \ln x_B} \quad . \tag{66}$$

Finally, using the definition of atom fraction, we obtain $\nabla \ln x_A$ in terms of ∇c_A using Eq. (34) to eliminate ∇c_B and using Eq. (33) to simplify. The result is

$$\nabla \ln x_A = (1/C\bar{v}_B)(\nabla c_A/c_A) \quad . \tag{67a}$$

Similarly

$$\nabla \ln x_B = (1/C\bar{v}_A)(\nabla c_B/c_B) \quad , \tag{67b}$$

where $C = c_A + c_B$. Using Eq. (64) for the chemical potentials and Eqs. (67), the diffusion flows given by Eq. (62) can be rewritten as

$$J_A = \left(L_{AA} - \frac{x_A}{x_B} L_{AB}\right)\left(\frac{kT}{Cc_A\bar{v}_B}\right)\left(1 + \frac{d \ln \gamma_A}{d \ln x_A}\right)\nabla c_A , \tag{68a}$$

$$J_B = \left(-\frac{x_B}{x_A} L_{BA} + L_{BB}\right)\left(\frac{kT}{Cc_B\bar{v}_A}\right)\left(1 + \frac{d \ln \gamma_B}{d \ln x_B}\right)\nabla c_B . \tag{68b}$$

Comparing Eqs. (68) with the definition of the intrinsic diffusion coefficients given by Eqs. (63) we see that

$$D_A^I = \left(L_{AA} - \frac{x_A}{x_B} L_{BA}\right)\left(\frac{kT}{Cc_A\bar{v}_B}\right)\left(1 + \frac{d \ln \gamma_A}{d \ln x_A}\right) , \tag{69a}$$

$$D_B^I = \left(-\frac{x_B}{x_A} L_{BA} + L_{BB}\right)\left(\frac{kT}{Cc_B\bar{v}_A}\right)\left(1 + \frac{d \ln \gamma_B}{d \ln x_B}\right) . \tag{69b}$$

Part of the composition dependence of the intrinsic diffusion coefficients, the last two factors in Eqs. (69a) and (69b), is now known and depends upon that of the atomic volumes and that of the activity coefficients. We note that, in view of Eq. (66), the factor $(1 + d \ln \gamma_i/d \ln x)$ is the same for both intrinsic diffusion coefficients. We can call

$$\frac{kT}{Cc_A\bar{v}_B}\left(1 + \frac{d \ln \gamma_A}{d \ln x_A}\right) \quad \text{and} \quad \frac{kT}{Cc_B\bar{v}_A}\left(1 + \frac{d \ln \gamma_B}{d \ln x_B}\right)$$

the thermodynamic factors of the intrinsic diffusion coefficients, since they depend entirely upon the equilibrium properties of the system and are independent of the diffusion mechanism. When A and B form an ideal solid solution, the activity coefficients are unity. What we have called the thermodynamic factor differs slightly from what has been called the same thing in the literature. Part of this difference is because we have defined our intrinsic diffusion coefficients in terms of concentration gradients rather than atom fraction gradients, part because we do not here assume that $\bar{v}_A$ and $\bar{v}_B$ are constant and equal. The composition dependence of that part of the intrinsic diffusion coefficients depending upon the phenomenological coefficients L_{ij} depends upon the details of the diffusion mechanism. If we define mobilities B_A^I and B_B^I by

$$c_A B_A^I = L_{AA} - \frac{x_A}{x_B} L_{AB} , \tag{70a}$$

$$c_B B_B^I = - \frac{x_B}{x_A} L_{BA} + L_{BB} , \tag{70b}$$

we can rewrite Eqs. (69) as

$$D_A^I = \frac{kTB_A^I}{C\bar{v}_B} \left(1 + \frac{d \ln \gamma_A}{d \ln x_A}\right) , \tag{71a}$$

$$D_B^I = \frac{kTB_B^I}{C\bar{v}_A} \left(1 + \frac{d \ln \gamma_B}{d \ln x_B}\right) , \tag{71b}$$

so that the chemical diffusion constant $\mathscr{D}$, which can be obtained from an experimental concentration-penetration curve by Eq. (47), can be written using the above equations and Eq. (39) as

$$\begin{aligned}\mathscr{D} &= c_B \bar{v}_B D_A^I + c_A \bar{v}_A D_B^I \\ &= (x_B B_A^I + x_A B_B^I) kT \left(1 + \frac{d \ln \gamma_A}{d \ln x_A}\right) \\ &= \left(\frac{x_B L_{AA}}{x_A} + \frac{x_A L_{BB}}{x_B} - 2L_{AB}\right)\left(\frac{kT}{C}\right)\left(1 + \frac{d \ln \gamma_A}{d \ln x_A}\right) .\end{aligned} \tag{72}$$

C. The Relationship between Tracer Self-Diffusion and Intrinsic Diffusion Coefficients

By the application of irreversible thermodynamics we have obtained part of the composition dependence of the intrinsic and chemical diffusion coefficients [Eqs. (71) and (72)] in a form that can be evaluated from experimental data. The next step would be to determine the rest of the composition dependence, which by Eq. (71) would amount to determining the composition dependence of the mobilities B_i^I. These will depend upon the concentration of the point defect involved in the diffusion mechanism, but the equations are complicated by the presence of cross coefficients L_{ij}, $i \neq j$, which may not be equal to zero, due at least in part to correlation effects. In the older literature, and still occasionally, the quantities kTB_i^I were referred to as self-diffusion coefficients and were erroneously assumed to be equal to the corresponding tracer self-diffusion coefficients. One can show that this is not true for diffusion via point defects in solids in two ways: in a general way by using irreversible thermodynamics, and for special cases but more explicitly from kinetic treatments of diffusion. The development is lengthy for both approaches so we cite only the results here and refer the interested reader to a review article by Howard and Lidiard [21] for the application of irreversible thermodynamics and the book by Manning [8] for a kinetic treatment.

Consider the simultaneous diffusion of species a, α (which are isotopes of A), b, and β (which are isotopes of B). There are then 4 diffusion flow equations of the form of Eqs. (60). Each flow depends linearly upon the chemical potential gradients of all 4 species, so there are 16 phenomenological coefficients. By the Onsager reciprocity relation, Eq. (56), only 6 of the 12 nondiagonal coefficients are independent, leaving 10 coefficients. However, by considering tracer self-diffusion of the A atoms, then of the B atoms, 5 independent relations are obtained, so that finally only 5 of the 16 phenomenological coefficients are independent. In principle, these could be determined by measuring $\mathscr{D}$, $D_A^* = D_\alpha$, $D_B^* = D_\beta$, the marker velocity u_{IL}, and finally, isotope drift in the presence of an overall gradient in the chemical composition. As a result of this sort of analysis, the intrinsic diffusion coefficients given by Eqs. (71) can be written as

$$D_A^I = \left(\frac{D_A^*}{kT} + \frac{c_A L_{a\alpha}}{c_a c_\alpha} - \frac{L_{ab}}{c_B} \right) \left(\frac{kT}{c\bar{v}_B} \right) \left(1 + \frac{d \ln \gamma_A}{d \ln x_A} \right), \tag{73a}$$

$$D_B^I = \left(\frac{D_B^*}{kT} + \frac{c_B L_{b\beta}}{c_b c_\beta} - \frac{L_{ab}}{c_A} \right) \left(\frac{kT}{C\bar{v}_A} \right) \left(1 + \frac{d \ln \gamma_A}{d \ln x_A} \right) . \qquad (73b)$$

By a comparison with Eqs. (71) we see the quantity kTB_i^I is equal to D_i^* plus additional terms that are not zero unless the cross coefficients $L_{a\alpha}$, $L_{b\beta}$, and L_{ab} are zero. The chemical diffusion coefficient and the marker velocity can be written in terms of the tracer self-diffusion coefficients and these cross coefficients by substituting Eqs. (73) into Eq. (72) and Eq. (38) respectively.

Manning [8, pp. 218-221] has given a kinetic treatment of diffusion via the vacancy mechanism in a nondilute, disordered, A-B solid solution in which the A atoms and B atoms jump into the same type of vacancy. Assuming the total atom concentration is constant [so that $\bar{v}_A = \bar{v}_B =$ constant and $C\bar{v}_A = C\bar{v}_B = 1$ in Eqs. (71)], the phenomenological coefficients of Eqs. (62) are

$$L_{AA} = [c_A D_A^*/kT][1 + 2x_A D_A^* F] \quad ,$$

$$L_{BB} = [c_B D_B^*/kT][1 + 2x_B D_B^* F] \quad , \qquad (74)$$

$$L_{AB} = L_{BA} = 2c_A c_B D_A^* D_B^* F/CkT \quad ,$$

where

$$F = [M_0 (x_A D_A^* + x_B D_B^*)]^{-1}$$

and M_0 is a constant which represents the effect of the crystal structure on the correlation coefficient and is 7.15 for a fcc structure, 5.33 for a bcc structure, 3.77 for a simple cubic structure, and 2 for a diamond structure. The larger the value of M_0, the smaller is the cross coefficient L_{AB} and the more closely Eqs. (70) for the mobilities approach the approximate relation $B_i^I = D_i^*/kT$. The intrinsic diffusion coefficients given by Eqs. (69) can be expressed in terms of the tracer self-diffusion coefficients using Eq. (74). The chemical diffusion coefficient given by

Eq. (72) and the marker velocity given by Eq. (38) can then be written, respectively,

$$\mathscr{D} = \left(x_B D_A^* + x_A D_B^* + 2x_A x_B F(D_A^* - D_B^*)^2\right)\left(1 + \frac{d \ln \gamma_A}{d \ln x_A}\right) \tag{75}$$

and

$$u_{IL} = (D_A^* - D_B^*)\left(1 + \frac{2}{M_0}\right)\left(\frac{1}{C}\right)\left(1 + \frac{d \ln \gamma_A}{d \ln x_A}\right)\frac{\partial c_A}{\partial z} \quad . \tag{76}$$

(The integral term in the general expression for u_{IL} disappears since we have assumed $\bar{v}_A = \bar{v}_B$ = constant.) Thus this kinetic treatment does not give the same expression for the chemical diffusion coefficient and marker velocity as would be obtained by making the assumption that $B_i^I = D_i^*/kT$. The marker velocity is greater by a factor $1 + 2/M_0$, i.e., 1.14 for a fcc structure. The chemical diffusion coefficient is greater by a factor that (1) depends on $(D_A^* - D_B^*)^2$ and goes to zero as this difference does and (2) is smaller than the corresponding factor for u_{IL}.

As a second example consider diffusion in an ordered compound AB in which the A and B atoms occupy sites in separate sublattices and do not move into sites in the other sublattice during the diffusion jumps. It is then assumed that L_{ab} in Eqs. (73) is zero. Kinetic treatments give $c_A L_{a\alpha}/c_a c_\alpha$ and $c_B L_{b\beta}/c_b c_\beta$ as proportional to D_A^* and D_B^*, respectively, so that by Eqs. (70)

$$D_A^*/kT = B_A^I f_A \quad , \tag{77a}$$

$$D_B^*/kT = B_B^I f_B \quad , \tag{77b}$$

where f_A and f_B are correlation factors dependent only upon the crystal structure and diffusion mechanism and are listed in Table 1. By comparison with Eq. (18), the mobilities for the vacancy mechanism are

$$B_A^I = k_A(V_A/S_A) \quad , \tag{78a}$$

$$B_B^I = k_B(V_B/S_B) \quad , \tag{78b}$$

where V_A and V_B are the concentrations of A vacancies and B vacancies, respectively, S_A and S_B are the concentrations of sites in each sublattice, and k_A and k_B are composition-independent factors depending upon the jump frequency, jump distance, and coordination number. Similar equations apply for other diffusion mechanisms, the concentration of the point defect essential for the mechanism substituting for that of the vacancies in Eqs. (78). Equations (72) and (38) for the chemical diffusion coefficient and marker velocity become, respectively,

$$\mathscr{D} = \left(\frac{x_B D_A^*}{f_A} + \frac{x_A D_B^*}{f_B}\right)\left(1 + \frac{d \ln \gamma_A}{d \ln x_A}\right) , \tag{79}$$

$$u_{IL} = \left(\frac{D_A^*}{\bar{v}_B f_A} - \frac{D_B^*}{\bar{v}_A f_B}\right) \frac{\bar{v}_A}{C} \frac{\partial c_A}{\partial z} + \int_{-\infty}^{z} \frac{\mathscr{D}}{c_A \bar{v}_B} \left(\frac{\partial \bar{v}_B}{\partial c_A}\right)\left(\frac{\partial c_A}{\partial z}\right)^2 dz' . \tag{80}$$

If $f_A = f_B$, both $\mathscr{D}$ and the first term for u_{IL} are larger by a factor $1/f$ than would be calculated assuming $B_i^I = D_i^*/kT$.

VII. INTERDIFFUSION IN SEMICONDUCTING COMPOUNDS

A. Introduction

We now make use of our previous discussion in a detailed development of diffusion in a semiconducting compound. Generally, such compounds are ordered, the more metallic and less metallic elements each occupying sites in their own sublattices and possessing some fractional positive or negative charge, respectively, as a consequence of the partial ionic character of the crystal bonding. Those compounds with band gaps of about 4-5 eV and higher conduct an electrical current almost entirely by ions, the electronic transport number being close to zero (e.g., alkali metal and silver halides). These are usually referred to as ionic compounds and do indeed include those compounds in which the binding is most nearly ionic, as might be judged from a comparison of calculated and measured cohesive energies. Those compounds with band gaps of about 2-3 eV or less conduct an electrical current almost entirely by electrons and/or holes and the electronic transport number is close to unity (e.g., III-V, IV-VI, II-VI binary compounds; InAs, PbSe, ZnTe). The composition corresponding to the perfect crystal structure of the compound is called the stoichiometric composition. The homogeneity range of the compound is generally close to the stoichiometric composition if it in fact does not include it. Generally these homogeneity ranges are narrow, 1 at. % or much less, but for some of the transition metal semiconductors they are about 5 at. % wide (FeO, CoO).

Changes in the composition of the pure compound from one point in the homogeneity range to another by diffusion is an example of chemical diffusion and is generally called interdiffusion. An example has already been discussed in Section I. As illustrated there, for those semiconducting compounds whose homogeneity ranges are sufficiently wide to entail a significant variation in physical properties, interdiffusion is vitally connected with the preparation of samples with uniform composition or with desirable composition gradients. Among others, these include the IV-VI and II-VI binary compounds and many transition metal oxides. On the other hand, for some classes of compounds, the homogeneity range seems to be narrow enough that the carrier concentrations do not vary significantly across it. Most of the III-V compounds fall in this category although there is evidence that GaSb has a measurable homogeneity range. (For InAs and GaAs the chemical potentials have been shown to vary significantly across the homogeneity range.) Although the alkali metal and silver halides show some indications of a measurable homogeneity range, such as a varying concentration of cation or anion vacancies (color centers) upon annealing under various halogen pressures, most diffusion measurements on these materials are interpreted on the assumption that, if pure, the compounds are stoichiometric.

Aside from being involved in the change of composition, interdiffusion is also important in tarnishing reactions and received its earliest discussion for this type of problem. Here a metal is attacked by an electronegative gas to form a layer of a metal-gas compound on the metal surface. If this layer is adhering, subsequent attack of the underlying metal depends upon the diffusion of the metal and/or the electronegative gas, usually as ions, through the layer. The width of the compound's homogeneity range is of minor importance then, in the sense that the greatest part of the matter transport is used to increase the thickness of the oxidation layer.

We shall illustrate the general features of the interdiffusion process with a specific example based upon a modification of the original treatment by Wagner [22]. The original derivation is reproduced in the book by Jost [23] and has been extended by the author [24]. Although we shall start out in a general way, we shall carry the analysis to completion only for semiconducting compounds whose electronic transport number is close to unity.

B. The Vacancy Model and the Flow Equations for Charged Particles

Let the stoichiometric composition be MN where the more metallic element M forms a cation C in the compound bearing a charge $+\delta e$ and the less metallic element N forms an anion bearing a charge $-\delta e$. (e is the magnitude of the free electron charge; δ is a fractional or integral positive number.) The predominant atomic point defects are vacancies in the cation sublattice which are acceptor levels capable of a single stage of ionization and vacancies in the anion sublattice which are donor levels also capable of a single stage of ionization.

Following conventional nomenclature the concentrations of ionized and un-ionized M-vacancy acceptors are written as V'_M and V^x_M, respectively. The x superscript indicates the un-ionized acceptor has the same charge as the cation normally occupying the site, i.e., zero relative charge. To avoid confusion we shall not refer to these as neutral vacancies, as is usually done. When the acceptor is ionized by placing an electron there, the relative charge is negative as indicated by the prime superscript. The concentrations of ionized and un-ionized N-vacancy donors are written as $V_N^{\cdot}$ and V_N^{x}, respectively. The un-ionized N-vacancy is neutral in a relative sense. It is assumed to have the same real charge as the anion. There are, of course, compounds, e.g., ZnTe, in which a vacancy is capable of two stages of ionization. There is no difficulty in extending the treatment to cover these. If S is the concentration of sites in each sublattice, we then have

$$S = C_M + V_M^x + V_M' = C_N + V_N^x + V_N^{\cdot} \quad , \tag{81}$$

where we allow the symbols identifying the various types of vacancies to also represent the concentrations of these species.

We assume that diffusion occurs by the vacancy mechanism, the cations and anions moving between sites in their own sublattices and in such a way that local thermodynamic equilibrium and local electrical neutrality are maintained.

The condition for electrical neutrality can be obtained by equating the total positive charge in the compound to the total negative charge:

$$\delta C_M + \delta V_M^x + (\delta - 1)V_M' + p = \delta C_N + \delta V_N^x + (\delta - 1)V_N^{\cdot} + n \, , \tag{82}$$

where n and p are the concentrations of electrons and holes.

Using Eq. (81) we can write Eq. (82) in the usual form:

$$n - p = V_N^{\cdot} - V_M' \quad . \tag{83}$$

The fact that the actual charge on the cations or ions does not appear in Eq. (83) is a result of the fact that the charge on an un-ionized vacancy was taken to be the same as that on an ion in the same sublattice. This is important since the actual charges on the components of a crystalline compound have generally not been determined.

Now when an ion moves, it jumps into either an un-ionized vacancy or into an ionized vacancy in the same sublattice, so that the total diffusion flows for the components of the compound, J_M and J_N, are related to those of the vacancies by

$$J_M + J_{V^x_M} + J_{V'_M} = 0 \quad , \tag{84a}$$

$$J_N + J_{V^x_N} + J_{V^{\cdot}_N} = 0 \quad . \tag{84b}$$

As can be seen from the charges on the various species given in Eq. (82), there is no net transfer of charge when an ion exchanges places with the corresponding un-ionized vacancy, V^x_M or V^x_N. However, there is a transfer of charge when an ion exchanges places with the corresponding ionized vacancy. Thus, independent of the actual charge on the ion, a charge e is transferred in the direction of the cation jump and a charge -e in the direction of the anion jump for an exchange with the appropriate ionized vacancy. It is to be expected that the jump frequencies will be different for the four kinds of vacancies considered in our model and indeed for $\delta \geq 1$, much smaller for an un-ionized vacancy than for the corresponding ionized vacancy because of coulombic repulsion. If we had considered the vacancies to be doubly ionizable we would have to consider six different jump frequencies.

Now from irreversible thermodynamics, when an electrostatic potential is present in an isothermal system, the force conjugate to the diffusion flow J_i is given by

$$X_i = -\nabla(\mu_i - q\phi) \quad , \tag{85}$$

where q is the charge on the ith species. (The electrostatic potential used here is the negative of the quantity as usually defined so that a negative charge is driven to lower potentials.) Following our previous discussion, we assume the electrostatic potential acts in an ion-vacancy exchange proportional to the net charge transported. The diffusion flows are to be measured relative to inert markers or, equivalently, relative to the local crystal lattice. However, we shall omit the superscript I usually attached to the flow symbols J_i and the diffusion coefficient symbols D_i to indicate this choice of flow reference plane. The diffusion flows are then:

$$J_M = -C_M B_C \nabla\mu_M - C_M b_C \nabla(\mu_C - e\phi) \quad , \tag{86a}$$

$$J_N = -C_N B_A \nabla\mu_N - C_N b_A \nabla(\mu_A + e\phi) \quad , \tag{86b}$$

$$J_n = -nb_n \nabla(\mu_e + e\phi) \quad , \tag{86c}$$

$$J_p = pb_p \nabla(\mu_e + e\phi) \quad , \tag{86d}$$

where B_C and B_A are the mobilities for, respectively, cation and anion jumps into the corresponding un-ionized vacancies; b_C and b_A are the corresponding mobilities for jumps into ionized vacancies; and b_n and b_p are the mobilities of electrons and holes. We assume the distribution of electrons and holes is rapid compared to the movement of the ions, so that local electronic equilibrium prevails. Therefore the chemical potential for holes is the negative of that for electrons. The chemical potentials in Eqs. (86) are related by the equations

$$\mu_M = \mu_C + \mu_e \quad , \tag{87a}$$

$$\mu_N = \mu_A - \mu_e \quad , \tag{87b}$$

and the Gibbs-Duhem relation

$$x_M \nabla\mu_M + x_N \nabla\mu_N = 0 \quad . \tag{88}$$

The flows for the ions consist of two terms, the first being the negative of the flow of un-ionized vacancies. We assume the electrostatic field does not influence this flow, which involves no net transfer of charge upon the exchange of ion and vacancy. The second term in

J_M and J_N gives the negative flow of the ionized vacancies which involves a transfer of charge of $\pm e$ upon the exchange of ion and vacancy and is influenced by the electrostatic field. We also note that the chemical potentials appearing in the flows of un-ionized and ionized vacancies are different. The chemical potentials μ_M and μ_N are for the M-component and N-component, respectively, and are thermodynamic quantities independent of the charge these components may take in forming the compound MN. A statistical mechanical treatment gives two expressions each for μ_M and μ_N. In one, the chemical potential is of the form of Eqs. (87) where μ_C or μ_A depends upon the concentration of ionized vacancies and μ_e is the chemical potential of an electron. In the other, the chemical potential depends upon the concentration of un-ionized vacancies and μ_e does not appear. Thus $-\mu_C$ and $-\mu_A$ can be considered to be chemical potentials of, respectively, the ionized M-vacancies and the ionized N-vacancies and are the appropriate potentials for the driving forces for the flow of ionized vacancies. On the other hand $-\mu_M$ and $-\mu_N$ can be considered to be the chemical potentials of, respectively, un-ionized M-vacancies and un-ionized N-vacancies and are the appropriate potentials for their flows. Equations (86a) and (86b) then follow from Eqs. (84a) and (84b).

Now the driving forces in Eq. (86) are of the form required by irreversible thermodynamics. We have assumed, however, that all interactions among the flows are entirely accounted for by the electrostatic field and Eqs. (87) and (88). This is a plausible assumption since the cations and anions diffuse on separate sublattices and the electronic flows occur by mechanisms independent of those for the atomic point defects. If we were to consider the diffusion of M or N tracers or that of a substitutional impurity, each ion diffusion flow would then have to depend on the driving forces for all the ionic species occupying the same sublattice. These points have been discussed in connection with Eq. (77) and Eq. (79). We have also used the product of a concentration and a mobility in the place of the phenomenological coefficients L_{ij} used in

Section VI. This of course is simply a change of nomenclature, the choice in Eqs. (86) coinciding with that commonly used in discussing ion and electronic transport.

If all the chemical potentials are constant in space, the current is obtained from the flow equations (86) as

$$I = (C_M b_C + C_N b_A + nb_n + pb_p) e \nabla(e\phi) \quad , \tag{89}$$

and the conductivity then is

$$\sigma = (C_M b_C + C_N b_A + nb_n + pb_p) e^2 \quad . \tag{90}$$

The transport numbers are defined as:

$$t_e = (nb_n + pb_p) e^2/\sigma \quad , \tag{91a}$$

$$t_C = C_M b_C e^2/\sigma \quad , \tag{91b}$$

$$t_A = C_N b_A e^2/\sigma \quad . \tag{91c}$$

It is to be noted that the contribution of the cations and the anions to current flow is as though they possessed unit positive and negative charges, respectively, rather than the actual charges $+\delta e$ and $-\delta e$. Because the ion flows are accompanied by opposite flows of charged vacancies, the ion current is not simply $\delta e(J_M - J_N)$. Such a simple result would follow if the ions had an integral charge of magnitude Ze, and if the vacancies of both kinds were entirely ionized Z times, the case usually considered in discussions of interdiffusion.

C. Tracer Self-Diffusion for the Vacancy Mechanism

In order to learn more concerning the composition dependence of the interdiffusion coefficient we consider self-diffusion of the cations by tracers. The electron, hole, and anion diffusion flows of Eqs. (86) are each zero. The gradient of the diffusion potential is also zero. Electrical neutrality is maintained by the flows of stable isotope and tracer cations, having the same effective mobilities, $B_C + b_C$, being opposite and equal. The problem then reduces to essentially that considered in Section VI except that here there are two kinds of vacancies for the cations to jump into. Analogous results follow from a consideration of anion self-diffusion by tracers. As a result, we can write modified versions of Eqs. (77) and (78):

$$D_M^* = kT(B_C + b_C)f_M \quad , \tag{92a}$$

$$D_N^* = kT(B_A + b_A)f_N \tag{92b}$$

and

$$B_C = K_C(V_M^x/S) \quad , \qquad b_C = k_C(V_M'/S) \quad , \tag{93a}$$

$$B_A = K_A(V_N^x/S) \quad , \qquad b_A = k_A(V_N^{\bullet}/S) \quad . \tag{93b}$$

At this point, we note that when an external electric field is applied to a uniform sample (no gradients in the chemical potentials) the current carried by the cations is obtained from Eq. (86) as

$$C_M b_C e^2 \nabla\phi = \sigma_C \nabla\phi = t_c \sigma \nabla\phi \quad . \tag{94}$$

Comparing this to Eq. (92) for the self-diffusion coefficient, we obtain a modified Nernst-Einstein equation

$$\frac{\sigma_C}{D_M^*} = \frac{C_M b_C e^2}{kT(B_C + b_C)f_M} \quad . \tag{95}$$

An analogous equation is obtained relating the anion partial conductivity and the self-diffusion coefficient for anions, D_N^*. Equation (95) differs from the relation given in the older literature because of the presence of the correlation factor. It also differs because cation jumps into un-ionized vacancies, characterized by a mobility B_C, contribute to diffusion flow but not to the current flow. As a result, a factor $b_C/(B_C + b_C)$ is introduced into the right-hand side of Eq. (95). The modified Nernst-Einstein relation is extremely useful in establishing the diffusion mechanism in very ionic compounds for which the electronic transport is near zero and ion conductivities can be directly measured.

D. Interdiffusion in the Absence of Externally Applied Fields

We now wish to specialize the flow equations given by Eqs. (86) to the case of interdiffusion when there is no externally applied electrostatic field. Because the different charged species differ in their mobilities, an internal diffusion potential must be set up in order to maintain electrical neutrality. The magnitude of this diffusion potential can be obtained and then eliminated from the flow equations. We start by applying the equation of continuity to the electroneutrality condition given by Eq. (83):

$$0 = \frac{\partial}{\partial t}(n - p + V'_M - V^{\bullet}_N)$$

$$= -\nabla \cdot [J_n - J_p + C_M b_C \nabla(\mu_C - e\phi) - C_N b_A \nabla(\mu_A + e\phi)]$$

$$+ G_n - G_p + G_{V'_M} - G_{V^{\bullet}_N} \quad , \tag{96}$$

where $\nabla \cdot J_i$ is the divergence of the vector flow J_i and G_j is the source strength or generation rate for the jth species. The source strength enters because the concentrations of electrons, holes, and ionized vacancies can change without diffusion, as required by local electronic equilibrium, by simply shifting electrons among the various energy levels. Since the total number of electrons is conserved, the source strengths are coupled by the equation

$$G_n + G_{V'_M} = G_p + G_{V^{\bullet}_N} \quad . \tag{97}$$

Thus they make no net contribution to Eq. (96).

The diffusion potential can now be obtained by substituting for J_n and J_p from Eqs. (86) into Eq. (96). We obtain

$$\nabla(e\phi) = t_C \nabla\mu_C - t_A \nabla\mu_A - t_e \nabla\mu_e \quad , \tag{98}$$

where the transport numbers t_j are defined by Eqs. (91). Using Eq. (98) to eliminate the gradient of the diffusion potential, and Eqs. (87) and (88) to express the gradients in terms of those in μ_M and μ_N, the ion flows are

$$J_M = -C_M B_C \nabla\mu_M - C_M b_C (1 - t_C - t_A x_M/x_N) \nabla\mu_M \tag{99a}$$

$$= -C_M B_C \nabla\mu_M - C_M b_C t_e \nabla\mu_M - C_M b_C t_A (1 - x_M/x_N) \nabla\mu_M, \tag{99b}$$

$$J_N = -C_N B_A \nabla\mu_N - C_N b_A (1 - t_A - t_C x_N/x_M) \nabla\mu_N \tag{100a}$$

$$= -C_N B_A \nabla\mu_N - C_N b_A t_e \nabla\mu_N - C_N b_A t_c (1 - x_N/x_M) \nabla\mu_N \ . \tag{100b}$$

The first term in each equation is the contribution made by the diffusion of un-ionized vacancies. As discussed previously, there is no net current flow associated with the exchange of positions between an ion and the corresponding un-ionized vacancy. The remaining term (or terms) in each equation is the contribution made by the movement of singly ionized vacancies. Equations (99b) and (100b) are obtained from (99a) and (100a), respectively, by utilizing the fact that the transport numbers, by definition, add to unity. The last term in both Eqs. (99b) and (100b) is negligible compared to the middle term whenever $(1 - x_M/x_N)t_A$ and $(1 - x_N/x_M)t_C$ are small compared to t_e. This is surely the case when t_e is near unity, but for a compound with a narrow homogeneity range, $x_M \approx x_N$, it is still true for somewhat smaller values of t_e. Equations (99) and (100) are almost the same as those originally obtained by Wagner. Here we have shown that the actual charge carried by the cation or anion is not involved in the flow equations, so that their validity is not confined to purely ionic compounds and we have considered ion flow via un-ionized as well as by singly ionized vacancies.

The assumption of local electrical neutrality made at the beginning of this subsection is generally regarded as a good one for the more ionic compounds. An even better assumption for those compounds for which $t_e \approx 1$ might be that the gradient of the Fermi level, $\nabla(\mu_e + e\phi)$, is zero. However, the consequences of this assumption have not yet been developed. Some discussions of interdiffusion have been given in which the assumption of local electrical neutrality is set aside but only when the chemical potentials are those of an ideal solution. In the author's opinion, the best approach at present is to accept the approximation of electrical neutrality and to recognize that the ordered compounds considered are far from ideal.

E. Interdiffusion in a Semiconducting Compound with $t_e \approx 1$

Equations (99) and (100) are applicable to semiconducting compounds regardless of the relative magnitudes of the transport numbers. We now consider the case in which the electronic transport number is near enough to unity that the last term in both Eqs. (99b) and (100b) may be neglected. These equations for the ion flows are then

$$J_M = -C_M(B_C + b_C)t_e \nabla\mu_M \quad , \tag{101a}$$

$$J_N = -C_N(B_A + b_A)t_e \nabla\mu_N \quad . \tag{101b}$$

We can define practical diffusion coefficients from the equations

$$J_M = -D_M \nabla C_M \quad , \tag{102a}$$

$$J_N = -D_N \nabla C_N \quad , \tag{102b}$$

so that, using Eqs. (64) and (67) to convert $\nabla\mu_i$ to ∇C_i, we obtain the analogs of Eqs. (71):

$$D_M = (B_C + b_C)\left(\frac{kT}{C\bar{v}_N}\right)\left(1 + \frac{d \ln \gamma_M}{d \ln x_M}\right)t_e \quad , \tag{103a}$$

$$D_N = (B_A + b_A)\left(\frac{kT}{C\bar{v}_M}\right)\left(1 + \frac{d \ln \gamma_N}{d \ln x_N}\right)t_e \quad , \tag{103b}$$

where $C = C_M + C_N$. We recall that, by Eq. (66), the last factor in Eq. (103a) is equal to that in Eq. (103b).

The deviation from the exact stoichiometric composition Δ is of more interest than the individual ion concentrations. By Eq. (81) we can write the deviation from stoichiometry in terms of the defect concentrations as

$$\Delta \equiv C_M - C_N = V_N^x - V_M^x + V_N^{\cdot} - V_M^{\prime} \quad , \tag{104a}$$

or, using the electroneutrality condition given by Eq. (83), as

$$\Delta = (V_N^x - V_M^x) + n - p \quad . \tag{104b}$$

In semiconducting compounds such as PbS, PbSe, and PbTe, the atomic point defects appear to be completely singly ionized, so that the deviation from stoichiometry is equal to the net electron concentration, $n - p$, which is readily obtainable from Hall measurements, for homogeneous samples at least. The equality of Δ and $n - p$ for atomic point defects that are completely singly ionized has been developed here in terms of a vacancy model but is true independent of the nature of the electrically active atomic point defects.

Since we are primarily interested in the deviation from stoichiometry, Δ, we are interested in the excess cation flow, $J_M - J_N$. In particular, we wish to obtain $\partial\Delta/\partial t$, applying the results of Section V. As mentioned earlier, the diffusion flows J_i are measured relative to inert markers or, equivalently, relative to the local crystal lattice. Thus, D_M and D_N are intrinsic diffusion coefficients. The partial derivative of C_M or of C_N with respect to time is then not given simply by the negative divergence of the corresponding flow. However, these derivatives can be obtained from Eq. (40) upon a change to the symbols used in this section. For one-dimensional or planar diffusion, we have

$$\frac{\partial \Delta}{\partial t} = \frac{\partial}{\partial z}\left(\mathscr{D}\frac{\partial \Delta}{\partial z} - C_M I_M + C_N I_N\right) , \tag{105}$$

where

$$\mathscr{D} = C_N \bar{v}_N D_M + C_M \bar{v}_M D_N \tag{106}$$

and where I_M is the integral of Eq. (40) when $C_i = C_M$ and I_N is the integral when $C_i = C_N$. For interdiffusion in a compound MN with a homogeneity range of 1 at. % or less, we might expect that the partial atomic volumes $\bar{v}_M$ and $\bar{v}_N$ are each constant. In this case the integrals I_M and I_N, which contain, respectively, $\partial \bar{v}_N/\partial C_M$ and $\partial \bar{v}_M/\partial C_N$ as a factor in the integrand, are each zero. As a result, Eq. (105) simplifies to Fick's second law:

$$\frac{\partial \Delta}{\partial t} = \frac{\partial}{\partial z}\left(\mathscr{D}\frac{\partial \Delta}{\partial z}\right) . \tag{107}$$

Using Eqs. (103) we can write Eq. (106) for the interdiffusion coefficient as

$$\mathscr{D} = kT[x_N(B_C + b_C) + x_M(B_A + b_A)][1 + d \ln \gamma_M/d \ln x_M]t_e . \tag{108}$$

To recapitulate, we have arrived at Eqs. (107) and (108) assuming: (1) six species are diffusing, ionized and un-ionized vacancies of each kind, electrons and holes; (2) local thermodynamic equilibrium and local electrical neutrality are maintained; (3) the electronic transport number is close enough to unity that Eqs. (99b) and (100b) may be replaced by Eqs. (101a) and (101b), respectively; and (4) the partial atomic volumes $\bar{v}_M$ and $\bar{v}_N$ are constant. The result is that the time rate of change for the deviation from stoichiometry is given by Fick's second law, Eq. (107), with an interdiffusion constant given by Eq. (108). We note that the interdiffusion constant is very similar to that obtained for the diffusion of uncharged A atoms and B atoms in the A-B system given by Eq. (72). The only difference is that in Eq. (108) a sum of two mobilities appears where only a single mobility appeared in Eq. (72), and Eq. (108) contains the electronic transport number as a factor. Using Eqs. (92) in

Eq. (108) we can write the interdiffusion coefficient in terms of the tracer diffusion coefficient. The result is the same as Eq. (79), obtained for the interdiffusion of atoms, if the right-hand side of Eq. (79) is multiplied by the electronic transport number t_e.

F. Complete Composition Dependence of the Interdiffusion Coefficient for the Special Case of Singly Ionized Vacancies and $t_e \approx 1$

As can be seen from Eq. (108), the composition dependence of the interdiffusion coefficient can be obtained using Eqs. (93) for the mobilities, if the chemical potential μ_M is known as a function of Δ. In general, the chemical potentials for semiconducting compounds with narrow homogeneity ranges are very strong functions of the atom fraction, so that the thermodynamic factor is large and the interdiffusion coefficient is large compared to the tracer diffusion coefficients. We shall show this for a special case. We further characterize our model of the semiconducting compound MN by assuming the vacancies are randomly distributed and require a fixed energy of creation. Then if the electronic distribution in the semiconducting compound is nondegenerate, the chemical potential of M in the pure compound is [25]:

$$\mu_M = kT \sinh^{-1}[(n - p)/2n_i] + kT \sinh^{-1}[(n - p)/2k_s^{\frac{1}{2}}] + \mu_M(\text{int}) , \qquad (109)$$

where the intrinsic carrier concentration $n_i = (np)^{\frac{1}{2}}$ and the Schottky constant $k_s = V_M' V_N^{\bullet}$ are functions of temperature only and where μ_M(int) is the chemical potential for the intrinsic compound (i. e., when $n - p = 0$). If the vacancies are all singly ionized, then by Eq. (104b), $\Delta = n - p = V_N^{\bullet} - V_M'$. From the definition of Δ we also have

$$\Delta = (x_M - x_N)(C_M + C_N) \approx 2S(x_M - x_N) \quad , \qquad (110)$$

where terms of the order of $(V_M'/S)^2$, etc., have been neglected in obtaining the last equality. Then in Eq. (108) we can set x_M and x_N, appearing in the first parentheses, equal to $\frac{1}{2}$ with negligible error and take $B_A = B_C = 0$ because there are no un-ionized vacancies. Equation (108) becomes

$$\mathscr{D} = kTt_e\left[\frac{k_C V'_M}{V'_M + V^{\bullet}_N} + \frac{k_A V^{\bullet}_N}{V'_M + V^{\bullet}_N}\right]\left[1 + \left(\frac{(\Delta/2)^2 + k_s}{(\Delta/2)^2 + n_i^2}\right)^{\frac{1}{2}}\right] \tag{111a}$$

$$= \frac{kTt_e}{2}\left[k_C\left(1 - \frac{\Delta}{2Q}\right) + k_A\left(1 + \frac{\Delta}{2Q}\right)\right]\left[1 + \left(\frac{(\Delta/2)^2 + k_s}{(\Delta/2)^2 + n_i^2}\right)^{\frac{1}{2}}\right] \tag{111b}$$

where

$$Q = [(\Delta/2)^2 + k_s]^{\frac{1}{2}} = (V'_M + V^{\bullet}_N)/2 \quad .$$

Changes in the deviation from stoichiometry by interdiffusion are then described by Eq. (107) for $\partial\Delta/\partial t$ and Eq. (111a) or (111b) for the interdiffusion coefficient. Assuming $t_e = 1$, the composition dependence of $\mathscr{D}$ is given by that of factors shown in parentheses in Eq. (111b). The first factor can vary from a value of k_C, the cation jump frequency, to a value of k_A, the anion jump frequency if the homogeneity range is wide enough. It would not be unexpected to find the ratio k_C/k_A varying between 1/10 and 100 for various semiconducting compounds. On the other hand, if the homogeneity range is relatively narrow and the Schottky constant relatively large, the first factor tends to remain nearly constant at a value of $k_C + k_A$. The second factor in parentheses in Eq. (111b) is symmetric about the stoichiometric composition, approaching a value of 2 for sufficiently large values of $|\Delta/2|$ and being $(1 + k_s/n_i^2)$ at $\Delta = 0$. Since it is not expected that a crystal will have an electronic transport number close to unity if the ratio k_s/n_i is too large, this second factor is perhaps limited to a maximum value of about 100. The behavior of the reduced interdiffusion coefficient $\mathscr{D}/kTk_C$ is shown as a function of $\Delta/2n_i$ in Figs. 8 and 9. In Fig. 8 the jump frequencies k_C and k_A are taken as equal so that the first factor in parentheses in Eq. (111b) is a constant and the composition dependence of $\mathscr{D}$ is that of the second factor, which is symmetric about

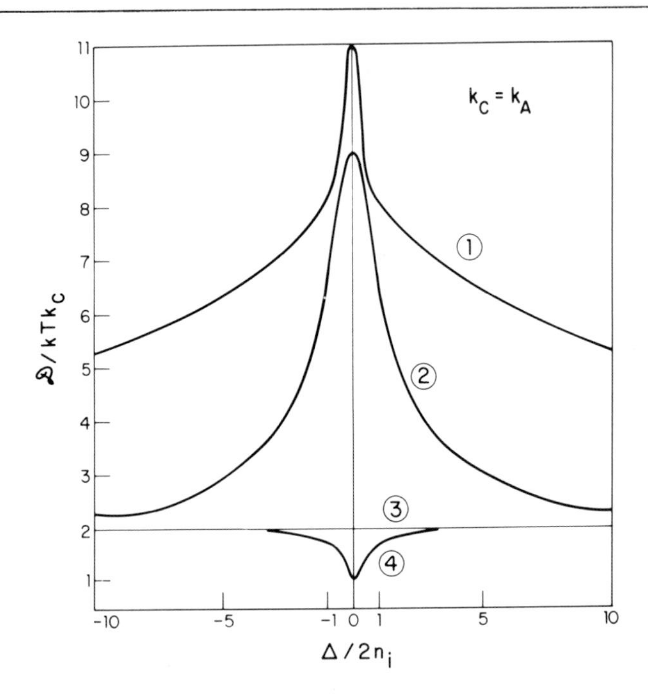

FIG. 8. Composition dependence of the reduced interdiffusion coefficient $\mathscr{D}/kTk_C$, where k_C is the cation jump frequency and k the Boltzmann constant, as a function of the reduced deviation from stoichiometry, $\Delta/2n_i$. The curves are obtained from Eq. (111) for a semiconductor compound with vacancies that are all singly ionized. For all the curves the cation and anion jump frequencies are equal, $k_C = k_A$, but the ratio of the Schottky constant for ionized vacancies to the square of the intrinsic carrier concentration, k_S/n_i^2, differs.

Curve	1	2	3	4
k_S/n_i^2	100	64	1	0.01

$\Delta/2n_i = 0$. When $k_S = n_i^2$ in addition to the $k_C = k_A$, $\mathscr{D}$ is independent of composition. In Fig. 9 the jump frequency for ionized M-vacancies, k_C, is taken as 10 times that for ionized N-vacancies, k_A, and $\mathscr{D}/kTk_C$ is shown for various values of the ratio k_S/n_i^2. This ratio might be termed an intrinsic disorder ratio; values larger than 1 indicate a greater intrinsic disorder of atomic point defects (vacancies) than of electrons. Values of k_S/n_i^2 less than 1 indicate the opposite. It is seen that $\mathscr{D}$ can be a strong function of composition.

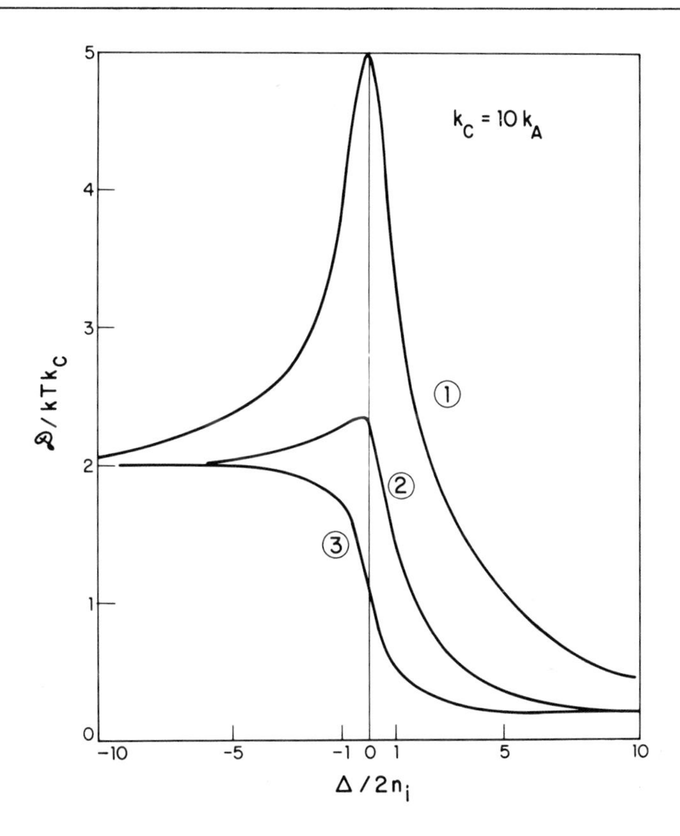

FIG. 9. The reduced interdiffusion coefficient as a function of the reduced deviation from stoichiometry from Eq. (111). For all the curves the cation jump frequency is ten times that for anions, $k_C = 10k_A$, but the intrinsic disorder ratio differs.

Curve	1	2	3
k_s/n_i^2	64	10	1

In general, the interdiffusion coefficient given by Eqs. (111) is larger than either tracer diffusion coefficient. At large negative values of Δ, where D_M^* is largest, we get, setting $t_e = 1$,

$$\frac{\mathscr{D}}{D_M^*} = \frac{2V_M'/(V_M' + V_N^\bullet)}{V_M'/S} + \frac{2k_A V_N^\bullet/(V_M' + V_N^\bullet)}{k_C V_M'/S}$$

$$\geq 2S/(V_M' + V_N^\bullet) \quad . \tag{112}$$

At the stoichiometric composition,

$$\frac{\mathscr{D}}{D_M^*} = \frac{[k_C + k_A]\{1 + [(1 + k_s)/(1 + n_i^2)]^{\frac{1}{2}}\}}{k_C(k_s^{\frac{1}{2}}/S)}$$

$$\geq S/k_s^{\frac{1}{2}} \quad , \tag{113}$$

where S is of the order of 10^{22} cm^{-3} and $k_s^{\frac{1}{2}}$ is of the order of 10^{15}-10^{20} cm^{-3}. The interdiffusion coefficient therefore can be 10^2 to 10^5 times as large as either tracer self-diffusion coefficient.

G. Application

Extensive experimental results for interdiffusion in semiconducting compounds with extremely narrow homogeneity ranges do not exist presently. Some of the major experimental difficulties are concerned with (1) keeping the concentrations of electrically active impurities well below those of the electrically active, native atomic point defects (at or below 10^{17} cm^{-3} for PbS, PbSe, and PbTe); (2) obtaining homogeneous crystals that are free from the internal microprecipitates that can occur when the solidus lines for the compound are retrograde and which significantly affect diffusion rates (experimental evidence indicates that these microprecipitates are difficult to avoid in some instances, e.g., Te-saturated PbTe, Se-saturated PbSe, Pb-saturated PbS, Te-saturated HgTe, and Hg-saturated HgSe); (3) developing experimental techniques to obtain concentration-penetration profiles. At present the crystal is usually cleaved or cut after diffusion and the p-n junction is located as the point at which the sign of the Seebeck coefficient changes sign using a fine probe.

For PbSe [26] the diffusion boundary conditions were established using initially homogeneous crystals and a two-phase vapor source giving a fixed surface concentration, as described in Section I. The p-n junction penetration depths were found to increase as $t^{\frac{1}{2}}$. The effective diffusion coefficients obtained by taking $\Delta = 0$ at the p-n

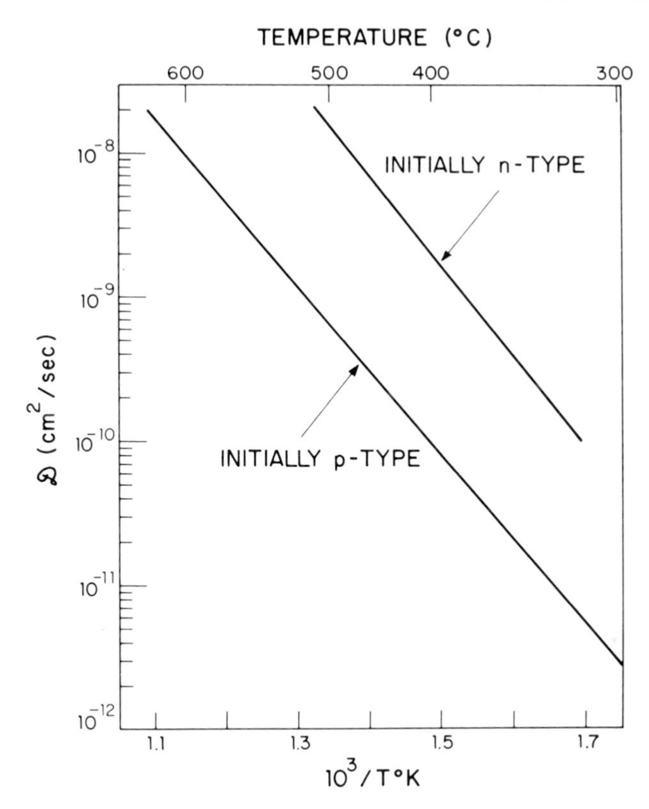

FIG. 10. Effective interdiffusion coefficient for PbSe plotted on a log scale against $10^3/T$ [27]. The coefficients were obtained from the location of the p-n junction and assuming the diffusion coefficient to be constant. The upper line is for the movement of a p-n junction into initially n-type PbSe, the lower for movement into initially p-type PbSe.

junction and assuming $\mathscr{D}$ is a constant in Eq. (107) are shown in Fig. 10. The fact that the diffusion constant is different depending upon whether initially p-type, Se-rich PbSe is converted to n-type, Pb-rich or whether an initially n-type crystal is converted to p-type is consistent with a composition-dependent interdiffusion coefficient. Approximating the interdiffusion coefficient by a step function which is 9 times greater for $\Delta < 0$ than for $\Delta > 0$ accounts fairly well for the observed effective diffusion constants and indicates the jump frequency for the atomic point defect predominant in p-type PbSe is 9 times greater than that for the predominant defect in n-type PbSe [27]. The effective interdiffusion coefficients are 2 to 3 orders of magnitude larger than the tracer diffusion coefficients [28]. The latter suggest Pb-vacancy acceptors and Pb-interstitial donors are the predominant atomic point defects in PbSe, although the results are not extensive.

Extensive tracer self-diffusion measurements have been made on semiconducting compounds. The physical situation is much simpler for tracer self-diffusion than for interdiffusion. Consequently the development of the appropriate equations is simpler and, involving fewer assumptions, is likely to be more generally valid. The appropriate equations are the analogs of Eqs. (92) and (93) which were given for the vacancy mechanism. The mobilities depend upon the concentrations of the appropriate point defects; and if the jump frequencies depend upon the temperature only, the variation of the tracer self-diffusion with composition is the same as the composition dependence of the atomic point defect concentrations. Often it is convenient to specify the composition indirectly by giving the temperature and the partial pressure of one elementary species for a binary system. The composition is then not necessarily known, but it is uniquely fixed, since by the Gibbs phase rule all intensive properties of a binary system are fixed at equilibrium when any two are. This convenience becomes a necessity when the homogeneity range is extremely narrow. Thus for the compound MN one obtains the tracer self-diffusion coefficients D_M^* and D_N^* as a function of T and of, say, p_{N_2}. If the predominant point defects are M-vacancies and N-vacancies, one expects D_M^* to increase and D_N^* to decrease with increasing p_{N_2}, since the M-vacancy concentration increases and N-vacancy concentration decreases with increasing p_{N_2}. In the range where the hole concentration is determined by that of M-vacancies, D_M^* will vary as $p_{N_2}^{\frac{1}{4}}$ or $p_M^{-\frac{1}{2}}$ if the M-vacancies are singly ionized acceptors or as $p_{N_2}^{\frac{1}{6}}$ or $p_M^{-\frac{1}{3}}$ if they are doubly ionized. If M-vacancies and M-interstitials are the major atomic point defects, the dependence of D_M^* upon p_{N_2} can show a minimum. The behavior of D_N^* will depend upon the type of point defect associated with the N-atoms. These will play a minor role in establishing the electrical properties.

The composition dependence of tracer self-diffusion coefficients and equilibrium properties is not always simple. For instance a predominant atomic point defect may establish the carrier concentration while a lower concentration defect, because of its higher mobility, determines the self-diffusion coefficient. Or some electrically inactive defects, undetected by electrical measurements, may determine the self-diffusion coefficient. The latter seems to be the case for II-VI compounds (the 1-1 compounds of Zn or Cd with S, Se, or Te) [29]. The tracer self-diffusion coefficient of the Group VI elements in general varies as the reciprocal of the metal partial pressure, consistent with the presence of neutral Group VI element interstitials. The tracer self-diffusion coefficient of the metallic component is more varied. Near metal saturation, D_M^* goes as $p_M^{\frac{1}{2}}$, consistent with a singly ionized metal interstitial (CdS, CdSe) or else is independent of p_M (ZnTe, CdTe) indicating that the point defect concentrations are essentially those set by thermal disorder. Near the Group VI homogeneity limit, a vacancy mechanism predominates in some cases.

Considerable diffusion data exist for FeO, MnO, and CoO which have homogeneity ranges wide enough to make traditional physico-chemical measurements feasible. Some of these will be discussed under Tarnishing Reactions. Recently the results in F have been successfully applied to CoO [37] and to PbTe [38].

H. Tarnishing Reactions

For many systems the thickness $\mathcal{E}_T$ of an adhering tarnish film changes with time according to

$$d\mathcal{E}_T/dt = k_1/\mathcal{E}_T \tag{114}$$

when the temperature and the partial pressure of the oxidizing gas over the metal being attacked are held constant. The integrated form of this so-called parabolic law is

$$\mathscr{E}_T^2 - \mathscr{E}_{T,0}^2 = 2k_1 t \quad , \tag{115}$$

where $\mathscr{E}_{T,0}$ is the thickness at $t = 0$. At the lowest temperatures and for very thin films, a logarithmic law is observed that has been explained in terms of the tunneling of electronic carriers through the film, the resulting electrostatic field increasing the diffusion rate of the ions [30]. The analysis of interdiffusion given in the previous sections, being based on the assumption of local electroneutrality, is not applicable in this case. However, it should lead to an understanding of the parabolic law assuming the slow step is diffusion through the tarnish film rather than a surface reaction. Unfortunately an exact solution of the appropriate boundary value problem is not known, due to the complication of the moving boundary at the oxidizing gas-tarnish film interface. For the special case of a composition-independent diffusion coefficient, it has been shown [31, 32] that the concentration gradient across the tarnish film is constant at any instant when the atomic concentration varies by only a small fraction across the film. (This is the case when the width of the homogeneity range of the compound forming the tarnish film is small.) Since the interdiffusion coefficient is generally not a constant, we shall not make use of this result.

Consider the tarnish film to be the compound MN for which the electronic transport number is close enough to unity that Eqs. (101a) and (101b) are a good approximation for Eqs. (99b) and (100b). The flow of the deviation from stoichiometry measured relative to a plane fixed in the laboratory coordinate system must be the quantity in parentheses on the right-hand side of Eqs. (107), $\mathscr{D}\, \partial\Delta/\partial z$. For our purpose here it is convenient to express this flow in terms of the gradient in μ_M. We can do so using Eqs. (64) and (67) after changing the subscripts A and B in those equations to M and N, respectively. The result is

$$(J_M^L - J_N^L) = -[x_N(B_C + b_C) + x_M(B_A + b_A)][CC_M][\bar{v}_N + \bar{v}_M]t_e \nabla\mu_M \quad . \tag{116}$$

For a narrow homogeneity range, C_M and C_N are almost equal, so that using Eq. (33) we can set $C_M(\bar{v}_N + \bar{v}_M)$ equal to unity with no significant error. Moreover the total atom concentration C is essentially equal to twice the concentration of sites in each sublattice, 2S, and x_M and x_N are essentially equal to $\frac{1}{2}$. Equation (116) then reduces to

$$(J_M^L - J_N^L) = -[(B_C + b_C) + (B_A + b_A)]St_e \nabla\mu_M \quad . \tag{117}$$

We now make the approximation that at any instant this flow is constant throughout the film, neglecting the divergence that must occur in order to adjust the composition of the tarnish film. Multiplying both sides of Eq. (117) by dz, the differential of the spatial coordinate, and integrating over the film gives

$$(J_M^L - J_N^L)\mathscr{E}_T = -\int_{\mu_M(z=0)}^{\mu_M(z=\mathscr{E}_T)} [(B_C + b_C) + (B_A + b_A)]St_e \, d\mu_M \quad . \tag{118}$$

If $v_{MN} = \frac{1}{2}(\bar{v}_M + \bar{v}_N)$ is the volume per atom of MN (assumed constant) then

$$(J_M^L - J_N^L)v_{MN} = d\mathscr{E}_T/dt \quad , \tag{119}$$

and from Eqs. (114)

$$k_1 = -v_{MN}\int_{\mu_M(z=0)}^{\mu_M(z=\mathscr{E}_T)} [(B_C + b_C + (B_A + b_A)]St_e \, d\mu_M \quad . \tag{120}$$

Now the integral is indeed a constant. This follows since the mobilities t_e and the chemical potential μ_M are all functions of Δ at a fixed temperature. Therefore the integral will be the difference of a function of Δ evaluated at $z = \mathscr{E}_T$ and $z = 0$. However the value of Δ is fixed at a constant value at $z = 0$ by the M–MN equilibrium and at another constant value at $z = \mathscr{E}_T$ by the MN–N (gas at fixed pressure and temperature) equilibrium. For a particular instant of time, the mean value theorem of calculus can be used to write the integral of Eq. (120) as the value of the mobility sum in parentheses evaluated at some value of Δ, say Δ''', intermediate between the fixed values at $z = 0$ and $z = \mathcal{E}_T$, times the integral of $d\mu_M$. This gives

$$k_1 = v_{MN}\{t_e[(B_C + b_C) + (B_A + b_A)]\}_{\Delta=\Delta'''}(\mu'_M - \mu''_M) \quad , \tag{121}$$

where μ''_M is the fixed value of the chemical potential of M at the N(gas)-MN interface and μ'_M is the fixed value at the M-MN interface. Since this can be done for every instant of time and μ''_M and μ'_M are independent of time, Δ''', and hence the effective value of the mobility sum, is independent of time also. The chemical potential difference is equal to $RT \ln(p'_M/p''_M)$. If the homogeneity range is of the order of 1 at. % wide or less, the sum of the chemical potentials, $\mu_M + \mu_N$, will be essentially independent of composition, so that Eq. (121) can be rewritten as

$$k_1 = \{v_{MN}t_e[(B_C + b_C) + (B_A + b_A)]\}_{\Delta=\Delta'''}(RT/2)\ln(p''_{N_2}/p'_{N_2}) \quad . \tag{122}$$

The value of Δ''' is unknown and in general it may be a function of p''_{N_2}. In the past, equations similar to Eqs. (121) and (122) have been derived by assuming $t_e[(B_C + b_C) + (B_A + b_A)]$ is constant. We have shown here that the integral in Eq. (120) is constant in the more general case and that the parabolic law follows if the flow $J^L_M - J^L_N$ is assumed to be constant throughout the film at any instant in time.

Before proceeding further, we should point out that the results of this subsection are independent of the microscopic diffusion mechanism assumed and apply equally well for the direct interstitial and interstitialcy mechanisms as well as for the vacancy mechanism. We have not used any particular expression for the mobilities or the chemical potentials as a function of Δ. We have, however, implicitly assumed that the space charge density is zero throughout the diffusion zone and local electroneutrality prevails in arguing that the integrand of Eq. (120) depends on Δ (or μ_M) alone at a fixed temperature.

Having shown that interdiffusion leads to the parabolic law for the time dependence of the tarnish film thickness, we would like to obtain the mobility sum from measured values of the parabolic rate constant k_1. Returning to Eq. (120) we note that, if the upper limit of the integral is changed to $\mathscr{E}$, where $0 < \mathscr{E} < \mathscr{E}_T$,

$$(J_M^L - J_N^L)\mathscr{E} = v_{MN} \int_{\mu_N(z=0)}^{\mu_N(z=\mathscr{E})} [(B_C + b_C) + (B_A + b_A)]St_e \, d\mu_N' , \quad (123)$$

where we have used $d\mu_M = -(x_N/x_M)\, d\mu_N \approx -d\mu_N$. Physically, Eq. (123) states that for $\mu_N = \mu_N''$ at the $MN-N_2(g)$ interface at $z = \mathscr{E}_T$, the flow times a distance $\mathscr{E}$ can be obtained by performing the specified integration from the MN–M interface at $z = 0$ to a distance $\mathscr{E}$ within the tarnish film. However, as stated before, the integral is a function of the limits of integration only. Its value therefore is the same as would be obtained upon integrating over the entire film thickness in another experiment in which the value of the chemical potential of N at the $MN-N_2(g)$ interface is μ_N. Writing the rate constant as $k_1(\mu_N)$ to indicate it is a function of μ_N at the $MN-N_2(g)$ interface at a fixed temperature, we can divide Eq. (123) by Eq. (120) and get

$$\frac{\mathscr{E}}{\mathscr{E}_T} = \frac{k_1(\mu_N)}{k_1(\mu_N'')} . \quad (124)$$

Thus if experiments have been performed at a given temperature to obtain the rate constant as a function of μ_N, one can use this information along with Eq. (124) to obtain μ_N as a function of $\mathscr{E}/\mathscr{E}_T$ for any value of μ_N at the $MN-N_2(g)$ interface within the measured range. The value of μ_N is then fixed and independent of time not only at the film surfaces but also at any prescribed fractional part of the film thickness (such as at $\mathscr{E}/\mathscr{E}_T = \frac{1}{2}$). Differentiating Eq. (120) with respect to the value of μ_N at $\mathscr{E}_T$,

$$dk_1/d\mu_N'' = v_{MN}[(B_C + b_C) + (B_A + b_A)]St_e . \quad (125)$$

Thus the slope of the experimental plot of k_1 as a function of μ''_N yields the right-hand side of Eq. (125) as a function of μ_N. This general procedure has been used [33] to obtain values of $RT[(B_C + b_C) + (B_A + b_A)]St_e$ across the oxide films CoO, MnO, and FeO. (In Ref. [33] the equations were written as though there were a single type of defect with a single, effective mobility. This is reasonable for these oxides whose homogeneity ranges lie on the oxygen-rich side of the 50 at. % stoichiometric composition. If b is this effective mobility, St_eRTb in our nomenclature was called the self-diffusion constant for the metal component.)

Finally, when matter transport across a tarnish film occurs primarily via one type of ion, e.g., Fe^{2+} in FeO, one expects a significant Kirkendahl effect. If the cations account for most of the matter transport, inert markers originally at the metal-tarnish film interface will tend to remain there. On the other hand, they will be progressively displaced into the tarnish film if the anions account for most of the matter transport.

VIII. DIFFUSION IN IONIC CRYSTALS

From the point of view of the discussion in Section VII, ionic crystals are those crystals for which the ion transport numbers are relatively large and the transport number for electronic carriers essentially zero. The presence of electrons and holes can then be neglected not only in diffusion but also in discussing equilibrium properties. Thus for energetic reasons, electrons and holes play no important role in maintaining electrical neutrality when aliovalent impurities are incorporated substitutionally in ionic crystals. For every Cd^{2+} ion occupying a Na^+ site in NaCl, one Na^+ site must be vacant. The addition of about 5 mol % of CaO to zirconia, ZrO_2, not only stabilizes the cubic CaF_2-type structure, but also introduces one oxygen vacancy per Ca. The high concentration of vacancies results in the transport number of oxygen being near one. The conductivity is still dependent on the oxygen pressure also. These facts, coupled with the fact that diffusion measurements in the alkali metal and silver halides are successfully interpreted assuming the compounds are exactly stoichiometric, make the discussion of diffusion in ionic crystals somewhat simpler than for semiconducting compounds in general. There is one complication, however, that we have not mentioned as yet. Namely, a significant fraction of an aliovalent impurity may be tightly bound to vacancies to form pairs. Most of the published literature is concerned with tracer self-diffusion in pure or doped crystals or with tracer diffusion of impurities. We note that the ion flows for interdiffusion given by Eqs. (99a) are likely to be very small for ionic compounds ($t_e = 0$) when the vacancy mechanism is predominant. The second term in each equation is near zero when $x_M = x_N$ because the sum of the ion transport numbers is near unity. On the other hand, the first term in each equation is small for the vacancy mechanism, since the free energy barrier for the exchange of an ion and a vacancy of the same charge is expected to be large, resulting in small values of the corresponding mobilities.

The possibility of determining the contribution of each ion to the electrical conductivity of ionic crystals makes the modified Nernst-Einstein equation, given by Eq. (95) for the vacancy mechanism, a valuable one in determining the diffusion mechanism or mechanisms. If the vacancy mechanism is operative, and the crystal is stoichiometric, the vacancies of each kind will be entirely ionized, so that B_C in Eq. (95) is zero. In this case the value of f_M calculated from Eq. (95) using the measured cation conductivity and self-diffusion coefficient should be 0.78 for a fcc lattice. More generally the effective correlation coefficient calculated from

$$f_M = \frac{C_M e^2 D_M^*}{kT\sigma_C} \tag{126}$$

allows the diffusion mechanism to be identified. If more than one mechanism is operating, f_M in general will show a dependence upon temperature. The introduction of substitutional divalent cation impurities into a uni-univalent crystal will increase the concentration of cation vacancies and repress that of cation interstitials, so that for such doped crystals the vacancy mechanism will be enhanced in importance and may be studied independent of the interstitials. The use of aliovalent substitutional impurities also allows the free energies of defect motion and formation to be determined separately if the impurity itself is relatively immobile. For pure crystals the activation energy for diffusion is the sum of these two quantities, as shown in Eq. (17). For a uni-univalent crystal doped with a divalent cation impurity, the cation vacancy concentration at low temperatures will be fixed at that of the impurity and the diffusion activation energy will be that for the free energy of defect motion. Combining the results on pure and doped crystals, both energies can be obtained.

A detailed illustration of these effects in AgCl, for which Ag interstitials and Ag vacancies are the predominant point defects, is offered in papers by Friauf and his students [2-5].

IX. MISCELLANEOUS

In our discussion of diffusion in solids we have chosen to develop two general topics in considerable detail: the phenomenological description of diffusion in an isothermal, binary system in Sections V and VI and interdiffusion in a pure semiconducting compound, MN, in Section VII. The former, inasmuch as it shows that Fick's second law is valid only when the partial atomic volumes can be considered constant and supplies a more generally valid equation (40), is basic and moreover has not been covered in textbooks. The latter topic has been treated previously and requires many assumptions which restrict the generality of the final equations obtained. Nevertheless the topic is of basic importance for a wide class of compounds, and we have felt it worthwhile to try to provide an adequate development, the framework of which, if not the details, is generally applicable. However, as a consequence, a number of topics of current research could only be briefly presented or had to be omitted entirely. Among these are kinetic treatments of diffusion, some of the results of which have been quoted, but the development of which has not been reproduced in the complexity necessary to obtain correlation coefficients for tracer self-diffusion or to discuss diffusion of two species via the same vacancies. The interested reader is referred to the recent book by Manning [8] on this subject. Another area is the diffusion of impurities in semiconductor compounds. In view of correlation effects, which would be important in all but the interstitial mechanism, an adequate treatment is complicated. A recent review of the experimental data for III-V compounds has been given by Kendall [34]. A somewhat older discussion of diffusion in semiconductors is contained in the book by Boltaks [35]. Finally, the diffusion of impurities in elemental semiconductors has not been covered. An approach similar to that used here for compound semiconductors has been applied [36]. In Ge and Si there is believed to be only one category of native atomic point defect, namely, vacancies that act as acceptor levels. As a consequence, when the Fermi level is raised so that a significant fraction of the vacancies become ionized, the jump frequency for an ionized acceptor impurity decreases while that for any ionized donor impurity increases since the latter has a charge opposite in sign to that of an ionized vacancy.

REFERENCES

1. J. Bardeen and C. Herring, Atom Movements (J. H. Holloman, ed.), Am. Soc. Metals, Cleveland, Ohio, 1951, p. 87.
2. R. J. Friauf, J. Appl. Phys., 33, 494 (1962).
3. M. D. Weber and R. J. Friauf, J. Phys. Chem. Solids, 30, 407 (1969).
4. J. P. Gracey and R. J. Friauf, J. Phys. Chem. Solids, 30, 421 (1969).
5. R. J. Friauf, J. Phys. Chem. Solids, 30, 429 (1969).
6. J. G. Mullen, Phys. Rev., 121, 1649 (1961).
7. A. D. LeClaire, Phil. Mag., 14, 1271 (1966).
8. J. R. Manning, Diffusion Kinetics for Atoms in Crystals, Van Nostrand, Princeton, New Jersey, 1968.
9. A. B. Lidiard, Phil. Mag., 46, 1218 (1955).
10. J. R. Manning, Phys. Rev., 116, 819 (1959).
11. A. H. Schoen, Phys. Rev. Lett., 1, 138 (1958).
12. J. G. Kirkwood, R. L. Baldwin, P. J. Dunlop, L. J. Gosting, and G. Kegeles, J. Chem. Phys., 33, 1505 (1960).
13. A. C. Smigelskas and E. O. Kirkendahl, Trans. AIME, 171, 130 (1947).
14. C. Robinson, Trans. Faraday Soc., 42B, 12 (1946).
15. D. Lazarus, Solid State Physics (F. Seitz and D. Turnbull, eds.), Vol. 10, Academic, New York, 1960.
16. R. W. Baluffi, Acta Met., 8, 871 (1960).
17. J. Crank, The Mathematics of Diffusion, Oxford Univ. Press, London and New York, 1956, pp. 232, 238.
18. L. S. Darken, Trans. AIME, 175, 184 (1948).
19. S. R. deGroot and P. Mazur, Non-Equilibrium Thermodynamics, North-Holland Publ., Amsterdam, 1962.
20. D. D. Fitts, Nonequilibrium Thermodynamics, McGraw-Hill, New York, 1962, pp. 44, 239.
21. R. E. Howard and A. B. Lidiard, Reports on Progress in Physics (A. C. Strickland, ed.), Vol. XXVII, The Institute of Physics and the Physical Society, London, 1964, p. 161.

22. C. Wagner, Z. Physik. Chem., B32, 447 (1936).
23. W. Jost, Diffusion in Solids, Liquids, and Gases, Academic, New York, 1952, p. 383.
24. R.F. Brebrick, J. Appl. Phys., 30, 811 (1959).
25. R.F. Brebrick, Progress in Solid State Chemistry (H. Reiss, ed.), Vol. 3, Pergamon, Oxford, 1966, p. 244.
26. A.R. Calawa, T.C. Harman, M. Finn, and P. Youtz, Trans. AIME, 242, 374 (1968).
27. R.W. Brodersen, J.N. Walpole, and A.R. Calawa, J. Appl. Phys., 41, 1484 (1970).
28. M.S. Seltzer and J.B. Wagner, Jr., J. Chem. Phys., 36, 130 (1962).
29. H.H. Woodbury, II-VI Semiconducting Compounds (D.G. Thomas, ed.), Benjamin, New York, 1967, p. 244.
30. N.F. Mott, Trans. Faraday Soc., 43, 429 (1947).
31. F. Booth, Trans. Faraday Soc., 44, 706 (1948).
32. W.J. Moore, J. Electrochem. Soc., 100, 302 (1953).
33. F.S. Pettit, J. Electrochem. Soc., 113, 1249 (1966).
34. D.N. Kendall, Semiconductors and Semimetals, Vol. 4: Physics of III-V Compounds (R.K. Willardson and A.C. Beer, eds.), Academic, New York, 1968, p. 163.
35. B.I. Boltaks, Diffusion in Semiconductors, Academic, New York, 1963.
36. H. Reiss and C.S. Fuller, Semiconductors (N.B. Hannay, ed.), Chap. 6, Reinhold, New York, 1959.
37. J.M. Wimmer, Ph.D. Thesis, Marquette University, Milwaukee (1972).
38. J.N. Walpole and R.L. Guldi, J. NonMetals, in press.

Chapter 10

SURFACE CHEMISTRY

Gabor A. Somorjai

Department of Chemistry
University of California
Berkeley, California

I. THERMODYNAMICS OF SURFACES

In order to cleave a solid along its rows of atoms to create a surface, work is required. In general, creation of a stable interface always has a positive free energy of formation. This "reluctance" of the solid or liquid to form a surface defines many of the interfacial properties of the condensed phases. Under conditions of constant temperature and pressure, the work dW required to increase the surface area A by an amount dA is given by $\delta W_{T,P} = \gamma\, dA$. Here γ is the two-dimensional analog of the pressure and is called the "surface tension," while the volume change is substituted by the change of surface area. This work is equal to the change in free energy for this process $\delta W_{T,P} = d(GA)$. Thus the increase of free energy with increasing surface area is

$$dG_{T,P} = \gamma\, dA \quad , \tag{1}$$

where γ is defined as the surface free energy per unit area (ergs/cm^2) that has to be expended to increase the surface. We may also consider γ as a pressure along the surface plane which opposes the creation of more surface. The pressure is force per unit area (dynes/cm^2), therefore the "surface pressure" γ is given by force per unit length (dynes/cm). (The customary units of surface tension, $ergs/cm^2$ or dynes/cm , are dimensionally identical.) As a consequence, liquids tend to minimize their surface area and therefore their surface free energy by assuming a spherical shape. For solids, surfaces of lowest surface free energy will form at the expense of other surfaces of higher free energy. Crystal faces that exhibit closest packing of atoms tend to be surfaces of lowest free energy of formation, and hence the most stable [1].

Since the free energy of formation of a surface is always positive, a particle which consists of surfaces only, i. e., platelets or droplets of atomic dimensions, would be thermodynamically unstable. There must be a stabilizing influence, however, which allows small particles of atomic dimensions to form and grow, a common occurrence in nature. This is, in fact, given by the free energy of formation of the condensed phase (liquid or solid) from the vapor at a given temperature:

$$\Delta G = -nRT \ln(P/P_0) \quad . \tag{2}$$

The condensed phase of n moles will grow if the vapor pressure P is larger than the equilibrium vapor pressure P_0, or it will vaporize if $P < P_0$. Under conditions of growth ($P > P_0$) the particle also possesses a surface free energy, which has to be taken into account. For a spherical droplet, the surface free energy is $4\pi r^2 \gamma$. Thus the total free energy of a growing particle is

$$\Delta G(\text{total}) = -nRT \ln(P/P_0) + 4\pi r^2 \gamma \quad . \tag{3}$$

The first term on the right is negative while the second term is always positive. In order to see the dependence of ΔG on the dimensions of the condensed particle, expressed through its radius r, let us substitute n by $n = \frac{4}{3}\pi r^3/\bar{V}$, where $\bar{V} = M/\rho$ is the molar volume of the condensed phase, M is its atomic weight, and ρ is the density. We have

$$\Delta G(\text{total}) = -(\tfrac{4}{3}\pi r^3/\bar{V}) RT \ln(P/P_0) + 4\pi r^2 \gamma \quad . \tag{4}$$

Initially, when the condensed particle is very small, the surface free energy term must be the larger of the two, and ΔG increases with r. In this range of sizes the particles are unstable. Above this critical size, however, the volumetric term becomes larger and dominates; hence, a particle of critical size or larger grows spontaneously. When ΔG is at a maximum, that is, $\partial \Delta G(\text{total})/\partial r = 0$, the particle reaches a critical size which it must have for growth to begin:

$$\frac{\partial \Delta G(\text{total})}{\partial r} = -\frac{4\pi}{\bar{V}} r^2 RT \ln\left(\frac{P}{P_0}\right) + 8\pi\gamma r = 0 \quad , \tag{5}$$

$$r(\text{critical}) = \frac{2\gamma\bar{V}}{RT \ln(P/P_0)} \quad . \tag{6}$$

This value is about 6-10Å for most materials, which means that the droplet of critical size contains between 50 and 100 atoms or molecules.

The free energy of a particle of critical size can be expressed by substituting Eq. (6) into Eq. (4). We obtain

$$\Delta G(\text{maximum}) = \frac{4\pi r^2_{\text{crit}}\gamma}{3} \quad . \tag{7}$$

Thus, the total free energy for a spherical particle of critical size is one-third of its surface free energy.

A condensed particle must be larger than a certain critical size for spontaneous growth to occur. "Nucleation" of the condensed phase by simultaneous clustering of many atoms, however, is very improbable. Growth of condensed phases generally occurs on solid surfaces already present, such as the walls of the reaction chamber or dust particles present in the atmosphere. The introduction of solid particles to induce the formation and growth of the condensed phase is frequently called "seeding." It is used to facilitate the growth of crystals or the condensation of water droplets from the atmosphere (rainmaking).

If there is equilibrium between the surface and the bulk of a condensed droplet, any small change in the equilibrium state of one will cause similar small changes in the other [1]. Equating the surface and volumetric free energy changes in Eq. (5), we have

$$\ln\left(\frac{P}{P_0}\right) = \frac{2\gamma\bar{V}}{rRT} \quad . \tag{8}$$

This is the well-known Kelvin equation which describes the dependence of the vapor pressure of a spherical particle on its size. Here, we can take P_0 to be the vapor pressure over a particle of infinite size. We can see that, according to Eq. (8), small particles have higher vapor pressures than larger ones. This explains why "supersaturated" vapor can exist, i.e., ambient conditions in which the vapor pressure of a condensible substance is larger than its equilibrium vapor pressure. The actual vapor pressure is higher in the presence of droplets of subcritical size. Similarly, very small particles of solids have greater solubility than large particles [2, 3]. If we have a distribution of particles of different sizes, we find that the larger particles grow at the expense of the smaller ones, as predicted by Eq. (8) [3, 4].

It should be noted that such differences in vapor pressure or solubility, which depend on particle size, can only be observed for particles ≤ 100Å. Assuming representative values [$\gamma(H_2O) = 73.4$ ergs/cm^2, $\bar{V} = 18$ cm^3/mole], P/P_0 approaches unity rapidly above this radius, i.e., $\ln(P/P_0)$ vanishes.

A. Surface Heat Capacity

It is well known that the heat capacity of a monatomic solid, which is due to atomic vibrations, can be well approximated by the equation [5]

$$C_v = 9R\left(\frac{T}{\Theta}\right)^3 \int_0^{x_m} \frac{x^4 e^x}{(e^x - 1)^2}\, dx \quad , \tag{9}$$

where $x = h\nu/kT$, k and h are the Boltzmann and Planck constants, respectively, $x_m = h\nu_m/kT = \Theta/T$, ν_m is the maximum frequency of lattice vibrations, Θ is the Debye temperature (defined by $\Theta = h\nu_m/k$), which is characteristic of a given solid. At high temperatures, $x \ll 1$, the exponential may be expanded, and Eq. (9) reduces to $C_v = 3R$; at low temperatures ($T \ll \Theta$), we have $C_v \propto T^3$. Thus, the lattice heat capacity approaches zero degrees Kelvin as T^3. Both limits have been established by experiments for many solids [6].

Montroll [7] and others have shown that one can calculate heat capacity due to surface atoms by considering and averaging over the frequencies of lattice vibrations of surface atoms separately from that of the atoms in the bulk. At low temperatures ($T \ll \Theta$), the surface heat capacity varies with temperature as $C_v(\text{surface}) \propto T^2$. This temperature dependence may be measured for samples of large surface-to-volume ratio (thin films for example). At higher temperatures at which surface heat capacity becomes temperature independent, its contribution to the total heat capacity (which is proportional to the mole fraction of surface atoms) should be negligible.

II. THE STRUCTURE OF SOLID SURFACES

A description of a surface using thermodynamic functions, which are derived by averaging over certain physical-chemical characteristics of many surface atoms, necessarily reflects the macroscopic properties of that surface. Surface tension and surface heat capacity are macroscopic properties, which should not markedly depend on small changes in the structure, i.e., atomic arrangements in the surface. We must know the structure of the surface, however, to be able to investigate and interpret physical-chemical processes on an atomic level. The description of the overall vaporization or growth rates of a crystal may not necessitate the knowledge of the surface structure. However, in order to uncover the important atomic positions from which atoms desorb or to which atoms are added, to pinpoint the atomic sites at which electronic rearrangements are taking place, which may be rate determining, we should know the structure of the surface. In order to describe the overall adsorption rate of gas atoms or their net rate of chemical rearrangement on the solid surface, structural information is often not necessary. Determination of the nature of the interaction and the configuration of the adsorbed gas, the scrutiny of the details of energy or electron transfer processes, however, require the availability of accurate structural information.

A macroscopic picture of a metal surface is shown in Fig. 1. This picture was taken using an electron microscope which is a valuable tool in the study of surfaces. A description of the principles of electron microscopy may be obtained from several reference sources [9, 10]. In Fig. 1(a) there are several steps distinguishable on the surface. Some of them are of atomic height; others are several tens of angstroms high. There are also pits of different sizes. Their shapes, however, are defined by the orientation of the crystal faces that constitute their boundaries. Fig. 1(b) shows that there are several steps visible within each pit. Thus, the electron microscopic picture clearly shows that the surface is heterogeneous. Atoms may occupy positions in which they are surrounded by different numbers of nearest neighbors. Since their binding energies can be assumed to be proportional to the number of neighbors, there are several atomic sites of different binding energy. Pits may form at the emergence of line defects (dislocations, etc.) to the surface. At a larger magnification we should see point defects, i. e., single atomic vacancies or aggregates of vacant sites.

What is the arrangement of the atoms that occupy different equilibrium surface sites on such a heterogeneous surface? Figure 2 shows a diffraction pattern of a (100) surface of a fcc crystal (nickel) obtained by low energy electron diffraction. Low energy electrons, which are backscattered from the surface of a single crystal without the loss of energy (elastic scattering), give us information about the arrangement of surface atoms and the structure of the surface [11, 12]. This very important technique, which provided most of the structural information about solid surfaces, will be discussed in more detail later. The

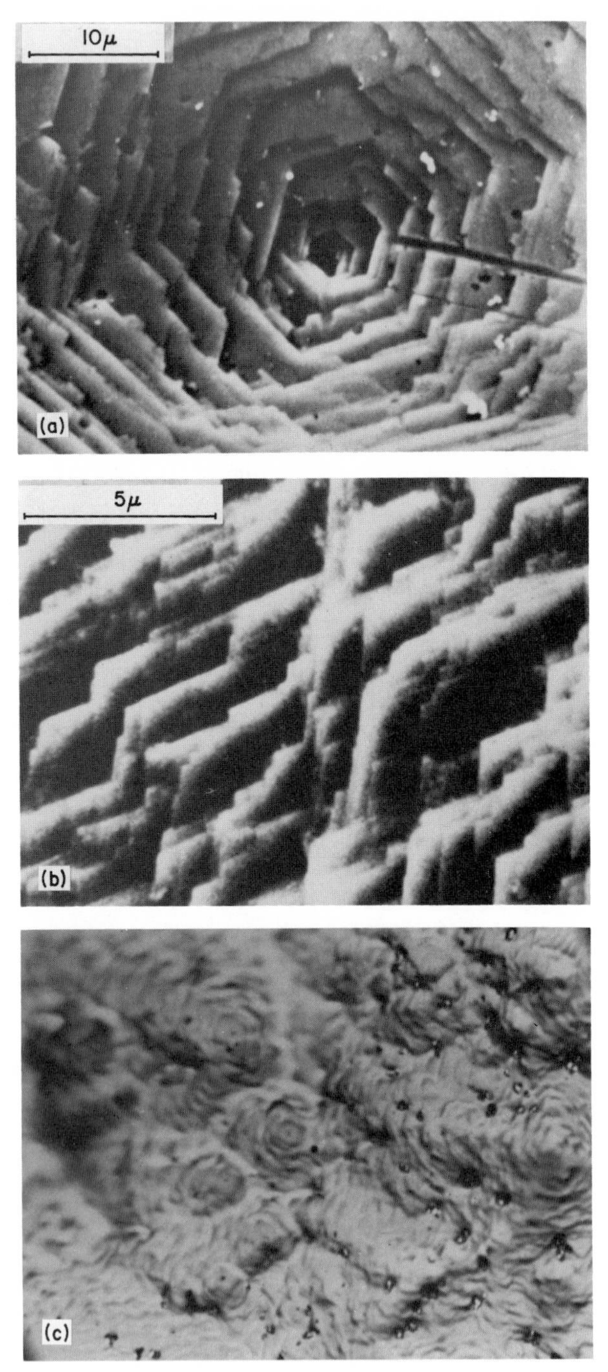

FIG. 1. Electron microscope picture of a zinc surface (magnification 1000X).

presence of such sharp spots of high intensity are indicative of well-ordered surface domains of at least 500Å in diameter in which most of the atoms are located in their equilibrium lattice position. Any loss of order would show up by the broadening of the diffraction spots, the appearance of other diffraction features such as streaks and an overall increase in the background intensity. For comparison, Fig. 3 shows the low energy electron diffraction (LEED) pattern which was obtained from the same surface, after ion bombardment using high energy (300 eV)

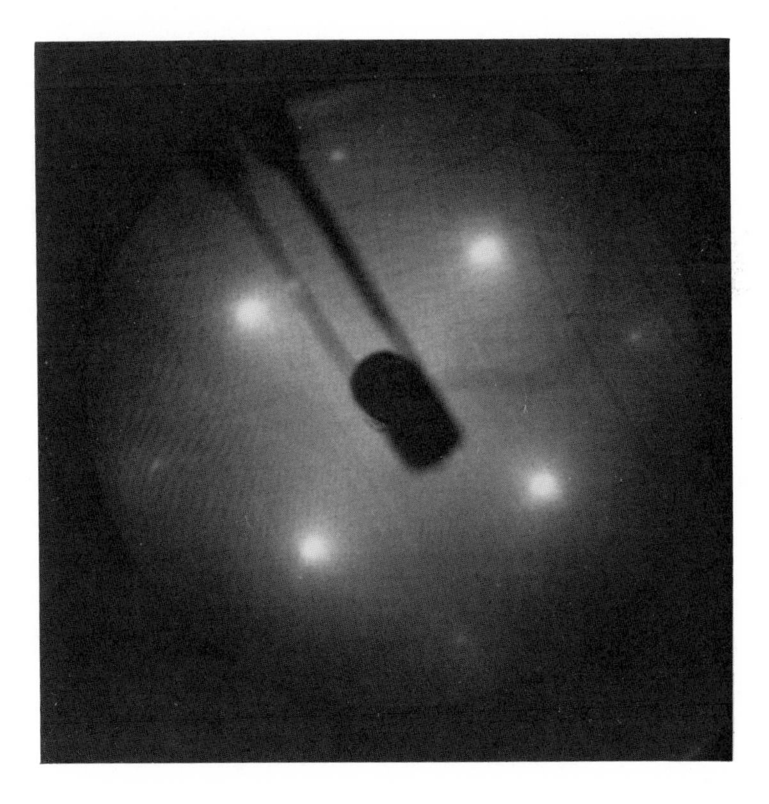

FIG. 2. Low energy electron diffraction pattern of the (100) face of a nickel single crystal.

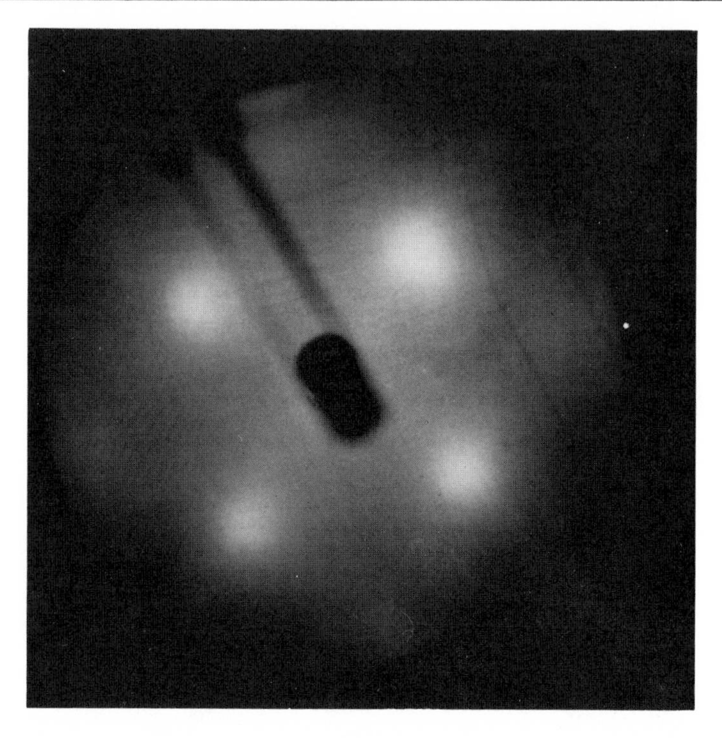

FIG. 3. Low energy electron diffraction pattern of the (100) face of a nickel single crystal after ion bombardment.

ions. Such a treatment dislocates a large fraction of the surface atoms, disrupts the periodicity, and introduces large concentrations of defects in the surface. The effect of increasing disorder is clearly documented in the diffraction pattern.

Thus, we find that the crystal surface is ordered on an atomic scale despite the macroscopic heterogeneity, the presence of imperfections, and many different atomic positions. Most of the surface atoms are in their equilibrium positions surrounded by neighbors of different number but which are at well-defined interatomic distances.

A. Atomic Displacements

Atoms do not stay rigidly in their equilibrium position but undergo thermal motion. The energy stored in lattice vibrations gives rise to the heat capacity of a solid. Due to atomic oscillations whose amplitudes increase with temperature, the conduction electrons are scattered and give rise to the observed increasing electrical resistance of metals with increasing temperature. The detailed nature of the atomic vibrations in solids, or as it is frequently called "lattice dynamics," is under thorough investigation, since it provides information about the forces that hold atoms together in the condensed phase and about the atomic potential. Several references [13, 14] on this subject would make valuable reading to introduce the unfamiliar reader to this subject.

Surface atoms, just like atoms in the bulk of the solid, undergo oscillations about their equilibrium positions. The amplitude of their mean displacement, however, is different from that for bulk atoms. Since the number of neighbors about an atom in the surface is, in general, smaller than the number of neighbors surrounding atoms in the interior of a crystal, it is expected that surface atoms may not be held as tightly as atoms in the bulk. This indeed was found experimentally [15] to be the case. The amplitude of lattice vibrations of surface atoms is roughly twice as large as for bulk atoms for several fcc crystals.

The method used to measure the mean displacement of surface atoms is called Debye-Waller factor measurement, and it employs low energy electrons. In such a study, one measures the intensity of the diffraction spots, obtained by low energy electron diffraction, as a function of temperature. It can be shown [16-18] that the intensity I of a given diffraction spot as a function of temperature can be expressed as

$$I(T) = I(T=0) \exp\left[-\left(\frac{16\pi^2 \cos^2\phi}{\lambda^2}\right)\langle u^2\rangle\right], \tag{10}$$

where $\langle u^2\rangle$ is the mean square displacement and ϕ is the scattering angle with respect to the surface normal. All other terms have been defined previously. Using the Debye approximation (1) in which the atoms are treated as harmonic oscillators, the mean square displacement can be expressed as

$$\langle u^2 \rangle = \frac{12N\hbar^2}{Mk\Theta^2} T \quad , \tag{11}$$

where M is the atomic weight of the solid and Θ is the Debye temperature. Using low energy electrons and measuring the scattered beam intensities as a function of temperature permits calculation of a "surface Debye temperature" or a "surface root mean square displacement," which is characteristic of atoms in the studied crystal surface, from the slopes of the ln I vs T curves. Table 1 lists values obtained for several fcc metals. The table contains only that component of the root mean square displacement perpendicular to the surface plane. It can be seen that for these solids the vibrational amplitude of surface atoms is markedly larger than that for atoms in the interior of the solid.

TABLE 1

The Surface and Bulk Root-Mean-Square Displacements and Debye Temperatures for Palladium, Lead, Platinum, and Silver

	Pd	Pb	Pt	Ag
$\langle u_\perp \rangle$ (surface) (Å)	0.144	0.263	0.135	0.129
$\langle u_\perp \rangle$ (bulk) (Å)	0.074	0.160	0.064	0.089
Θ_D (surface) (°K)	140 ± 10	55 ± 10	110 ± 10	155 ± 10
Θ_D (bulk) (°K)	273.4	90.3	234	225

B. Surface Rearrangements

The atomic environment of surface atoms is markedly different from the environment of atoms in the bulk. Surface atoms are surrounded by neighbors in the surface plane and from one side under the surface, while there is a vacuum (i.e., no neighbors) on the other side. The outermost layer of atoms is always in an anisotropic environment as compared to bulk atoms. This drastic change of environment is expected to have an effect on many of the physical-chemical properties of surface atoms. We have seen that the mean square displacements of surface atoms greatly differ from those in the interior of the solid. Calculations on molecular [19], ionic [20], and metal [21] crystal surfaces indicate that there should also be a "net" displacement of atoms in the surface plane. Thus, not only may the amplitude of vibration be different, but surface atoms should occupy equilibrium positions which are displaced outward (away from the underlying atomic plane) or inward (toward the atomic plane below) by a few percent (~ 2-8%) of the interplanar distance. Having such different atomic properties, it would not be surprising for surface atoms to be arranged in a different periodicity or structure than that which prevails in the bulk. It is, in fact, common to find monatomic or diatomic solids that have crystal faces showing different periodicity or repeating unit cells from the corresponding bulk structures. These rearranged crystal faces are characterized by unit cells that are integral multiples of the bulk unit cell. For example, the (111) face of silicon (diamond structure) has a surface structure indicating periodic arrangement of surface atoms that are twice the interatomic distance from each other as are atoms in the bulk. This is called the (2 × 2) surface structure. If the surface atoms are located in the positions expected from the projection of the bulk atomic structure to the surface, this arrangement is called a (1 × 1) surface structure. Most insulators and semiconductors studied so far [12, 22, 23] show surface rearrangements on all, or on at least one, of the crystal faces. In general, the more "open" the crystal face (the lower its atomic density), the more likely it is that rearrangement will occur. Some of the surface structures have well-defined temperature ranges of stability. At temperatures above and below this range, the surface undergoes a transformation into another surface structure. Rearranged surface structures have also been found on several metal surfaces [24-26]. Here, the

TABLE 2

Surface Structures Detected on Several Monatomic Solid Surfaces

Solid	Crystal face	Surface structure
Silicon	(111)	(2 × 1)
	(111)	(7 × 7)
	(110)	(5 × 2)
	(100)	(2 × 2)
Germanium	(111)	(2 × 1)
	(111)	(8 × 2)
	(100)	(2 × 2)
Platinum	(100)	(5 × 1)
Gold	(100)	(5 × 1)
	(110)	(1 × 2)
Bismuth	$(11\bar{2}0)$	(2 × 10)
Antimony	$(11\bar{2}0)$	(6 × 3)
Tellurium	(0001)	(2 × 1)

surface structures appeared to be stable throughout the studied temperature ranges up to the melting point of the metal. Several other metals show no sign of surface rearrangements but maintain their surface structure as characterized by the bulk unit cell. Table 2 lists some of the solids that have exhibited surface rearrangements and the surface structures formed.

What are the mechanisms of surface rearrangements? Our discussion will be restricted to two types of possible mechanisms.

1. Surface Relaxation

This mechanism may best be illustrated by considering what happens to the atoms in a solid in the neighborhood of a vacancy, i. e., a vacant lattice position. If we remove an atom from its equilibrium position in the bulk to the gas phase, the atoms surrounding the now vacant site "relax," i. e., will be displaced slightly toward the vacancy. They are no longer restrained from larger displacement in the direction of the empty site by the strong, repulsive atomic potential. Therefore, the free energy of removing an atom from its bulk equilibrium position to the gas phase is partially offset by the lattice "relaxation" about the vacancy. The free energy of vacancy formation from a rigid lattice that is not allowed to relax can be approximated by the cohesive energy, the energy necessary to break a solid into single atoms infinitely separated from each other. We find that, where these quantities have been measured, the free energy of forming vacancies is appreciably smaller than the cohesive energy for most solids. For example, for copper, the cohesive energy is 81.0 kcal/mole, while the free energy of vacancy formation is 32.2 kcal/mole. This leaves over 48 kcal/mole, a very large relaxation energy.

Surface atoms are in an anisotropic environment, as though they were surrounded by atoms on one side and by vacancies on the other. The atoms can "relax" out of plane, perpendicular to the surface which is not allowed for the bulk atoms. Depending on the bonding characteristics of the solid, atoms may be displaced out of plane in a periodic manner. One type of periodic relaxation is shown schematically in Fig. 4. The appearance of any new surface periodicity will be reflected in the diffraction pattern obtained by low energy electron diffraction. It is likely that surface structural rearrangements in germanium, silicon, and other semiconductor surfaces occur by this mechanism. Another alternative is that there may be a periodic arrangement of vacancies in the surface, which may also propagate through the bulk crystal.

2. Surface Phase Transformation

The crystal structure adopted by a solid has been shown [27, 28] to depend on the number of unpaired s and p valence electrons per

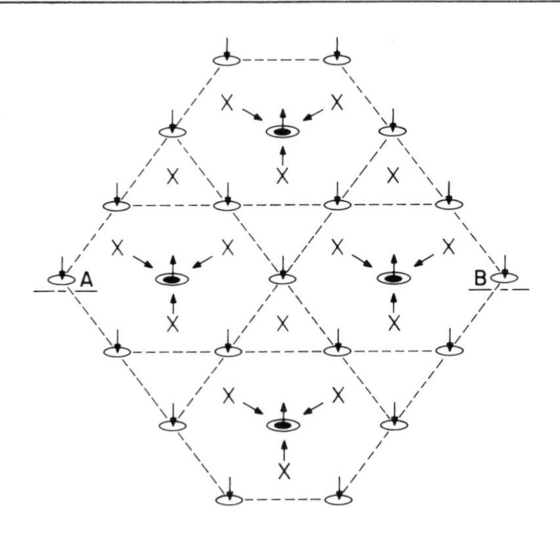

FIG. 4. Schematic representation of one type of surface relaxation which can give rise to a (2 × 2) surface structure.

atom which are available for binding. For example, atoms that have one unpaired s or p electron form a bcc structure when condensed into a solid. Atoms with two unpaired s and/or p electrons will have a hcp structure; three unpaired valence electrons will give an fcc structure; and four unpaired valence electrons a diamond crystal structure. A theory based on this concept, when extended to include the contribution of unpaired d electrons to the binding, can explain and predict the structure and stability range of alloys [29].

Surface atoms, in addition to being in an asymmetric environment, have fewer neighbors than atoms in the bulk of the solid. Since their electron density distribution should be different from that in the bulk, they may have less or more valence electrons available for binding than bulk atoms. Thus, they may undergo phase transformations in the surface plane with respect to their crystal structure in the bulk [30]. A

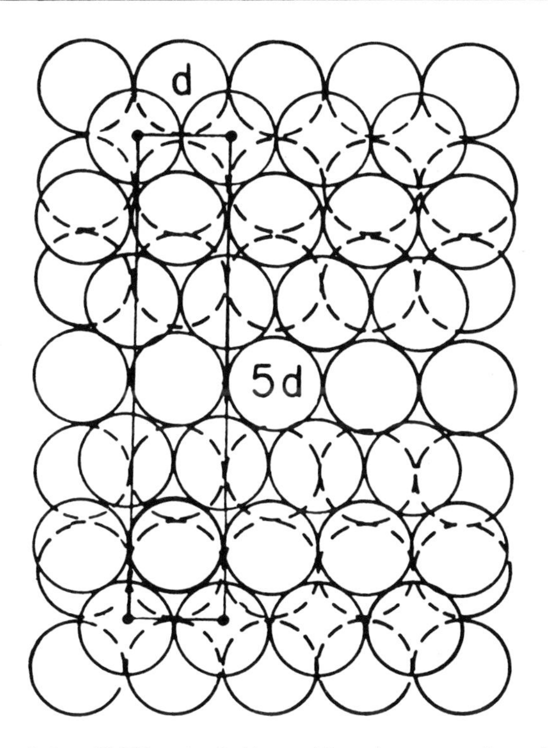

FIG. 5. A fcc (100) orientation with a hexagonal surface layer.

surface phase transformation of this type is shown schematically in Fig. 5. A hexagonal surface layer is formed on the top of a (100) face of a fcc crystal, which is characterized by the square unit mesh. Thus, a fcc → hcp surface phase transformation has occurred. It is likely that this mechanism governs the phase transformations of the (100) surfaces of gold and platinum.

It should be noted that impurity atoms with different numbers of unpaired valence electrons per atom may cause or accelerate surface phase transformations of this type in crystal surfaces.

III. SURFACE DIFFUSION

Surface atoms can migrate from one lattice position to another by way of diffusion along the surface. If we were to label one surface atom (for example, by using a radioactive isotope) its square migration distance $\langle x^2 \rangle$ at a given temperature would be given by

$$\langle x^2 \rangle = 2Dt \tag{12}$$

from the theorem of random walk [30] in one dimension, where t is the diffusion time and D is the diffusion coefficient, $D = \frac{1}{2} Pa^2\nu$. Here a is the interatomic distance, ν is the vibrational frequency, and P is the probability of the atomic jump. For P equal to unity, an atom will make a jump to a neighboring position at every lattice vibration (one jump per 10^{-13} sec). The probability of atomic jumps is much smaller, however, as it depends on the number of vacant sites available, the energy necessary to break away from the surface site, and the temperature. Thus, the temperature dependence of the diffusion coefficient (due to the temperature-dependent probability) is given by

$$D = D_0 \exp\left(-\frac{\Delta E}{RT}\right) , \tag{13}$$

where ΔE is the activation energy for surface diffusion and D_0 is the diffusion constant which contains all the temperature-independent terms used in defining the diffusion coefficient. In general, the diffusion coefficient or diffusion constant is given in units of cm^2/sec. Returning to Eq. (12), with a diffusion coefficient of $D = 10^{-5}$ cm^2/sec, the radioactive atoms may move, on the average, 1 mm in 1000 sec. The activation energy for diffusion along the surface will reflect the net energy necessary to move atoms from one atomic position to another. Its determination is of primary importance in diffusion studies since it reveals the mechanism of atomic transport along the surface.

If the surface were homogeneous, i.e., all atoms located in lattice positions with identical binding energy, the activation energy for

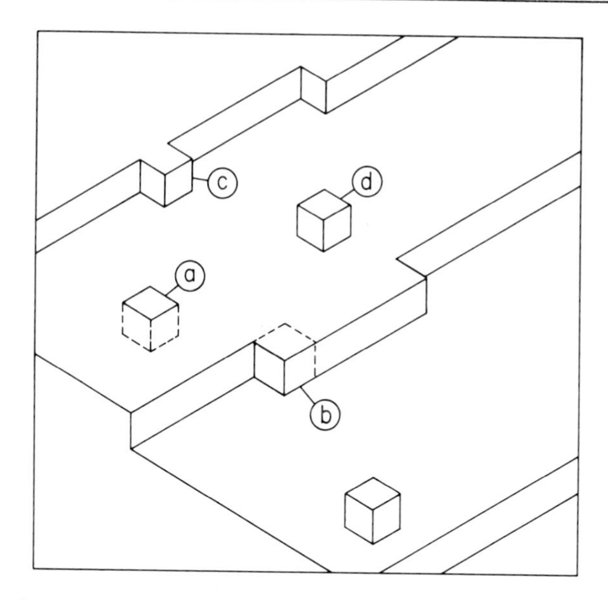

FIG. 6. A model of a metal surface depicting atoms in the following positions: (a) in surface, (b) kink, (c) at ledge, and (d) adsorbed on the surface.

surface diffusion would be a well-defined and, in fact, predictable quantity. We could count the number of bonds which have to be broken for an atom to break away from its neighbors and could even compute the effect on the bonding of those atoms further removed from the diffusing species (long range interactions)[31]. However, surfaces are heterogeneous, and the measured ΔE depends on the dominant types of surface sites between which atomic transport takes place. Figure 6 depicts some of the different surface sites on a heterogeneous surface. There are atoms in ledges, at kinks in the surface, and adsorbed on the surface. In order to move an atom from a kink position in a ledge to join another ledge, the atom has to break away from the ledge and diffuse along the surface as an adsorbed atom. This should require a greater activation energy than moving atoms already in adsorbed positions to a ledge. Thus, in any surface diffusion experiment, the relative abundance of atoms in the different surface sites can markedly influence the results.

Surface diffusion rates are, in general, appreciably larger at a given temperature than atomic diffusion rates in the bulk of the solid. The activation energies of diffusion on surfaces of fcc solids are approximately one-half of the heats of sublimation ($\sim 0.54\ \Delta H_{subl}$) [31]. These values are believed to be characteristic of the diffusion process by which atoms are moved from ledge to ledge. The activation energies for surface diffusion by a surface vacancy mechanism are much smaller, approximately $0.24\ \Delta H_{subl}$. Techniques able to distinguish between different atomic transport mechanisms have become applicable only recently. These include field emission microscopy, low energy electron diffraction, laser interferometry, and radiotracer measurements [32, 33].

IV. VAPORIZATION

Studies of the mechanisms of vaporization reveal the reaction steps by which atoms break away from their neighbors in the crystal lattice and are removed into the gas phase [34]. Detailed understanding of these steps should allow us to control vaporization and to increase or suppress its rate by suitable adjustment of the conditions of sublimation [35].

Vaporization studies may be carried out (a) near equilibrium between the solid and the vapor or (b) far from equilibrium ("free" or vacuum vaporization). Equilibrium studies are carried out to obtain vapor pressure and other thermodynamic data (enthalpy and free energy of vaporization). Vaporization studies of this type do not yield information about the reaction path. Vaporization rate measurements should be carried out far from equilibrium in order to determine the mechanism of sublimation.

Consider the vaporization of one crystal face of a monatomic solid A. If we assume that the overall vaporization reaction is A(solid) → A(vapor), the net rate of vaporization, J, may be expressed as

$$J(\text{mole/cm}^2\ \text{sec}) = k(A)_s - k'(A)_v \quad , \tag{14}$$

where k and k' are the rate constants for vaporization and condensation, respectively, $(A)_s$ is the concentration of atoms in the surface sites from which vaporization proceeds, and $(A)_v$ is the vapor density. For studies of the kinetics of vaporization, the condensation rate $k'(A)_v$ must be smaller than the rate of sublimation, $k(A)_s$. In such investigations, the evaporation rate of the solid is measured under nonequilibrium conditions, most frequently in a vacuum. For sublimation in a vacuum (free vaporization) the rate of condensation may be taken as zero, and Eq. (14) can be simplified to

$$J_v(\text{mole/cm}^2\ \text{sec}) = k(A)_s = k_0 (A)_s \exp(-E^*/RT) \quad , \tag{15}$$

where k_0 is a constant related to the frequency of attempted motion of vaporizing molecules over the energy barrier E^* (see the surface diffusion section).

The maximum theoretical rate of vaporization from the surface at a given temperature, J_{max}, would be attained if the solid were in dynamic equilibrium with the vapor $[k(A)_s = k(A)_v]$. The vacuum sublimation rate may have any value depending on the mechanism of vaporization, but its upper limit is that which is obtainable under conditions of equilibrium at a given temperature. It is customary to express the deviation of the vacuum evaporation rate J_v from the maximum equilibrium rate J_{max} in terms of the evaporation coefficient α which is given by

$$\alpha(T) = \frac{J_v(T)}{J_{max}(T)} \quad . \tag{16}$$

For most metals, for example, the vacuum evaporation rate equals the maximum rate $[\alpha(T) \simeq 1]$, whereas for other substances (for example, As, CdS, GaN, Al_2O_3) it may be orders of magnitude smaller than the maximum rate $[\alpha(T) << 1]$.

In studies of the kinetics of vaporization the evaporation rate of one face of a single crystal is measured under a variety of experimental conditions (as a function of temperature, in the presence of impurities or adsorbed gases, under irradiation, etc.). The use of a microbalance allows the measurement of the total weight loss of the sample and the determination of the absolute evaporation rate. The vapor composition and its temperature dependence can also be monitored with a mass spectrometer. This technique should certainly be used if the vapor is composed of more than one species. In this way the vapor constituents are identified and their activation energies of vaporization are determined separately.

For detailed descriptions of studies of vaporization mechanisms of solids and liquids the reader is referred to recent reviews [34, 36]. Here, we shall briefly mention some of the interesting results of recent vaporization measurements. Cadmium sulfide, CdS, dissociates upon vaporization according to the dominant net reaction, CdS(solid) = Cd(vapor) + $\frac{1}{2}S_2$(vapor). It was found that the concentration of free charge carriers (electrons and holes) at the vaporizing surface plays an important role in determining the sublimation rate. The evaporation rate can be increased by illumination of the vaporizing surface by light of suitable energy which increases the free carrier concentration (photocurrent). On the other hand, impurities (copper) or off-stoichiometry (excess cadmium or sulfur) in the crystals were found to markedly decrease the evaporation rate since they tend to decrease the concentration of free electrons or holes.

Sodium chloride vapor is composed mostly of monomer (NaCl) and dimer (Na_2Cl_2) molecules [37]. It was found that the concentration of dislocations in the crystal affect the evaporation rate [37]. The higher the dislocation density, the higher the evaporation rate until the maximum rate [$\alpha(T) = 1$] is reached (at a dislocation density $\sim 10^7$ cm^{-2}). Divalent ion impurities (such as Ca^{2+}) in the crystal lattice decrease the evaporation rate markedly. This is the result of small off-stoichiometry (excess Cl^- or Na^+ vacancies are present), which is introduced by doping. Monovalent ion impurities have no effect on the evaporation rates.

Arsenic vaporizes to form dominantly tetramer molecules (As_4) [38]. The vacuum evaporation rates are five to six orders of magnitude smaller than the maximum rates [$\alpha(T) \simeq 10^{-5}$] [38]. It was found that in the presence of a liquid metal (thallium, for example) the evaporation rate can be increased by several orders of magnitude. It appears that the formation of As_4 molecules is catalyzed by the presence of the liquid metal.

V. GROWTH

When a saturated solution of a salt is cooled from the saturation temperature T_2 to another temperature T_1, the salt concentration c corresponding to the solubility difference at these two temperatures, $\Delta c = c(T_2) - c(T_1)$, will precipitate [39]. Similarly, after achieving solid-vapor equilibrium at T_2, if the temperature of the system is decreased to T_1, the vapor molecules corresponding to the pressure difference $\Delta P = P_{eq}(T_2) - P_{eq}(T_1)$ will condense. For growth, which is addition of atoms or molecules to a surface, some degree of "supersaturation" is needed. The supersaturation σ is defined as

$$\sigma = \frac{P}{P_{eq}} - 1 \quad ,$$

where P is the vapor pressure and P_{eq} is the equilibrium vapor pressure of the solid. If $P = P_{eq}$, $\sigma = 0$ and there is no growth. For $P > P_{eq}$, σ is always larger than zero and growth may occur. (The condition $P < P_{eq}$ when $\sigma < 0$ is called "subsaturation." Under this condition vaporization occurs.)

It has frequently been observed that only very small supersaturation is needed for growth when crystals of the growing substance are already present at the onset of the crystal growth experiment [40]. It is well known that in the presence of a crystalline "seed," as it is often called,

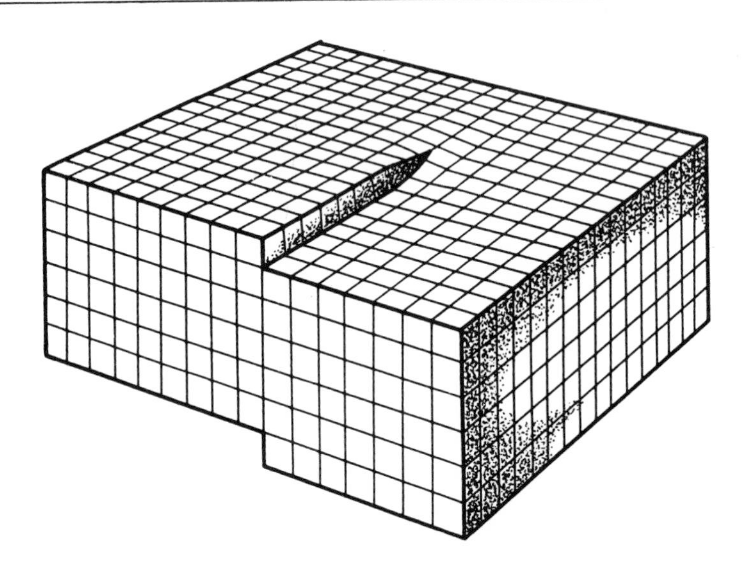

FIG. 7. Schematic representation of one type of dislocation.

growth commences rapidly. On the other hand, in the absence of a crystal surface, very large pressure differences ($P >> P_{eq}$) or solubility differences [$c >> c$(solubility)] often have to be established before growth begins. Since most frequently these large pressure or solubility differences are produced by lowering the temperature of the system, "undercooling" by many tens of degrees is necessary for growth to begin. For example, liquid copper may be cooled 236° below its melting point (1356 °K) in the absence of a "seed." As soon as the "undercooled" system is "seeded" (by introducing a crystal surface), rapid growth occurs.

The reason for the large difference between the growth characteristics on a crystal surface and in the absence of such a surface is that nucleation may not be needed in one case while nucleation is a prerequisite for growth in the other. Atoms can apparently add to the already existing surface to continue to build up the crystal. The presence of atomic aggregates of critical size or larger are needed (see Section IX of this chapter) before their growth as a condensed phase can occur.

In order to form such large particles, which finally become stable nuclei, a large supersaturation ($\sigma >> 0$) is needed [39].

If we had a perfect surface, addition of atoms would be very difficult without nucleation. It appears that a surface must always provide ledges where growth can continue without nucleation. A dislocation that results from a small mismatch of a row of atoms can provide such a continuous source of ledges. Figure 7 shows schematically a crystal with a dislocation emerging at the surface. The presence of growth spirals, i. e., growth along the dislocation, has been well documented on electron microscope pictures of growing crystals of many types. Lattice imperfections have to be present at the growing surface to have growth without nucleation. Since the surface is heterogeneous (Fig. 6) and imperfect, growth sites where atoms can be added to the surface are abundant [39].

Rate of Growth

Consider the growth of one face of a monatomic crystal from the vapor if the face is well provided with ledges and lattice imperfections at which condensation of atoms can occur without nucleation. Atoms will strike the surface; a fraction of the atoms adsorb while the rest reevaporate. The adsorbed atoms diffuse on the surface, and a certain fraction will reach a ledge or some other atomic site where growth can occur, before desorption. The fraction of the adsorbed atoms reaching the growth sites will effectively lose their excess energy and condense. The rate of the slowest reaction step in this series of consecutive reactions will determine the overall growth rate, under steady state conditions [34]. Vapor atoms strike the surface at a rate [34]

$$J(\text{vapor})\ (\text{moles/cm}^2\ \text{sec}) = \tfrac{1}{4}(A)_V \bar{v} \quad , \tag{17}$$

where $(A)_V$ is the vapor density (moles/cm^3) and $\bar{v}$ is the average velocity (cm/sec) of the vapor atoms. From the ideal gas law we have $(A) = P/RT$ and from the kinetic theory of gases, $\bar{v} = (8RT/\pi M)^{\frac{1}{2}}$, so that we can rewrite Eq. (17) as

$$J(\text{vapor})\left(\frac{\text{moles}}{\text{cm}^2\ \text{sec}}\right) = P_{eq}(2\pi MRT)^{-\frac{1}{2}} \quad . \tag{18}$$

A fraction of these atoms will not immediately reevaporate but will remain adsorbed on the surface and will move about. The rate of surface diffusion, J(diffusion), is given by

$$J(\text{diffusion}) = -D\left(\frac{\partial c}{\partial x} + \frac{\partial c}{\partial y}\right) \quad , \tag{19}$$

where D is the diffusion coefficient and $\partial c/\partial x$, $\partial c/\partial y$ are the surface concentration gradients toward the growth sites along the x and y directions on the surface. A fraction of the adsorbed atoms will reach the ledges and condense. The condensation rate is given by

$$J(\text{condensation}) = k''(C)(A')_s \quad , \tag{20}$$

where (C) is the concentration of adsorbed atoms at the growth sites, $(A')_s$ is the surface concentration of the growth sites, and k'' is the rate constant. Assuming that growth occurs via these consecutive reaction steps, any one of these steps may be rate limiting. For low supersaturations, Eq. (18) may control the overall rate, while at low temperatures the atoms have very small diffusion rates on the growing surface (D has exponential temperature dependence). Under these conditions, diffusing atoms may never reach a growth site but either desorb or condense on the surface randomly. It is well known that condensation on a cold surface on which atomic mobility is low produces amorphous and disordered layers instead of ordered single-crystal surfaces. Finally, Eq. (20) can be rate controlling if the adsorbed atoms must undergo electronic or structural rearrangements in order to condense at a growth site. Association [for example, $Cd(vapor) + \frac{1}{2}S_2(vapor) \rightarrow CdS(solid)$ to form CdS crystals] or dissociation [for example $As_4 \rightarrow 4As$ to form the arsenic lattice] reactions, which must take place upon condensation, are likely to make Eq. (20) the slow step in crystal growth.

Crystal growth phenomena are very important both in nature and in modern technology. The reader is referred to textbooks on this subject for further study [39, 40]. Growth at the solid-liquid interface differs considerably from growth at the solid-vapor interface, and impurities at the surface often play a significant role in catalyzing or inhibiting growth. The growth rates of different crystal faces can vary considerably and can also be competing processes.

VI. THE DENSITY AND DISTRIBUTION OF CHARGES AT SOLID SURFACES

A solid of atomic weight $M = 60$ g/mole and specific gravity of $\rho = 6$ g/cm^3 has an atomic density $(C) = \rho N/M \simeq 6 \times 10^{22}$ atoms/cm^3, where N is Avogadro's number. The surface density of atoms (S) is approximately given by $(S) \simeq C^{\frac{2}{3}} = 1.5 \times 10^{15}$ atoms/cm^2. The atomic density varies somewhat from one crystal face to another, but certainly less than by an order of magnitude.

The density of free electrons in a metal is almost as large as the atomic density. This can be determined experimentally by electrical conductivity and Hall effect measurements. If we trap all the free electrons about surface atoms or somehow remove them from the surface, they may immediately be replenished from the neighboring atoms under the surface. Thus, the mobile charge density at the metal surface should remain high even after the removal of free carriers from all of the surface atoms ($\sim 10^{15}$ electrons/cm^2).

The situation is entirely different in the case of insulating or semiconducting solids which have small free carrier concentrations. These substances have charge carrier densities around 10^{14}-10^{17} cm^{-3} at room temperature. Thus, there may be only one free electron per 10^7 atoms. If we now would like to trap $\approx 10^{15}$ free carriers at the surface, or remove them somehow, we have to transfer electrons from 10^5 atomic layers below the surface [41]. That is, removal of 10^{15} electrons at the surface will affect the charge distribution of an insulator in a layer of thickness 10^{-4} to 10^{-3} cm. Consequently, any change in the electron density distribution at an insulator or semiconductor surface will have a large effect on the charge distribution in the bulk. The layer near the surface whose charge distribution is changed upon removal of charges to the surface is called the space charge layer. The thickness of the space charge layer normal to the surface is commonly referred to as the "Debye length" [41, 42].

Because surface atoms have a different environment from that for the bulk atoms, they have localized electronic states in which electrons or holes may be trapped. The concentration of these surface states is of the same order of magnitude as the surface density of atoms ($\approx 10^{15}$ cm^{-2}). For a metal surface the presence of these surface states has virtually no effect on their charge distribution. For insulators, however, trapping of electrons or holes in surface states changes the carrier concentrations markedly since this process involves a large fraction of the total concentration of free carriers. As the surface becomes richer in charge carriers with respect to the bulk, a potential develops between the surface and the bulk which opposes further charge transfer to the surface [41]. The space charge layer which develops at the surface of an n-type semiconductor, due to trapping of electrons in surface states, is shown schematically in Fig. 8. The density of charges, ρ_+, in the space charge layer can be calculated from the Poisson equation:

$$\frac{d^2V}{dz^2} = -\rho_+/\kappa\varepsilon_0 \quad , \tag{21}$$

where κ is the dielectric constant of the solid and ε_0 is the permittivity of free space (8.85×10^{-12} farad/m). Assuming that electrons transferred to the surface states all come from one type of ionized donor center, N_D^+ ($\rho_+ = eN_D^+$ where e is the unit charge), we have

$$\frac{d^2V}{dz^2} = \frac{eN_D^+}{\kappa\varepsilon_0} \quad . \tag{22}$$

Integration of Eq. (22) gives the relationship

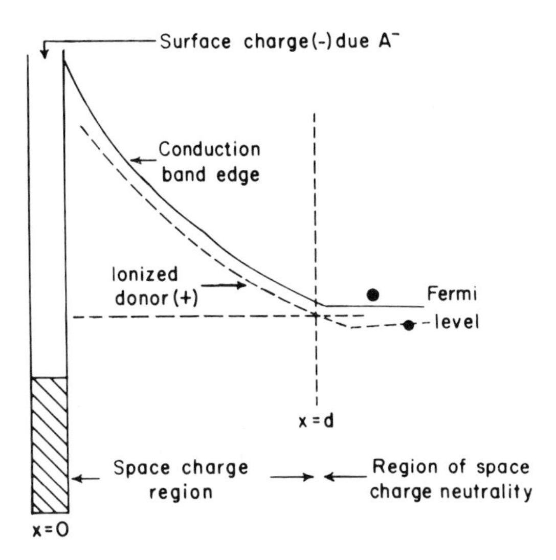

FIG. 8. Schematic representation of an n-type space charge layer at the surface.

$$V(z) = -\frac{eN_D^+}{2K\varepsilon_0}(z - d)^2 , \tag{23}$$

where at $z = d$, $V(z) = 0$. Thus, we see that the height of the potential barrier V and the thickness of the space charge layer z (Debye length) both depend on the density of charges, i.e., the concentration of trapped electrons [42].

Ionic solids (NaCl, for example), for which the mobile charge carriers are ions, form space charge layers at their surface similar to those observed in insulators, for which electrons are the charge carriers.

The space charge layer on insulating surfaces plays an important role in gas adsorption and chemical surface reactions. Since charge transfer is one of the reaction steps in gas-surface interactions, charge transfer over the space charge potential barrier can control the rate of adsorption or chemical reaction at the surface.

VII. EMISSION PROPERTIES OF SOLID SURFACES

A. Work Function

When light of suitable energy strikes a metal surface, electrons are emitted. The minimum energy necessary to remove an electron from the Fermi level in the metal into vacuum is called the "work function" Φ [43]. A photoelectric work function may be measured by varying the wavelength of incident light and detecting the threshold energy at which photoexcited electrons appear. Electrons are also emitted in a vacuum from a heated filament that is negatively charged (used as a cathode) with respect to a collector plate placed a short distance away. The electron current density depends on the filament temperature, and at a given accelerating potential, large enough to provide an electron current at its saturation value [33], it is given by

$$i = AT^2 \exp[-e\Phi/kT] \quad , \tag{24}$$

where Φ is the work function and A is a constant, $A = 120\,A\ sec/cm^2\ deg^2$. Thus, from the temperature dependence of the emission current the work function can be obtained [43].

The work function is always positive; the electron in the solid has to overcome an attractive potential in order to be emitted. The magnitude of this attractive potential varies somewhat from crystal face to crystal face. In general, low index (high atomic density) crystal faces have higher work functions than high index (lower atomic density) faces. Table 3 lists representative values of work functions of different crystal faces of tungsten. Since the work function may markedly change in the presence of gases adsorbed on the emitting surface or in the presence of impurities, care must be taken in these experiments to keep the surface clean. In turn, changes in the measured work function due to the adsorption of gases can be used to study the nature of gas-solid interactions and the quantity of the adsorbed gas. Investigations of this type are discussed later.

TABLE 3

The Work Function of Different Crystal Faces of Tungsten

Solid	Crystal face	Work function (eV)
Tungsten	(110)	4.68
	(112)	4.69
	(111)	4.39
	(001)	4.56
	(116)	4.39

B. Surface Ionization

A high electron density solid surface (e.g., a metal) may act as an acceptor or donor of electrons for atoms adsorbed on the surface. The metal surface may strip electrons from adsorbed atoms, forming positive surface ions which can then be emitted if the surface temperature is high enough to give rise to an appreciable desorption rate [43].

Consider a beam of atoms of flux J and of ionization potential I ($M \xrightarrow{I} M^+ + e^-$) as they strike a hot metal surface. The ratio of ion flux to the flux of neutral atoms in the reemitted beam is given by the Saha-Langmuir equation [43]:

$$\frac{J^+}{J_0} = A' \exp\left[\frac{e(\Phi - I)}{kT}\right] \quad . \tag{25}$$

Here e is the unit charge and A' is a constant which includes the following: the ratio of the statistical weights of the ionic and atomic states (for alkali atoms $\frac{1}{2}$, because of two spin states for the valence electron), correction factors to account for the possible lack of thermal equilibration between the atoms and the surface, and the ratio of probabilities for the ions and atoms adsorbed on the surface that have enough energy to vaporize (desorption probability). The larger the work function of the metal with respect to the ionization potential, the larger is the

TABLE 4

The Ionization Potentials of Alkali Atoms and the Average Work Functions of Tungsten and Platinum

	Li	Na	K	Rb	Cs	Pt	W
Ionization potential (eV)	5.40	5.12	4.32	4.10	3.88		
Average work function						6.35	4.52

positive ion flux that may be obtained. Table 4 lists the ionization potentials of the alkali metals and the average work function for tungsten and platinum (averaged over the work functions for the different crystal orientations). These metals are used most frequently in surface ionization studies. For cesium $I < \Phi$; thus $(J^+/J_0) > 1$, and the incident neutral atoms will desorb dominantly as ions. For sodium $I > \Phi$ and $(J^+/J_0) < 1$; a larger fraction of the incident flux should remain un-ionized.

Negative ions may form by surface ionization if neutral atoms with large electron affinity S $(X + e^- \xrightarrow{-S} X^-)$ strike a metal surface. As in Eq. (25), the ratio of negative ion flux to the flux of neutral particles in the reemitted beam is given by

$$\frac{J^-}{J_0} = A'' \exp\left[\frac{e(S - \Phi)}{kT}\right] . \qquad (26)$$

Here A'' is a constant consisting of the same terms as in A for this electron donation process [33]. For larger values of the electron affinity

TABLE 5

Electron Affinities of Several Atoms

	F	Cl	Br	I	O	S
Electron affinity (eV)	3.6	3.7	3.5	3.2	3.1 (2.3)	2.4

with respect to the work function, larger negative ion fluxes are obtained. Table 5 lists some of the atoms with large electron affinities. Even for chlorine or fluorine, however, $S < \Phi$ and, therefore, the ratio J^-/J_0 is smaller than unity.

C. Field Emission

The application of a large electric field at the surface, about 10^7-10^8 V/cm, distorts the potential barrier and permits the electrons to tunnel through it and be emitted. Such an intense electric field can be obtained by applying a negative potential of 10^3-10^4 V across a small cathode tip that has a radius of curvature 10^{-5}-10^{-4} cm. The current density emitted from the tip is given by the Fowler-Nordheim equation [43]

$$J = \frac{a\mathcal{E}^2}{\Phi} \exp\left[-\frac{b\Phi^{\frac{3}{2}}}{\mathcal{E}}\right] , \qquad (27)$$

where $\mathcal{E}$ (V/cm) and Φ (eV) are the field intensity and work function of the emitter, respectively, and a and b can be taken as constants [43, 44]. Using this effect, Muller [45] has constructed a field emission microscope which allows one to study the field emission distribution from solid surfaces. The electrons emitted from the cathode tip are accelerated onto a fluorescent screen, where the emission distribution of the tip is displayed with a magnification proportional to the ratio of the screen surface area and to the area of the cathode tip ($\approx 10^5$) [46]. Since the field emission current is very sensitive to small changes of the work function [see Eq. (27)], the different crystal faces (which all have different work functions) are easily distinguishable [35]. The adsorption of gases on the emitter tip can also be monitored since it gives rise to a work function change at the surface. This technique has become very important in studies of surface diffusion and gas adsorption [46]. Its application is limited, however, to those solids which can be used to fabricate cathode tips of a very small radius of curvature and which are not destroyed by high electric fields (heating effect, field evaporation).

In the field ion microscope, helium atoms are ionized at the field emission tip (now at a positive potential with respect to ground) in the presence of the large electric field [47]. The positive ions are then repelled and accelerated onto the fluorescent screen, where the greatly magnified image of the tip surface is displayed. The use of ions instead of electrons to form the image of the tip markedly increases the resolution. Due to the larger ionic mass (as compared with the electrons) there is little displacement of ions along the surface of the tip during the emission process, which could introduce uncertainty in the ion position and, therefore, blurring of the image. Thus, single atoms can be identified on the different crystal faces of the tip, and their movement may be monitored [47]. Adsorbed atoms of different kinds (Be, Cu, Fe, etc.) have been ionized by a large electric field applied to the surface with the production of several species with unusual valence states, for example, BeH^+, CuH_2^+, FeH_2^+ [48].

VIII. SCATTERING PROPERTIES OF SURFACES

A. Electron Scattering

When low energy electrons, 5-500 eV, impinge on a surface, they are backscattered without penetrating more than a few atomic layers [12, 15]. Therefore, their scattering properties provide considerable information about the properties of atoms in or near the surface. Higher energy electrons (> 1000 eV) penetrate farther below the surface and are scattered primarily by bulk atoms, unless directed at the solid at a grazing angle [49].

A small fraction of the incident low energy electrons (5-15%) backscatter without losing any energy in the collision (elastic scattering). Since the wavelength of these electrons, calculated from the de Broglie equation, $\lambda = h/p = (150/eV)^{\frac{1}{2}}$, is in the range of 0.5-5Å, diffraction from the rows of surface atoms occurs. Thus, low energy electron

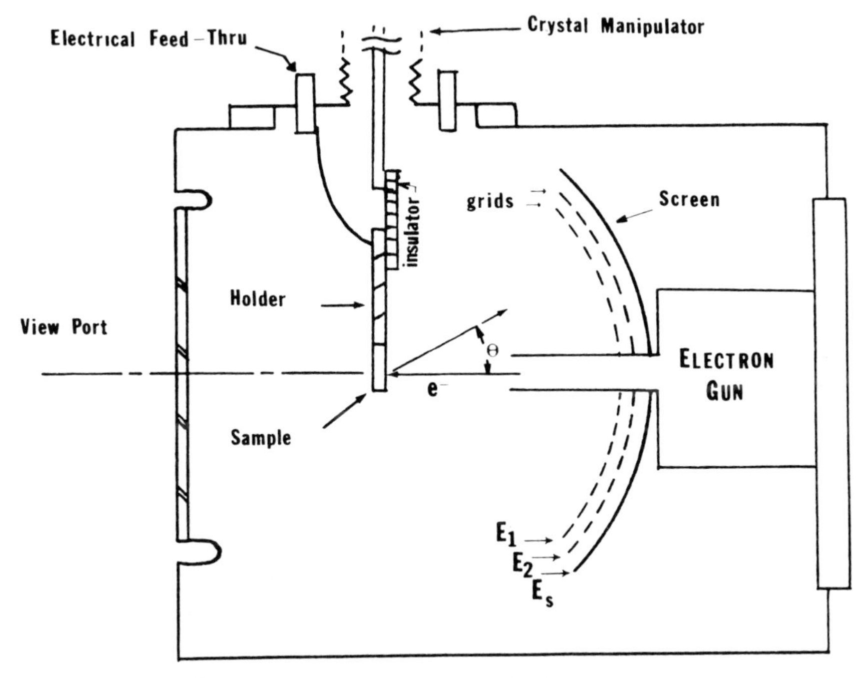

FIG. 9. Schematic representation of a low energy electron diffraction experiment.

diffraction allows us to investigate the structure of surfaces on an atomic scale. The other 85-95% of the incident low energy electrons undergo inelastic collisions with the surface, i.e., they lose energy in the scattering process. These electrons, when energy-analyzed, give information about the electronic structure of surface atoms [50] (electron binding energies) and permit identification of surface impurities [51].

B. Low Energy Electron Diffraction (LEED)

In LEED experiments a monochromatic beam of electrons (5-500 ± 0.2 eV) is focused on the surface of a single-crystal sample in an apparatus as illustrated in Fig. 9. (The electron wavelength can be varied easily by changing the acceleration potential.) The back-reflected electrons, after traveling a short field-free path (~7 cm) are separated;

the inelastically scattered electrons are retarded while the elastically scattered fraction, which contains the diffraction information, is post-accelerated onto a spherical fluorescent screen where the diffraction pattern is displayed. For accelerating potentials smaller than 75 eV, most of the electrons are backscattered from the surface plane. Thus, the diffraction pattern reflects the arrangements of atoms in the surface as the two-dimensional character of the diffraction essentially dominates. At higher electron energies the fraction of electrons that penetrate a few atomic planes below the surface increases. Thus, at higher beam voltages (> 150 eV) the diffraction pattern is more characteristic of the bulk structure near the surface, i. e., the three-dimensional character of the diffraction becomes more pronounced. The diffraction pattern can be viewed and photographed through a window. In the diffraction chamber pressures of the order of 10^{-10} to 10^{-9} Torr can be easily maintained using modern vacuum techniques (vacuum ionization and sorption pumps). Ultrahigh vacuum is necessary to avoid possible contamination of the sample surface by the adsorption of gases from the ambient [11].

Unlike x-rays, which are scattered by the atomic electrons, the incident electrons are scattered by the crystal potential [51]. Thus, hydrogen atoms can be as effective scatterers as heavier elements. Due to their large scattering cross section, which is several orders of magnitude larger than for x-rays, low energy electrons can be used to obtain high intensity diffraction patterns from ordered surface structures composed of no more than 5-10% of the total number of surface atoms ($\approx 10^{14}$ atoms/cm^2).

The preparation of the single-crystal surfaces is an important part of the LEED experiment. Careless cutting and polishing can produce an amorphous surface layer several microns thick ($1\mu = 10^{-4}$ cm). In order to obtain the diffraction features characteristic of the single-crystal surface, the damaged surface layer is removed by ion-bombardment [12], chemical etching [12], or by cleavage inside the diffraction chamber. It should be noted that, while bombardment using ions of 200-500 eV energy disrupts the surface or dislodges and removes atoms in the solid, electrons of the same energy rarely cause surface damage. This difference is caused by the large ion masses (at least 10^3 times larger than the electronic mass) which give the incident ions large momenta.

We have briefly mentioned some of the important discoveries made using low energy electron diffraction. It was found that surface atoms reside in different kinds of surface structures. These surface structures are characterized by lattice parameters that are integral multiples of the unit-cell dimensions of the bulk crystal. It was also found that atoms adsorbed on the surface can form ordered surface structures. It has also been discovered that, during a strongly exothermic surface reaction (oxidation for example), surface atoms can be dislodged from their equilibrium positions resulting in the formation of a "reconstructed" ordered surface layer composed of both solid and adsorbed atoms. Several investigations of the deposition of condensible vapors on single-crystal surfaces have been carried out using LEED. Again, deposition and growth of one solid on another take place via the formation of ordered surface structures [53].

Low energy electron diffraction studies of surfaces reveal their structure and variations of their structure under a variety of conditions (adsorption, phase transformation, etc.). This structural information is necessary for the investigation of the chemistry of surfaces on an atomic scale.

1. Inelastic Electron Scattering

When an incident electron beam strikes a surface, the larger fraction of the backscattered electrons (85-95%) loses energy in the collision process. This energy is used in several different electronic processes by the surface atoms. If the energy is imparted to the valence electrons, "secondary electron emission" occurs. Thus, for every incident electron one or more additional valence electrons may leave the solid. These electrons have a broad energy distribution and their yield depends on the incident beam energy. At higher electron beam energies (> 400 eV), electrons from lower lying electronic energy bands (inner shells) can be removed by electron impact. As electrons from outer levels drop into these vacancies, energy is released. This energy may appear as a photon or may be transferred to another bound electron which is then ejected with a characteristic energy, that is, by a so-called Auger process [50]. The energy distribution of the Auger electrons changes very little with the incident electron energy. Auger electrons can give information about the energy of deeper lying electronic bound states. Since these energy losses due to Auger processes are characteristic of the atoms from which the electrons come, Auger electrons from oxygen, for example, will have different energies than Auger electrons from copper atoms. Thus, the detection of Auger electrons allows chemical analysis of the surface and the detection of minute impurities [50].

Measurement of the energy distribution of inelastically scattered electrons can be carried out in an apparatus very similar to that shown in Fig. 9. By changing the negative potential on one of the metal grids, electrons with various energies can be collected and their distribution accurately analyzed [51].

2. Scattering of Molecular Beams from Surfaces

When a beam of atoms or molecules with well-defined kinetic energy strikes a solid single-crystal surface, part of the beam is scattered elastically. The wavelength associated with such a beam can be calculated from the de Broglie equation. A helium atom beam, for example, having thermal energy $E = 0.02$ eV for $T \simeq 300\,°K$ has a wavelength $\lambda = 1\,Å$. Thus, a helium beam can be diffracted by the surface, and the diffraction of He from a lithium fluoride surface has been observed [15].

The largest fraction of the beam, however (just as for electrons), undergoes inelastic collision with the surface. By analyzing the time of flight of the backscattered molecules one can determine their residence time on the surface. Measurements of the velocity and angular distribution of the scattered beam reveal the energy exchange and the nature of the interaction between the gas and surface atoms. The extent of the energy exchange between the incident beam and the surface can be expressed in terms of an "energy accommodation coefficient," $\alpha_E(\theta)$:

$$\alpha_E(\theta) = \frac{E(\text{incident}) - \bar{E}(\text{final})}{E(\text{incident})} \quad . \tag{28}$$

Here, E(incident) and $\bar{E}$(final) are the energies of the incident molecules and reflected molecules at angle θ [54]. Both $\alpha_E(\theta)$ and the residence time of the molecules depend on the surface temperature and the beam velocity. Experiments indicate that the molecular beam, which strikes the solid surface at a specified angle of incidence, is reemitted in a lobular pattern, which also shows a sharp maximum [54]. The angle of reflection at which this maximum appears is related in a reproducible way to the angle of incidence and to the surface temperature. The appearance of a lobe distribution instead of a single maximum is caused mostly by surface roughness.

There are two limiting cases of beam scattering mechanisms: (a) diffraction and (b) diffuse scattering. The dominance of elastic scattering (a) indicates the lack of energy exchange between surface atoms and the atomic beam. Diffuse scattering (b) is characterized by a scattered beam distribution, which varies as the cosine of the angle between the scattering direction and the surface normal. If the incident atoms reside long enough on the surface to equilibrate with it, diffuse scattering results.

In most molecular beam scattering experiments, the scattered beam has been found to have a spatial distribution which lies between the two limiting cases. The spatial distribution tends to be more specular for clean surfaces and more diffuse in the presence of adsorbed gases or for roughened surfaces. The spatial distribution also depends strongly on the angle of incidence and the mass of the incident molecules. The maximum in the scattered beam intensity shifts toward the surface normal as the surface temperature increases or as the energy of the incident beam decreases. Thus, as the attractive interaction between the beam and the surface increases, scattering becomes more diffuse, as expected.

A theory that successfully explains most of the experimental results has been developed recently [55-57]. It considers the interaction of gas atoms with symmetrically placed mass points, which are connected by harmonic springs to each other and to a fixed site. A Lennard-Jones interaction potential [48] is used in the calculations.

Molecular beam scattering studies from surfaces allow one to investigate energy transfer in some of the elementary steps of adsorption. Their extension to studies of reactive scattering will permit the study of energy transfer in surface reactions.

3. Ion Beam Scattering from Surfaces

When an ion beam strikes a solid surface, only a very small fraction of the ions are back-reflected (1 in $\sim 10^4$). Most of the ions become trapped, exchange energy and charge with the solid surface, and desorb as neutral atoms [43]. The ion energy is converted into electronic processes which are similar to those that take place during inelastic electron scattering. Secondary electron emission and Auger electron emission occur, their relative yields depending on incident ion energy and mass. At higher ion energies (> 100 eV) dislocation and removal of surface atoms occur, as mentioned previously. This treatment is frequently employed to prepare or clean solid surfaces. A recent review of ion impact phenomena on surfaces discusses in detail the electronic processes that take place as a consequence of ion-surface interactions [43].

C. Adsorption

When atoms or molecules strike a solid surface the largest fraction will lose energy in the collision, as clearly indicated by molecular beam scattering experiments. Most of the atoms that reside on the surface for a long time (10^{-6}-10^{-3} sec) with respect to the time of atomic vibration ($\approx 10^{-12}$ sec) will lose enough of their kinetic energy so that they are no longer able to leave the surface and are, therefore, adsorbed [58]. How long a residence time is necessary for an incident molecule to adsorb on the surface depends on the nature of the interaction between the molecule and the surface atoms and the temperature of the surface. The desorption probability $\mathscr{P}$ of any atom on the solid surface depends exponentially on the binding energy ΔE and the surface temperature, $\mathscr{P} \propto \exp[-\Delta E/RT]$. For weak interaction between the adsorbed atom and the surface, lower surface temperatures and longer residence times are needed for adsorption to occur. The weak interactions, which are due to dipole-dipole, induced dipole, and other short-range forces (the interaction energy varies as the inverse third power or the inverse sixth power of the distance between the surface and the gas atom), result in "physical adsorption." Atoms or molecules which are held on the surface this way have heats of adsorption in the range 1-10 kcal/mole, the magnitude of the heats of condensation [2, 3, 58].

If the molecule is held by strong electrostatic forces due to partial or complete electron transfer between the surface atoms and the adsorbed species (the interaction energy varies as the inverse first power of the distance between the surface and the gas atom), there is strong "chemisorption." The heats of chemisorption are of the order of chemical binding energies, 15-150 kcal/mole. Molecules capable of such strong interaction with the surface adsorb after a very short residence time, and even at elevated temperatures.

There are few systems that exhibit clearly defined physical adsorption or chemisorption. Possibly some of the noble gases (He, Kr) belong to the former, and the most reactive gases (O_2, F_2 on W or Ni) belong to the latter. For most molecules that have relatively larger dipole moments or polarizabilities, the molecule-surface interactions lie between those defined by physical adsorption and chemisorption. For purposes of demonstrating the principles of adsorption, however, we shall discuss the two limiting cases, physical adsorption and chemisorption, separately.

1. Physical Adsorption

Physical adsorption studies are carried out at low temperatures. The structure of weakly adsorbed species has been investigated by low energy electron diffraction only recently. Surface structures due to adsorbed xenon and bromine have been reported on graphite single-crystal surfaces [59]. No ordered surface structures could be found, however, during the physical adsorption of several gases (Ar, Kr, Xe, O_2, C_2H_2, etc.) on silver single-crystal surfaces [60]. The presence and importance of ordered surface structures in physical adsorption awaits further investigations.

In measurements of adsorption the amount of gas adsorbed on the surface is monitored as a function of pressure at different equilibrium pressures and at one temperature over the surface. A representative plot of the average thickness of the adsorbed layer (proportional to the amount adsorbed) as a function of pressure, the so-called "adsorption isotherm," is shown in Fig. 10. The data were obtained by using an

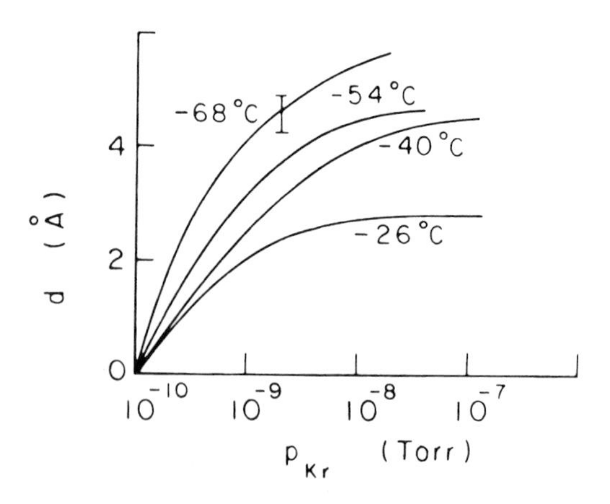

FIG. 10. Adsorption isotherms for krypton on the (110) crystal face of silver at various temperatures.

ellipsometer to monitor the adsorption of krypton on one face of a silver single crystal [60]. The adsorption isotherms in this case show an initial rapid rise with increasing equilibrium pressure until the coverage reaches a maximum, when the thickness becomes virtually independent of pressure. Isotherms of this shape are often called Langmuir isotherms [58, 61]. When adsorption isotherms are determined at different surface temperatures, one can plot the logarithm of the equilibrium pressures, necessary to obtain the same coverage at different temperatures, as a function of the reciprocal temperature. From the slope of these curves the isosteric heats of adsorption (the heat of adsorption at a constant surface coverage), ΔH_{st}, can be calculated using the Clausius-Clapeyron equation. Since ΔH_{st} can be obtained at several different surface coverages, one may plot the isosteric heats as a function of the amount of gas adsorbed on the surface. This is shown in Fig. 11 for the physical adsorption of several gases on a silver surface. The isosteric heat of adsorption first increases with coverage and then, in some cases, it decreases [60].

What is the meaning of these curves? First, the extrapolation of ΔH_{st} to zero surface coverage, $\Delta H_{st}(f = 0)$, gives the interaction energy between the adsorbed gas molecule and the surface atoms. As more molecules adsorb on the surface there is a lateral interaction, i.e., an interaction between the adsorbed molecules in addition to their interaction with the silver surface. Since ΔH_{st} increases with increasing coverage, the lateral interaction appears to be strong, of the same magnitude as the interaction with the metal surface [60]. As the coverage increases, ΔH_{st} should approach the heat of condensation of the gas, which is indicated by the arrow.

Isotherms of the Langmuir type (shown in Fig. 10) are generally obtained when adsorption terminates at about monolayer coverage. The adsorption isotherm can be described by the equation [2, 3, 58]

$$f = bP/(1 + bP) \quad , \tag{29}$$

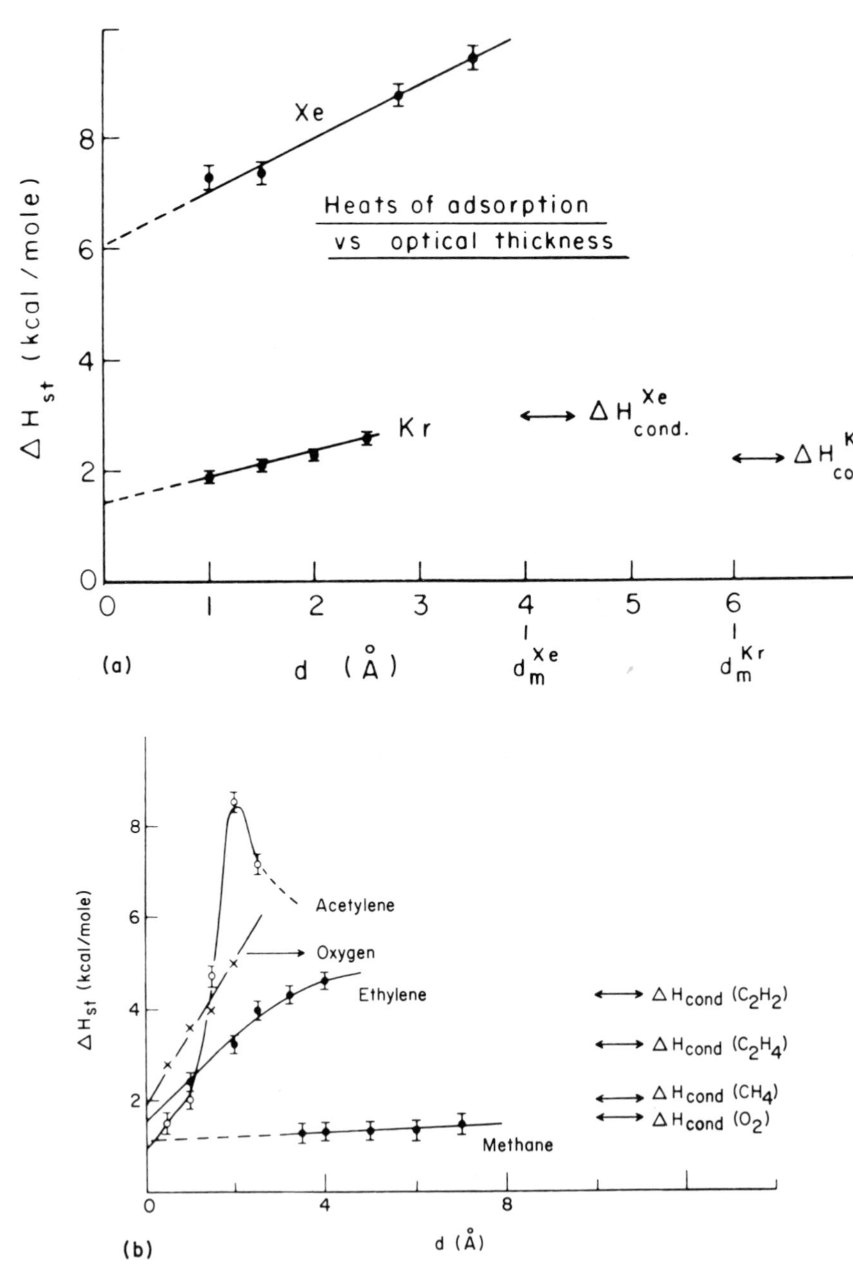

FIG. 11. The isosteric heats of adsorption vs optical thickness (i. e., average coverage) for several gases on the (110) crystal face of silver.

where f is the coverage and b is related to the equilibrium constant for the solid-gas equilibrium [54]. If adsorption occurs in multilayers, i.e., further gas adsorption takes place on top of the adsorbed monolayer, adsorption isotherms of different shapes are obtained. Their analysis, although somewhat more dubious since it involves the use of at least two adjustable parameters [58, 61], can also be carried out.

Most reactive surfaces of technological importance are polycrystalline. Adsorption isotherms are measured for these systems to determine their surface area [2, 3] and other adsorption characteristics. The analysis of adsorption data is complicated by the availability of several sites for adsorption at which the gas molecules may have different interaction energies [2, 3]. Physical adsorption on heterogeneous, polycrystalline surfaces has been the subject of many detailed experimental and theoretical investigations. The interested reader is referred to recent reviews [62, 63].

2. Chemisorption

The structures of chemisorbed gases on many metal and semiconductor surfaces have been studied by low energy electron diffraction. In most cases, ordered surface structures form during chemisorption. The type of surface structure produced depends on the chemistry and size of the gas molecules, the crystal orientation, and the amount of adsorbate [64]. It is often found that the gas structure converts into other surface structures as a function of coverage. The ordering process can be accelerated if one assures suitable mobility of the adsorbed atoms by selecting the proper surface temperature [64]. Increasing the surface temperature generally increases the surface mobility. This is especially important during the chemisorption of large molecules such as butadiene or isobutylene [64]. The heats of chemisorption are markedly larger than for physical adsorption. Therefore, chemisorption takes place easily at elevated temperatures (above room temperature).

Since chemisorption can be considered a chemical reaction between the gas molecules and the surface, it is often the precursor of other chemical changes that occur subsequently at the solid surface. The chemisorption of diatomic molecules is often followed by their dissociation on the solid surface. Hydrocarbon molecules can break apart and lose hydrogen or isomerize on the surface. Following chemisorption, reactive gases such as oxygen or hydrogen may diffuse from the surface into the bulk of the solid. Thus, chemisorption is not an isolated charge transfer process but may be one that initiates other chemical surface reactions.

When gas molecules chemisorb at a semiconductor surface, the space charge layer undergoes marked changes [65]. If the adsorbed gas atoms are electron acceptors (for example, oxygen) at an n-type semiconductor surface, more electrons are trapped and thus are no longer available for conduction. Therefore, the conductivity of the solid decreases during chemisorption. As more electrons are transferred to the surface and used up for chemisorption, the height of the space charge potential barrier increases. It becomes more difficult for the electrons in the bulk to pass over this barrier to the surface and, therefore, the chemisorption rate decreases. The rate of chemisorption, which is controlled by charge transfer over a potential barrier, can be expressed by the equation [65]

$$\frac{dn}{dt} = \frac{d(X^-)}{dt} = kn_0(X) \exp\left[-\frac{eV}{kT}\right] , \tag{30}$$

where dn/dt is the rate of electron transfer, $d(X^-)/dt$ is the rate of chemisorption, k is the rate constant, X is the concentration of adsorbed neutral atoms available for charge transfer, n_0 is the electron density in the bulk of the solid, and eV is the height of the space charge potential barrier. From Eq. (23), which was obtained by integrating the Poisson equation, we have [41, 42]

$$eV = c'n^2 \tag{31}$$

since $N_D^+ = n$ and $n = N_D^+ z$ and where c' is the constant of proportionality. Substitution of Eq. (31) into Eq. (30) gives

$$\frac{d(X^-)}{dt} = kn_0(X) \exp\left[-\frac{c'n^2}{kT}\right] \quad . \tag{32}$$

It can be seen from Eq. (32) that as charge transfer progresses the chemisorption slows down. The integrated rate equation has the form

$$Bn = \ln(t + c'') \quad , \tag{33}$$

where B and c'' are constants. Thus, plotting the logarithm of time against the change in the carrier concentration, which can be obtained by monitoring the conductivity of the solid, gives a straight line. This type of rate equation (frequently called the Elovich equation) has often been obtained for chemisorption on insulator or semiconductor surfaces [66].

The work function of the solid undergoes marked changes during chemisorption. This is not surprising since electron transfer to or from the solid changes the charge density at the surface [67]. If electrons are donated to the surface atoms by the adsorbed gas, the work function decreases, i.e., it becomes easier to remove an electron from the solid surface to infinity. If the adsorbed gas is an electron acceptor, the work function increases. Typical changes of work function, upon the monolayer adsorption of different gases on platinum and nickel surfaces, are given in Table 6. It has been established [67] that for many systems the work function change is linearly proportional to the concentration of adsorbed gases. Therefore, by monitoring the work function change as a function of time during adsorption (after suitable calibration) one may determine the amount of adsorbed gas at any temperature.

TABLE 6

Work Function Changes ($\Delta\Phi$) of Platinum and Nickel Surfaces upon Chemisorption

Gas	$\Delta\Phi$ (in eV)	
	Nickel	Platinum
H_2	+0.35	+0.15
CO	+1.35	+0.18
O_2	+1.6	+1.2
C_2H_4	-0.89	-1.11
C_2H_2	-	-1.4
C_3H_6	-	-1.36
C_6H_6	-1.3	-1.6
Xe	-0.85	-
N_2	-0.21	-

IX. CHEMICAL SURFACE REACTIONS

Many chemical reactions that take place at solid surfaces are of great technological importance. The surface provides a large density of sites on which atomic interactions can occur. More importantly, the surface provides alternate reaction paths by forming chemical intermediates with the reactant gases [68]. If the reaction rates are higher via these intermediates, the surface serves as a "catalyst" in the chemical reaction. The surface structure of some of the reactive intermediates has been studies only recently [12]. We shall list several surface catalyzed reactions of great industrial significance which may involve only two types of reactants:

(1) $4NH_3 + 5O_2 \xrightarrow[850\,°C]{Pt} 4NO + 6H_2O$,

$$(2)\quad 2C_2H_4 + O_2 \xrightarrow[260\,^\circ C]{Ag} 2CH_2 \overset{O}{\frown} CH_2 \ ,$$

$$(3)\quad N_2 + 3H_2 \xrightarrow[450\,^\circ C]{Fe} 2NH_3 \ ,$$

$$(4)\quad nCO + (2n+1)H_2 \xrightarrow[150\text{-}300\,^\circ C]{Co(Fe, Ru)} C_nH_{2n+2} + nH_2O \ ,$$

$$(5)\quad C_nH_{2n} + H_2 \xrightarrow[100\text{-}400\,^\circ C]{Pt(Pd)} C_nH_{2n+2} \ .$$

The first reaction is used to produce nitric acid via the catalytic oxidation of ammonia. Silver surfaces appear to be efficient catalysts for the preparation of ethylene oxide, which is the intermediate compound in producing polyethylene. Reaction (3) gives the well-known ammonia synthesis which is so important in fertilizer production. Reaction (4) is the synthetic process for converting inorganic gases ("water gas," which is 2:1 molar ratio H_2 and CO) to organic compounds. Finally, Reaction (5) is the catalytic industrial process for the hydrogenation of olefins to produce saturated hydrocarbons (for example, margarine from corn or vegetable oils).

There are several conditions that can be proven to be essential for a catalytic surface reaction to occur [68, 69]:

(1) Both reactants should be able to chemisorb on the surface. In case of reactions involving diatomic molecules of large binding energy (O_2, H_2, CO, N_2) the molecule should be able to dissociate either on the catalyst surface or in its bulk. The ability of the catalyst to atomize the molecules determines, to a large extent, its reactivity. Iron is known to form nitrides and to chemisorb nitrogen in atomic form. Cobalt is likely to form surface carbides which act as reaction intermediates in the Fischer-Tropsch [68] synthesis of hydrocarbons from carbon monoxide and hydrogen. Oxygen has a large solubility in silver and diffuses rapidly in the bulk in the atomic state. Palladium and

platinum dissolve hydrogen atoms effectively and are ready sources or sinks of atomic hydrogen in chemical surface reactions. However, the other reactant should chemisorb on the surface equally well, at least in a finite temperature range.

(2) The reactants and products should not be adsorbed too strongly. It is important that the products of the surface reaction desorb rapidly so that the surface remain available for continued reaction. This applies also for the reactants. Although they should be able to chemisorb, their removal rate should also be rapid. The bond energies of the M—O, M—H, M—N bonds for the active catalysts for reactions involving oxygen, hydrogen, and nitrogen have intermediate values between the strongest and weakest metal oxide, hydride, and nitride bonds. For example, both platinum and silver oxides dissociate upon vaporization at relatively low temperatures.

(3) The reaction temperature and pressure should be adjusted to minimize the decomposition of the products.

These catalytic surface reactions are complex, and the individual atomic steps have not been studied in sufficient detail to allow the proposal of a unique mechanism [68, 69]. It is well established, however, that under high hydrogen pressure the dissociative adsorption of N_2 on iron is the rate-limiting step in the ammonia synthesis. Investigations of the structure of the reactive intermediates will allow us to distinguish among several probable reaction mechanisms.

For critical reviews of catalytic surface reactions and for mechanistic interpretations of the observed surface reaction rates, the reader is referred to recent publications [68, 69].

REFERENCES

1. R. Defay, I. Prigogine, A. Bellemans and D. H. Everett, Surface Tension and Adsorption, Longmans, London, 1966.
2. G. A. Somorjai, Principles of Surface Chemistry, Prentice-Hall, New York, 1972.
3. A. W. Adamson, Physical Chemistry of Surfaces, Wiley-Interscience, New York, 1960.
4. G. A. Somorjai, J. Chem. Phys., 35, 655 (1961).
5. See, for example, F. C. Brown, The Physics of Solids, Benjamin, New York, 1967.
6. G. N. Lewis and M. Randall, Thermodynamics (revised by K. S. Pitzer and L. Brewer), McGraw-Hill, New York, 1961.
7. E. W. Montroll, J. Chem. Phys., 18, 183 (1950).
8. M. Dupuis, R. Mazo, and L. Onsager, J. Chem. Phys., 33, 1452 (1960).
9. D. Kay, Techniques for Electron Microscopy, Blackwell, London, 1965.
10. P. B. Hirsch et al., Electron Microscopy of Thin Crystals, Butterworth, London, 1965.
11. J. M. Morabito and G. A. Somorjai, J. Metals, 20, 5 (1968).
12. J. J. Lander, Progr. Solid State Chem., 2, 26 (1965).
13. R. F. Wallis, Lattice Dynamics, North-Holland Publ., Amsterdam, 1964.
14. H. Boutin and H. Prask, Surface Sci., 2, 261 (1967).
15. G. A. Somorjai, Ann. Rev. Phys. Chem., 19, 251 (1968).
16. R. W. James, The Optical Principles of the Diffraction of X-rays, Bell & Sons, London, 1954.
17. R. M. Goodman, H. H. Farrell, and G. A. Somorjai, J. Chem. Phys., 48, 1076 (1968).
18. B. C. Clark, R. Herman, and R. F. Wallis, Phys. Rev., 139, A860 (1965).

19. B.J. Alder, J.R. Vaisnys, and G. Jura, J. Phys. Chem. Solids, 11, 182 (1959).
20. G.C. Benson, P.I. Freeman, and E. Dempsey, J. Chem. Phys., 39, 302 (1963).
21. J.J. Burton and G. Jura, Fundamentals of Gas-Surface Interactions, Academic, New York, 1967, p. 75.
22. A.H. MacRae and G.W. Gobeli, J. Appl. Phys., 35, 1629 (1964).
23. J.M. Charig, Appl. Phys. Lett., 10, 139 (1967).
24. A.E. Morgan and G.A. Somorjai, Surface Sci., 12, 705 (1968).
25. P.W. Palmberg and T.N. Rhodin, Phys. Rev., 161, 586 (1967).
26. F. Jona, Surface Sci., 8, 57 (1967).
27. A.J. Dekker, Solid State Physics, Macmillan, New York, 1958.
28. W. Hume-Rothery, The Structure of Metals and Alloys, Institute of Metals, London, 1936.
29. L. Brewer, II, Intern. Materials Symp., Berkeley, 1967, Wiley, New York, 1967.
30. H.B. Lyon and G.A. Somorjai, J. Chem. Phys., 46, 2539 (1967).
31. N.A. Gjostein and W.L. Winterbottom, Fundamentals of Gas-Surface Interactions, Academic, New York, 1967.
32. J.M. Blakely, Progr. Material Sci., 10, 395 (1963).
33. H.P. Bonzel and N.A. Gjostein, Appl. Phys. Lett., 10, 258 (1967).
34. G.A. Somorjai and J.E. Lester, Progr. Solid State Chem., 4, 1 (1967).
35. G.A. Somorjai, Science, 162, 755 (1968).
36. J.P. Hirth and G.M. Pound, Condensation and Evaporation, Pergamon, London, 1963.
37. J.E. Lester and G.A. Somorjai, J. Chem. Phys., 49, 2950 (1968).
38. G.M. Rosenblatt and Pang-Kai Lee, J. Chem. Phys., 49, 2995 (1968).
39. R.F. Strickland-Constable, Crystallization, Academic, New York, 1968.
40. W.D. Lawson and S. Nielsen, Preparation of Single Crystals, Butterworth, London, 1958.

41. E. Spenke, Electronic Semiconductors, McGraw-Hill, New York, 1958.
42. A. Many, Y. Goldberg, and N. B. Grover, Semiconductor Surfaces, North-Holland Publ., Amsterdam, 1965.
43. M. Kaminsky, Atomic and Ionic Impact Phenomena on Metal Surfaces, Springer-Verlag, Berlin, 1965.
44. R. H. Good and E. W. Muller, Handbuch der Physik, 21, 190 (1956).
45. E. W. Muller, J. Appl. Phys., 26, 732 (1955).
46. R. Gomer, Adv. Catalysis, 7, 93 (1955).
47. E. W. Muller, Adv. Electronics and Electron Phys., 13, 83 (1960).
48. D. F. Barofsky and E. W. Muller, Surface Sci., 10, 177 (1968).
49. G. W. Simmons, D. F. Mitchell, and K. R. Lawless, Surface Sci., 8, 130 (1967).
50. L. A. Harris, J. Appl. Phys., 39, 1419 (1968).
51. G. A. Somorjai and F. J. Szalkowski, Adv. High Temp. Chem., 4, 137 (1971).
52. B. K. Vainshtein, Structure Analysis by Electron Diffraction, Pergamon, London, 1964.
53. N. J. Taylor, Surface Sci., 4, 161 (1966).
54. L. M. Raff, J. Lorenzen, and B. C. McCoy, J. Chem. Phys., 46, 4265 (1967).
55. R. M. Logan and R. E. Stickney, J. Chem. Phys., 44, 195 (1966).
56. F. O. Goodman and H. V. Wachman, J. Chem. Phys., 46, 2376 (1967).
57. R. A. Oman, A. Bogan, and C. H. Li, Proc. Intern. Symp. Rarefied Gas Dyn., 4th, 1965.
58. J. H. de Boer, The Dynamical Character of Adsorption, Oxford Univ. Press, London and New York, 1953.
59. J. J. Lander and J. Morrison, Surface Sci., 6, 1 (1967).
60. R. F. Steiger, J. M. Morabito, G. A. Somorjai, and R. H. Muller, Surface Sci., 14, 279 (1969).
61. S. Brunauer, The Adsorption of Gases and Vapors, Oxford Univ. Press, London and New York, 1943.

62. W. A. Steele, The Solid-Gas Interface, Vol. 1, Dekker, New York, 1967.

63. K. W. Adolph and R. B. McQuistan, Surface Sci., 12, 27 (1968).

64. A. E. Morgan and G. A. Somorjai, J. Chem. Phys., 51; 3309 (1969).

65. G. A. Somorjai and R. R. Haering, J. Phys. Chem., 67, 1150 (1963).

66. D. A. Melnick, J. Chem. Phys., 26, 1136 (1957).

67. R. V. Culver and F. C. Tompkins, Adv. Catalysis, 11, 68 (1959).

68. G. C. Bond, Catalysis of Metals, Academic, New York, 1962.

69. E. K. Rideal, Concepts in Catalysis, Academic, New York, 1968.

PART III

PURIFICATION AND CRYSTAL GROWTH

Chapter 11

PHASE EQUILIBRIA AND MATERIALS PREPARATION

A. Reisman

IBM Thomas J. Watson Research Center
Yorktown Heights, New York

I. INTRODUCTION

To the casual observer, problems in materials synthesis, purification, and crystal growth seem to require unique solutions for each identifiable need. While the mechanics of addressing specific problems require a degree of individuality, the underlying principles governing materials behavior are quite general and have extensive application to widely varied situations.

The synthesis and purification of substances, the growth of single crystals of same, and often the examination of interesting properties possessed by them are effected in systems comprised of two or more phases, i.e., solids, liquids, and vapor, coexisting in a state of stable or metastable equilibrium. Such being the case, we might anticipate that the behavior of materials is in some way regulated by the theorems of the body of science known as Chemical Thermodynamics. More specifically, the pertinent area of Chemical Thermodynamics is that concerned with heterogeneous equilibria.

The chemical and physical attributes of a system could be characterized in many ways. Whatever the set of parameters chosen to characterize a given state of a system uniquely are, it should be evident intuitively that, if a sufficient number of specifications are imposed, the values of all other possible parameters will be fixed, that is, no longer subject to independent variation. This sufficient (and necessary) number of independently variable specifications are known as the degrees of freedom or variance V possessed by the system.

In a system which is of chemical interest, its unique characteristic is an intensive property known as the chemical potential or partial molar free energy μ. If the chemical potential of each of the chemical components present (a term we have yet to define) is known, all other intensive properties of the system are fixed, even if these are properties we have no immediate interest in. Those intensive properties which are useful for fixing the state of the chemical potential, simply because the mathematical relationships among them have been developed, are the temperature T, the pressure p, and the chemical composition X of the system.

Based on a distillation of the relationships among them, at the heart of which is the fact that the chemical potential of a component in coexisting phases for a system in equilibrium is the same in each phase, J.W. Gibbs derived the Phase Rule. This relationship, Eq. (1), enables estimation of the number of degrees of freedom possessed by the system. Alternately, it enables one to determine the number of variables that must be specified in order to completely define the state of a system.

$$V^{\dagger} = C - P + 2 \quad . \tag{1}$$

In Eq. (1), $V^{\dagger}$ is the variance of a system comprised of C components distributed among P phases. A component is defined as an independent chemical variable, namely, one whose mass (or mole number based on an arbitrarily assumed molecularity) in the combined or uncombined state is independent of the masses of other components present. As a first approximation, each chemical element present is a component, but possible reductions in the number of apparent components present may be deduced from experiment. A phase is an agglomeration of matter, independent of the latter's level of subdivision, having identifiable characteristics, i.e., a single crystal or a powder of NaCl constitutes only a single phase.

The elegance of the Phase Rule is that it is independent of the chemical nature of the system or the purposes for which the experimentalist intends the system. Thus, the limits of widely varying processes, such as the melting of ice by the application of salt, the precipitation of an insoluble sulfate in an analytical procedure, the growth of a single crystal from a pure melt or a solution, the steam distillation of an organic compound, and the filling of a cavity with a mercury amalgam, are constrained by it. For example, if we have a one-component system distributed among two phases, i. e., solid and liquid copper, Eq. (1) tells us that the system possesses a single degree of freedom (is univariant). Similarly, a two-component system present in three phases is univariant, and a twenty-component system present in twenty-one phases is also univariant. In each of these systems if we specify the temperature, all other properties (pressure, composition, density, thermal coefficient of expansion, etc.) become fixed (even though we might not know what these fixed values are and have to determine them experimentally or by calculation). Except for a one-component system, where the composition starts out fixed, we may choose from among the parameters T, p, and X as the independent variables. Further inspection of Eq. (1) reveals that if the number of phases exceeds the number of components by two, the system possesses no degrees of freedom (is invariant). Equation (1) tells us also that if experiments are conducted at fixed pressure, fixed temperature, or fixed composition, or two or more of these fixed, we reduce the variance by that number, with the limitation that we cannot reduce the variance to less than zero.

Finally, it might be noted that a one-component system cannot exist in more than three phases or a ten-component system in more than 12 phases, each such situation being one of invariance. This restriction helps in determining the number of components present. Thus, if a system under examination never exhibits more than three coexistent phases over the range of experimental interest, its behavior is unary (one component) even though it may contain more than a single element. In such an event, the system is termed pseudounary. As a consequence of the preceding, we may state that an attribute of a chemical component is that a description of its stoichiometry in each coexisting phase is possible via the use of a single chemical symbol. In cases where too low a component number has been chosen, i.e., where it is found experimentally that the assumed component stoichiometry varies in coexisting phases, an inconsistency will be found between the number of phases present and the assumed component number, i.e., too many coexistent phases will be detected. Experimentally, conditions are frequently sought which restrict the system to a condition of univariance, since such systems are more amenable to analytical description. This is accomplished by fixing temperature, pressure, and one or more composition ratios or simply by examining the system generally to uncover univariant situations.

While the Phase Rule can define the variance of a system, it cannot in the case of univariance, for example, provide the analytical relationship describing this univariance. Since the sophistication of theoretical description of univariance via analytical equations is limited, the experimentalist has made use of another technique, graphical in nature, to describe the phase relationships in systems of one or more components. Such graphical representations depict the physical state of the system, namely, the conditions under which single- or multiple-phase existence occurs. Consequently, the representations are called diagrams of state or phase diagrams. Given such a diagram, univariant situations can then be fitted phenomenologically to simple relationships, or as is often done, each graph may be used without attempts at analytical description as a future aid for application to practical or scientific problems.

II. ONE-COMPONENT PHASE DIAGRAMS

A. Solids-Liquids-Vapor

Because composition is not variable in a one-component system, unary phase diagrams are depicted via pressure-temperature relationships and have the general appearance shown in Fig. 1. In a one-component

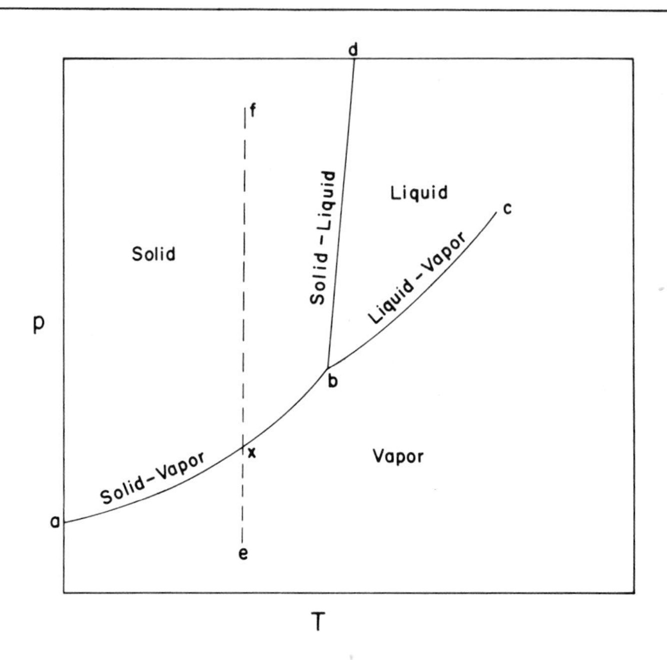

FIG. 1. A one-component pressure-temperature diagram.

system, a condition of univariance obtains only when two phases coexist. A condition of bivariance obtains when only one phase is present and the system is invariant when three phases coexist. In Fig. 1, the line a-b shows the pressure-temperature, p-T, relationship when solid coexists with vapor. The line b-c depicts this relationship when liquid coexists with vapor, and the line b-d defines the two-phase solid-liquid coexistence. Given the temperature along any of these lines, the pressure acquires a unique value. Given the pressure along any of these lines, the temperature acquires a unique value. Where the three lines intersect in a point we have an invariant three-phase situation. Neither the temperature nor the pressure is variable at this "triple point" without an attendant decrease in the number of phases present. The lines a-b, b-c, and b-d act as boundaries between the bivariant fields for pure solid, pure liquid, or pure vapor. In each bivariant field, both the pressure and temperature can be varied without changing the physical state of the system. Indeed, both must be specified to define the system uniquely.

Analytically, the univariant behavior along any line is described by the exact relationship, known as the Clapeyron equation,

$$dp/dT = \Delta H_p/T\Delta v \quad . \tag{2}$$

In Eq. (2), the general term ΔH_p represents the molar heat of sublimation, ΔH_s, or the molar heat of vaporization, ΔH_v, or the molar heat of fusion, ΔH_f. The term Δv represents the difference in molar volumes of the component in the coexisting phases. As the molar volume of vapor is always greater than that of solid or liquid, the solid-vapor, S-V, and liquid-vapor, L-V, curves always exhibit positive slope. Furthermore, ΔH_s always bears the relation to ΔH_v and ΔH_f given by

$$\Delta H_s = \Delta H_f + \Delta H_v \quad . \tag{3}$$

The molar volume of liquid is not necessarily greater than that of its coexisting solid, i.e., ice is less dense than water. Consequently, the S-L curve may exhibit negative, positive, or even infinite slope. A simplification of Eq. (1) known as the Clausius-Clapeyron equation is

$$dp/p = \Delta H/RT^2 \quad . \tag{4}$$

It makes use of the ideal gas equation and the fact that $v_{vapor} >> v_{condensed\ phase}$ to substitute for the term Δv in Eq. (1). Equation (4) in integrated form is generally employed for data fitting of unary condensed-vapor phase equilibria.

A further feature of interest in Fig. 1 is the termination of the L-V curve at point c. Point c is termed a critical point and represents that temperature above which vapor and liquid densities become identical, or the temperature above which the application of pressure fails to generate a liquid-vapor interface. This characteristic of one-component systems has been utilized for materials synthesis and notably for the growth of single crystals of very slightly water soluble materials such as quartz. The quasiliquid nature of the system above the critical point plus its bivariant nature enables one to vary both the temperature and pressure simultaneously over wide ranges. This can have the effect of increasing the solubility of "insoluble" materials sufficiently so that they can be recrystallized controllably from solution, particularly aqueous solution. As a matter of interest, it is to be noted that no phenomenon analogous to the critical point occurs along the solid-liquid curve b-d. The latter is called the melting point curve of a component and shows the effect of pressure on the triple point.

In Fig. 1, the curve a-b is termed the sublimation curve of the component in question, the curve b-c is called the vaporization curve, and b-d the melting point curve. The first two are the common vapor pressure curves of a material. If the material in the solid state achieves a vapor pressure equal to one atmosphere below its triple point, it will, when exposed to air, exhibit the phenomenon of complete sublimation without subsequent melting.* Dry ice (solid CO_2) is an example we are all familiar with. The temperature at which atmospheric pressure is achieved is called the "normal sublimation point." Analogously, when the liquid upon being heated in an open vessel achieves a temperature at which its pressure is one atmosphere, it will boil isothermally and never achieve the critical point temperature, hence the term "normal boiling point." Under its equilibrium pressure, however, the sublimation curve will always terminate in a S-L-V triple point and the L-V curve will terminate in a critical point.

Finally, the full significance of Fig. 1 can be better understood by following a hypothetical isothermal compression path such as the path e-f of Fig. 1. If we start with a defined mass of vapor at the

*Note that, with a constant pressure constraint, a one-component two-phase equilibrium is isobarically invariant. Consequently, neither the pressure nor temperature can vary unless one of the two phases disappears.

pressure and temperature of point e, and compress it, its pressure-temperature-volume relationship is given to a first approximation by the ideal gas relationship:

$$pv = nRT \quad . \tag{5}$$

Equation (5) is one describing bivariance. Thus, if both p and T for the number of moles, n, present are specified, the volume occupied becomes fixed. The equation, in fact, relates the dependent parameter volume v to the independent parameters T and p. Obviously, we could have chosen p and v or T and v as independent variables to fix the value of the remaining one. By considering the isothermal path e-f in Fig. 1, we have constrained the freedom of choice possessed by the system and thereby reduced its degrees of freedom by one. Thus, the compression line e-f is "isothermally univariant." Along e-x, the p-v relationship is given by Eq. (5) with T and n both constant. When the pressure achieves the value at point x, the vapor becomes saturated with respect to solid and the latter precipitates. As long as vapor and solid continue to coexist, the system is isothermally invariant and the pressure cannot change further until all of the vapor disappears. When it does, the system again becomes isothermally univariant and application of additional pressure causes the state of the system to move along the isothermally univariant line x-f. From point x on, no vapor phase is possible.

B. Temperature-Induced Phase Transformations in One-Component Systems

Many unary systems show solid-solid phase transformations in which at a solid-solid-vapor triple point the structure of a lower temperature-stable solid phase undergoes a thermally reversible transformation to a higher temperature-stable solid phase. In fact, systems are known in which several solid-solid transformations occur. Each of the solid phases is known as a polymorph or enantiomorph or allotrope,

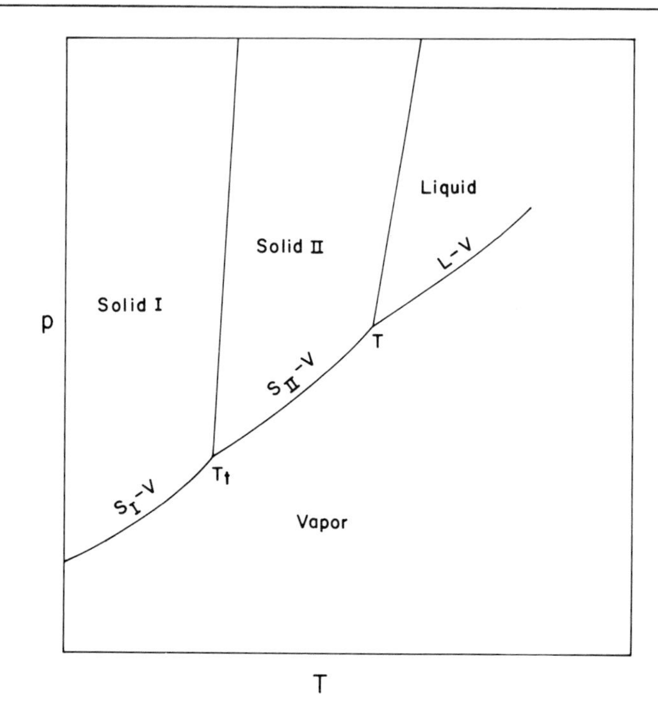

FIG. 2. A one-component system showing a phase change.

and the phenomenon is termed polymorphism, enantiomorphism, or allotropism. This is depicted schematically in Fig. 2. At the temperature T_t, a solid having a unique set of properties transforms reversibly to form a new solid having its own set of identifiable characteristics. Accompanying this transformation there is associated a latent heat of transformation with the relationship given in Eq. (3) being obeyed, at least in the vicinity of the triple point, e.g.,

$$\Delta H_s^{I} = \Delta H_t^{I-II} + \Delta H_s^{II} \quad . \tag{6}$$

ΔH_s^{I} represents the heat of sublimation of solid I, ΔH_t^{I-II} represents the heat of transformation of solid I → solid II, and ΔH_s^{II} represents the heat of sublimation of solid II. At the transformation temperature T, the relationship given in Eq. (3) of course applies to the heats of sublimation, fusion, and vaporization of solid II and its associated

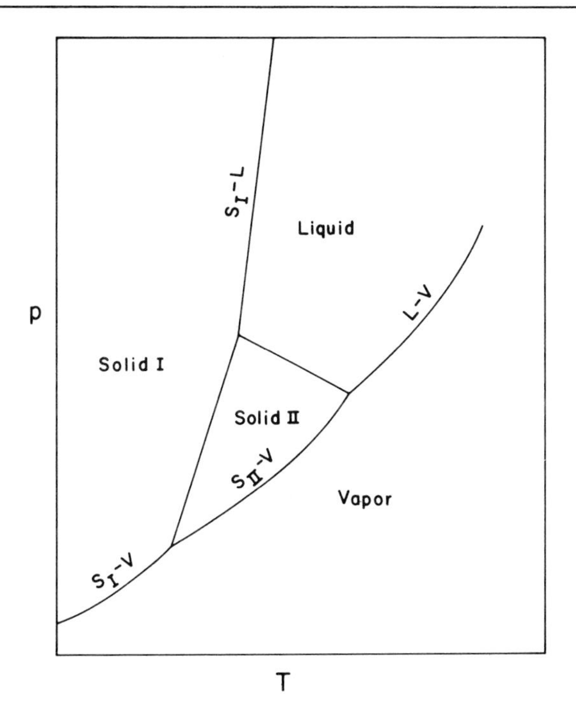

FIG. 3. Stability intervals as a function of pressure in a one-component system.

liquid phase. As with the slope of the solid-liquid curve, the slope of the solid-solid curve can be negative, positive, or infinite. When the S_I-S_{II} slope is positive and the S_{II}-L slope is negative or even when both are positive or negative, they may intersect giving rise to phase diagrams of the type shown in Fig. 3. We shall discuss the significance of the curve S_I-L in Section D.

There exist phase transformations in which the latent heats attending the transformations tend toward zero. Arguments have been advanced that such transformations are different from the kind we have considered and that no triple point occurs. These arguments contend that transformations of the second order, as they are sometimes termed, do not occur isothermally, a gradual change taking place over a finite temperature interval. Others have argued that transformations of the second order are not a separate case and that purported differences are in degree rather than in kind. This still remains an open question, with the author holding the second viewpoint. It is unquestionably the case that certain transformations exhibit vanishingly small latent heats of transformation. This is not unexpected, since one can certainly anticipate structural perturbations which are minor and which would not involve large energy changes. From a phase rule point of view, one-component second-order transformation phenomena would of course be excluded since equilibrium three-phase univariant occurrences are not allowed. Consequently, the physical concept of a second-order phenomenon appears to involve two considerations: (1) a temperature range of metastability (hysteresis) over which the free energy differences between the phases are vanishingly small, and (2) a kinetic barrier to transformation induced by this chemical potential equivalency.

It is interesting that many transformations are known which occur very slowly and which for this reason have not been observed with attendant latent heat anomalies. In this class of sluggish transformations, the kinetic limitations are caused by the major structural reorganizations associated with the phase change and the freezing-in of existent structures by low atom mobilities at the transformation temperature. When large structural reconstitutions are involved, however, it is always assumed that significant latent heats must be involved and that such transformations are first order. Experimentally, they exhibit most of the characteristics associated with the so-called second-order phenomenon.

C. Pressure-Induced Phase Transformations

Many systems are known in which solid-solid phase transformations are observed at pressures in excess of those at which a vapor phase is present. This phenomenon has been used for the preparation of materials not readily synthesized under normal ambient pressures, but which exhibit stability once prepared. The notable example of such a

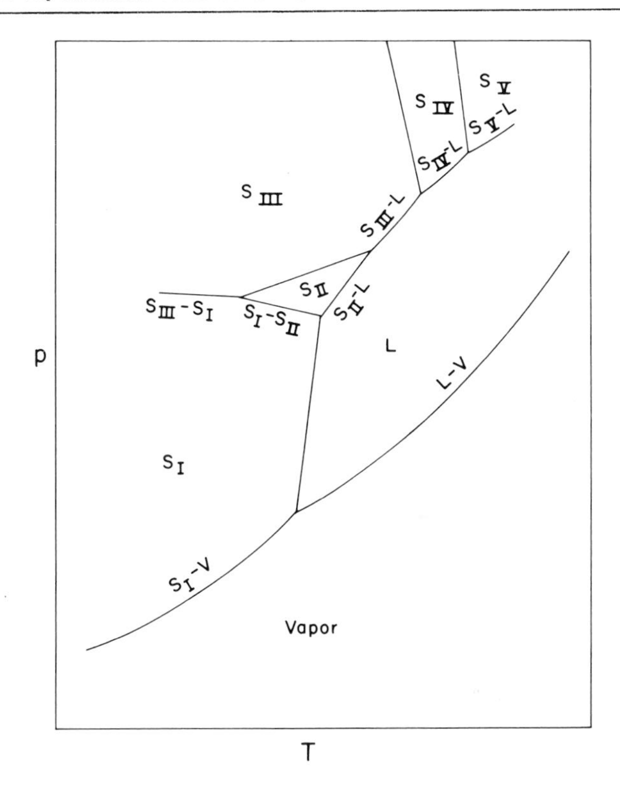

FIG. 4. High pressure effects in a one-component system.

high pressure phase is the diamond phase of carbon. A hypothetical example of solid-solid phase transformations occurring in the absence of a vapor phase is given in Fig. 4. It is always the case in systems such as those shown in Figs. 1-4 that when an isothermal compression path, of the type depicted in Fig. 1, cuts across a phase boundary with increase in pressure, it passes from a phase of lower to one of higher density.

D. Metastability in One-Component Systems

Many instances are known in which, upon cooling a liquid below the S-L-V triple point, the transformation L→S does not occur. In the temperature interval below the triple point, the supercooled (or undercooled) liquid is metastable with respect to some other state of the system, i. e.,

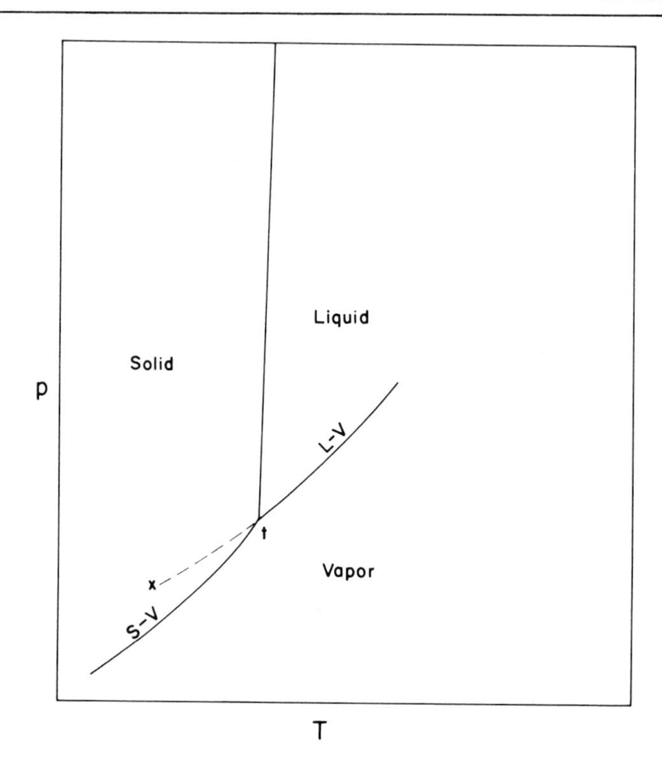

FIG. 5. Metastable extension of the liquid-vapor curve in a one-component system.

a solid phase. Thermodynamically, this implies that there exists a state of the system which possesses a lower Gibbs' free energy than is possessed by the liquid. Consequently, a negative free energy change is to be expected when the system transforms from a metastable to a more stable state, such negative free energy change being indicative of a spontaneously possible process. For condensed-vapor phase free energy changes, the relationship in Eq. (7) is applicable to a first approximation:

$$\Delta G = RT \ln(p_2/p_1) \quad . \tag{7}$$

In order for ΔG, the free energy change accompanying the process, to be negative, it is seen from Eq. (7) that the final pressure of the system, p_2, must be lower than the initial pressure p_1. The consequence of this is that the vapor pressure of a metastable phase must be greater than the vapor pressure of a more stable phase at the same

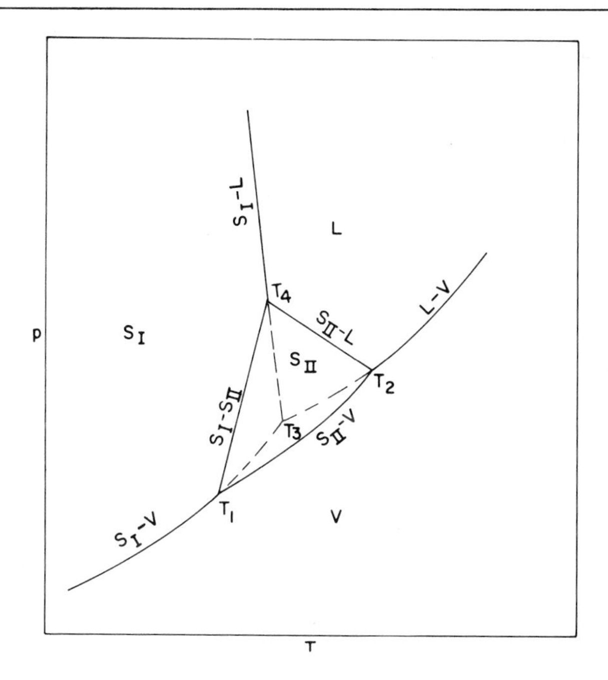

FIG. 6. Metastable extension of liquid and solid curves in a one-component system.

temperature. Graphically, this is seen in the phase diagram shown in Fig. 5. If the curve L-V is extended metastably below the triple point t, it continues to obey the analytical relation that it does in its stable temperature range. The requirement that its vapor pressure be greater than that of the S-V equilibrium below t results in the curve t-x lying above the curve S-V. Actually, the conclusion we have arrived at could have been anticipated from Eq. (3). Since the magnitudes of the slopes of the S-V and L-V curves are determined by ΔH_S and ΔH_V, respectively, and ΔH_S is always greater than ΔH_V, i.e., it contains the term ΔH_f also, it follows that the S-V curve has a greater positive slope than the L-V curve.

In solid-solid transformations, metastable extensions are possible above and below the triple point. This can lead to the situation shown in Fig. 6 where the metastable curve T_1-T_3 representing the S_1-V equilibria intersects the metastable L-V curve, T_3-T_2, at the metastable triple point T_3. At T_3, the low temperature solid S_I coexists metastably with liquid and vapor. Because of the slopes of the S_1-S_2 and S_2-L equilibria, the triple point T_4 is generated for the stable invariant equilibrium S_1-S_2-L. The dashed curve T_3-T_4 represents the metastable S_1-L curve which then continues at pressures greater than T_4 in a stable mode.

The determination of p-T diagrams of state involves pressure measurements. The commonly used methods are known as Knudsen effusion, transpiration, and Bourdon gauge techniques. In the Knudsen method the sample is heated in a chamber containing a small orifice. Within the chamber, equilibrium is assumed to obtain and knowing the orifice area, weight loss, and assumed or experimentally determined vapor phase species molecularity, and the temperature, vapor pressures can be calculated:

$$p = [\ell/a_0 t]\sqrt{2\pi RT/MW} \quad , \tag{8}$$

where p is the pressure, ℓ the weight loss, t the time, a_0 the orifice area, R is the ideal gas constant, and MW is the assumed molecular weight. This technique is most usually employed for low vapor pressure materials, and frequently in conjunction with a mass spectrometer for elucidating the nature of the effusing species.

The transpiration technique involves saturation of a flowing gas with the vapor of the specimen in question and calculating vapor pressures from a knowledge of the gas volume and the quantity of vaporized material carried by it:

$$p(\text{Torr}) = 760\left[n\Big/\left(\frac{V}{RT} + n\right)\right] \quad , \tag{9}$$

where p is the vapor pressure of material, n is the number of moles of this material based on an assumed molecular weight carried in a volume V measured at a temperature T(°K), and R is the ideal gas constant. The technique is sometimes referred to as the gas saturation method for obvious reasons. It has a range of use from very low vapor pressures (the micron range) up to the high mm Hg range.

In the range 1 Torr and up, the most powerful approach, since it requires no assumptions as to the molecularity of vapor phase species or the ideality of the vapor phase, is the Bourdon gauge method. The sample here is contained in a compartment (quartz or other suitable material) which has on one of its sides a distensible diaphragm. The distension of the diaphragm due to a pressure buildup within the compartment is counterbalanced by an externally applied and measured counterpressure. In its most sophisticated form, where it is employed to elucidate vaporization mechanism models, the latter are reconciled with observed p-T variations.

III. TWO-COMPONENT SYSTEMS

A. Introduction

When the component number rises to two, the composition becomes a variable, independent or dependent, depending on choice and/or the number of phases present. Since the component mole fraction can vary between 0 and 1, a convenient means of defining two-component system compositions is via the use of the mole fraction concept where, by definition,

$$M_A = \frac{m_A}{m_A + m_B} , \tag{10}$$

$$M_B = \frac{m_B}{m_A + m_B} , \tag{11}$$

$$M_A + M_B = 1 , \tag{12}$$

or

$$M_A = 1 - M_B . \tag{13}$$

In Eqs. (10)-(13), M_A and M_B represent the mole fractions of the two components while m_A and m_B represent the number of moles of each present.

In two-component systems, simplifications in their study or description can be effected by invoking a "reduced phase rule." Thus, if the system is not particularly volatile, i. e., its condensed mass does not change under free evaporation conditions in an open container, the system can be examined exposed to air or some other constant total pressure inert atmosphere. Because p is constant, the condensed system thereby loses a degree of freedom and the variables are T and X. Similarly, if we study the system under its equilibrium pressure under limited vapor available space conditions, the mass of the condensed phases will not change drastically and we can again invoke the reduced phase rule. In both cases, the composition of the vapor phase is ignored, all considerations then being confined to the condensed phases. Because the behavior of two-component systems is too varied to even attempt a detailed overview in this chapter, we will choose selectively those aspects which address most immediately problems liable to be encountered in the solid state.

When two components interact, they can do so in several ways. The interaction can be primarily physical in the sense that the phenomenon of interaction is one involving solubilization of the components one in the other, or chemical in that new compounds are formed which in turn then interact physically. The interactions can specifically follow one of two extreme routes or an infinite number of intermediate paths. Thus, liquid phases may show continuous solubility of the components one in the other or no measurable solubility. Coexisting solid phases may also exhibit these boundary states and the intermediate situations. All permutations of these cases between liquids and solids may be observed. In different situations, the resultant physical and chemical interactions may be put to significant practical use. For example, the semiconductor industry makes extensive use of the minuscule solubility of "impurity" components in the semiconductor to impart desired electrical characteristics. Based on the properties possessed by systems exhibiting varying degrees of solid solubility of the components in each other, a very powerful dynamic technique for purifying organic and inorganic materials has evolved known as zone refining. Based on the existence of systems in which the components are essentially insoluble in each other in the solid state, but miscible in all proportions in the liquid state, a powerful exploratory method for single-crystal growth known as flux or solvent growth has been perfected. Based on variations in equilibrium vapor pressures over vanishingly small composition intervals in systems exhibiting very limited solubility of the components in each other, and the effect on the stoichiometry of compounds due to these pressure-composition effects, single-crystal growth of controlled stoichiometry binary compounds is possible using pressurized systems. These practical implications will become more evident as we delve further.

B. Systems Exhibiting Complete Solid Solubility

A limiting case exists where the components (end members) are completely miscible in one another in both the solid and liquid states. If we determine experimentally the compositions of coexisting phases as a function of temperature at either constant pressure, or the equilibrium pressure of the system, the phase diagram would have the general characteristics shown in Fig. 7. The composition axis is divided into ten equal parts. By dint of Eq. (12), one end of the graph represents pure A, while the other end represents pure B. Thus, moving to the right in Fig. 7 represents increasing B (decreasing A) while moving to the left represents increasing A (decreasing B), the sum $M_B + M_A$ adding up to unity at all compositions.

The points T_A° and T_B° represent the melting points of the components A and B, respectively, at the pressure in question (either the constant applied pressure or the equilibrium pressure). The top curve (known as the liquidus) represents the composition of liquid in

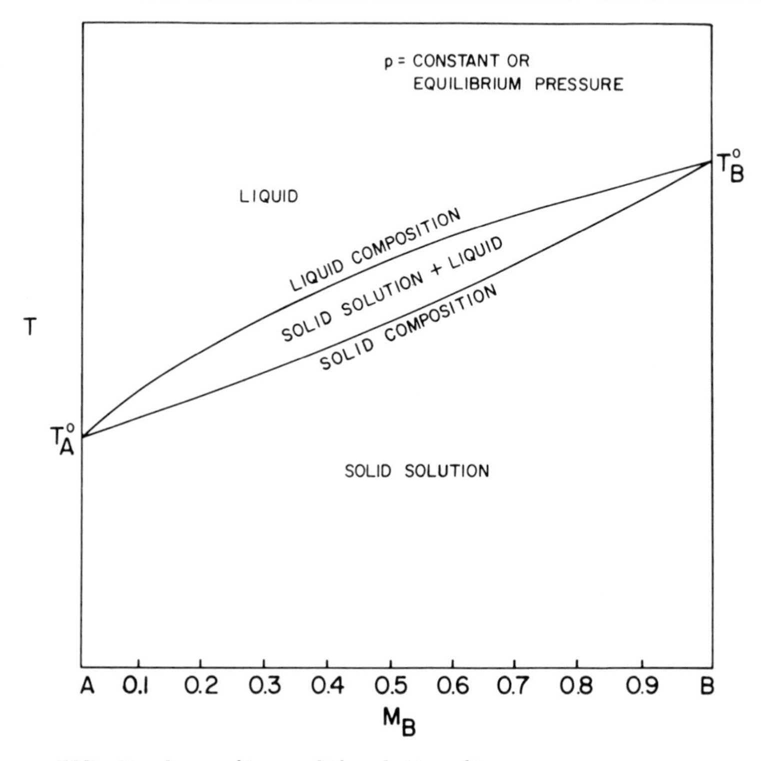

FIG. 7. Ascending solid solution diagram.

equilibrium with a solid solution whose composition varies continuously from pure A to pure B and whose composition at any temperature is given by the bottom curve (known as the solidus). From the Phase Rule, we note that in a two-component two-phase equilibrium where the pressure variable has been eliminated, the system is univariant. The significance of Fig. 7 is best understood by considering the behavior of a sample of specified composition in a cooling or heating cycle, Fig. 8.

Assume that the starting composition has a value a. If this sample is heated to the temperature T_0, it will become completely molten. If this melt is cooled, it will upon reaching the temperature T_1 become saturated with respect to solid solution having the composition c, and a minute amount of the solid precipitates. The liquid at this point, b, still has essentially the composition "a." If we proceed to

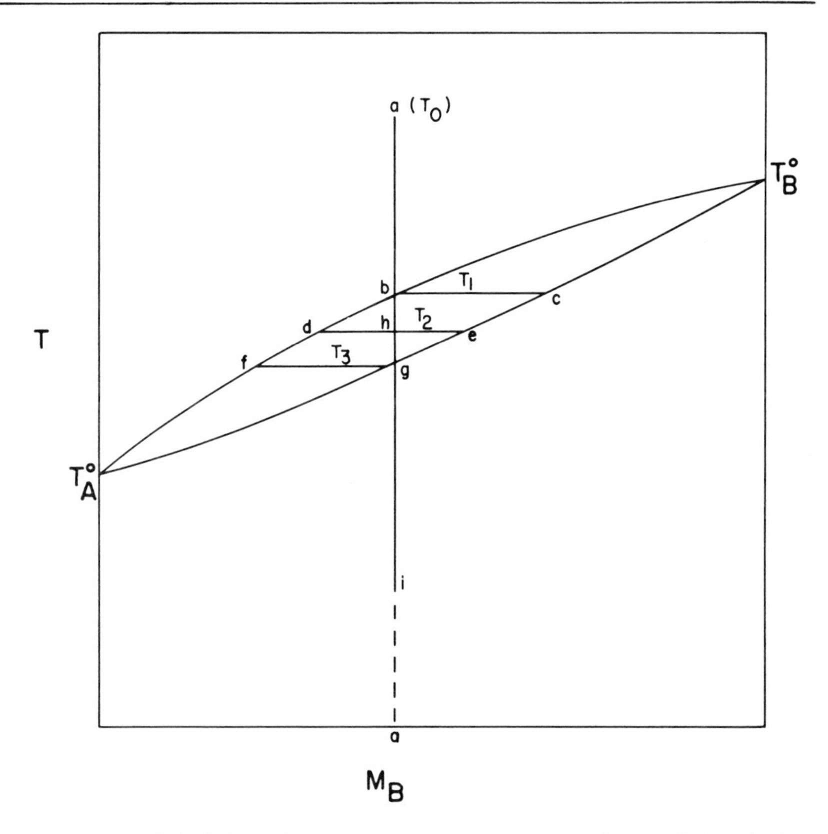

FIG. 8. Solid-liquid compositions in an ascending solid solution.

cool the sample further, solid continues to precipitate, the overall composition of the solid moving along curve c-e toward "e," and the composition of liquid in equilibrium with this solid moving from b toward d. It is to be noted that during this cooling process the liquid divests itself of the component having the higher melting point. The net result is that the solid is always richer in this component than the liquid with which it equilibrates, the liquid becoming richer in the lower melting end member. During this precipitation process, latent heat of fusion of both end members is being evolved. When the temperature T_2 is achieved, the liquid has the composition d and the total deposited solid has the composition e.

If the sample is cooled still further to temperature T_3, the solid acquires the composition g and the liquid, the composition f. At this point, very little liquid remains and cooling a little further results in complete solidification. When complete solidification has occurred, the solid now has the composition "a." Since the solidification process involves equilibration of the two components in a single lattice and the overall composition of the solid varies from a starting value "c" to a final value equal to the starting composition "a," sufficient time must be allotted for this changing solid composition to equilibrate. This is for most cases not easy to achieve in a dynamic cooling process, and if equilibrated alloys (as the solid solutions are termed) are desired, it is necessary frequently to anneal the solid samples at as high a temperature as is possible for long periods of time. An interesting aspect of the continuous solid solution system is that the end members must be isomorphic (possess the same crystal structure), since only a single solid phase is ever present. Thus, the end members must be capable of substituting in the lattice sites occupied by the other end member as the solid composition varies. A rule of thumb is that in the case of isomorphic substances, if the lattice constants are within ten percent of each other, continuous solid solubility will probably be observed.

A geometrically determined principle known as the lever arm or fulcrum principle may be applied to the isotherms in a T-X phase diagram such as that depicted in Fig. 8 to determine the quantity of precipitate at any temperature in the two-phase solid-liquid region. Thus, in Fig. 8, if the line a-a is thought of as a fulcrum and the isothermal arms b-c, d-h, e-h, and f-g are thought of as levers, the ratio, moles of both components in the liquid phase to moles in the solid phase, is given by the length of the right arm over the length of the left arm, respectively. At T_1, this ratio approaches infinity, as it must, since the liquid is just saturated at this temperature, and only a minute quantity of solid is present. At the temperature T_2, the ratio, total number of moles in the liquid/total number in the solid, is given by the lengths $|e\text{-}h|/|d\text{-}h|$.

At T_3 the ratio tends toward zero since the last trace of liquid is present. Isothermal lines such as b-c, d-e, and f-g are referred to as tie arms since their intersection with the liquidus and solidus curves tie together the compositions of the coexisting phases at the temperature in question. In addition to liquid-solid and solid-liquid ratios, the lever arm principle can be employed to determine the fraction of the initial number of moles of A + B that has either precipitated or still remains in the liquid phase. Thus, the ratio $|e-h|/|e-d|$ gives the fraction of the starting number of component moles still present in the liquid at T_2 and the ratio $|d-h|/|e-d|$ gives the fraction of the starting number of component moles precipitated out at T_2. While application of the lever arm principle to the mole fraction diagram gives answers in fractions of total moles, the latter can be converted to mass ratios in obvious ways, since we know the molecular weights of the components based on the molecularities assigned to them. Frequently, when phase diagrams are to be utilized for just this type of calculation, in purification or crystal growth processes, it is more convenient to express compositions in mass fraction terms so that, starting with a known sample mass, direct mass answers are obtained.

For an ideal system the solidus and liquidus curves may be adequately described by Eqs. (14) and (15):

$$M_A^{(\text{liquid})} = \frac{\exp\left[-\frac{\Delta H_A^\circ}{R}\left(\frac{1}{T} - \frac{1}{T_A^\circ}\right)\right]\left\{\exp\left[-\frac{\Delta H_B^\circ}{R}\left(\frac{1}{T} - \frac{1}{T_B^\circ}\right)\right] - 1\right\}}{\exp\left[-\frac{\Delta H_B^\circ}{R}\left(\frac{1}{T} - \frac{1}{T_B^\circ}\right)\right] - \exp\left[-\frac{\Delta H_A^\circ}{R}\left(\frac{1}{T} - \frac{1}{T_A^\circ}\right)\right]}, \tag{14}$$

$$M_A^{(\text{solid})} = \frac{\exp\left[-\frac{\Delta H_B^\circ}{R}\left(\frac{1}{T} - \frac{1}{T_B^\circ}\right)\right] - 1}{\exp\left[-\frac{\Delta H_B^\circ}{R}\left(\frac{1}{T} - \frac{1}{T_B^\circ}\right)\right] - \exp\left[-\frac{\Delta H_A^\circ}{R}\left(\frac{1}{T} - \frac{1}{T_A^\circ}\right)\right]}. \tag{15}$$

In Eqs. (14) and (15), M_A represents the mole fraction of A in either solid or liquid, H_A° and H_B° represent the standard molar heats of fusion of the components A and B, respectively, T_A° and T_B° represent the melting points of the pure components, and T represents the temperature in question. It is a simple matter to demonstrate that T always lies between T_A° and T_B° in the case in question. The values for $M_B^{(liquid)}$ and $M_B^{(solid)}$ can be evaluated by dint of Eq. (13). It is also to be noted that as a consequence of thermodynamic conventions, heats of fusion are positive in sign, i. e., melting is always accompanied by absorption of heat (endothermic).

C. Systems Exhibiting No Solid Solubility

It is of course to be expected that all components exhibit some solid solubility in all other components, even if this solubility is

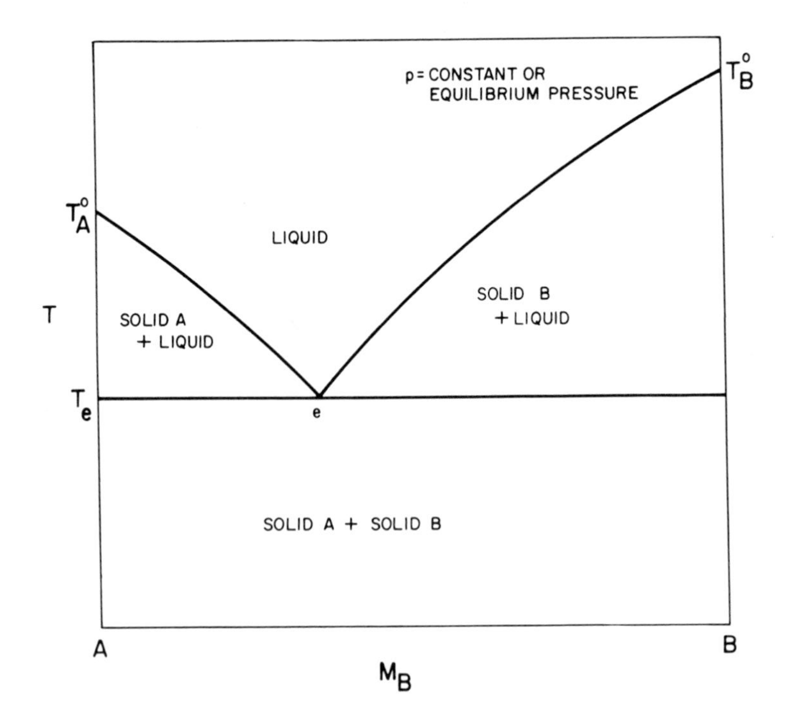

FIG. 9. Simple eutectic binary diagram.

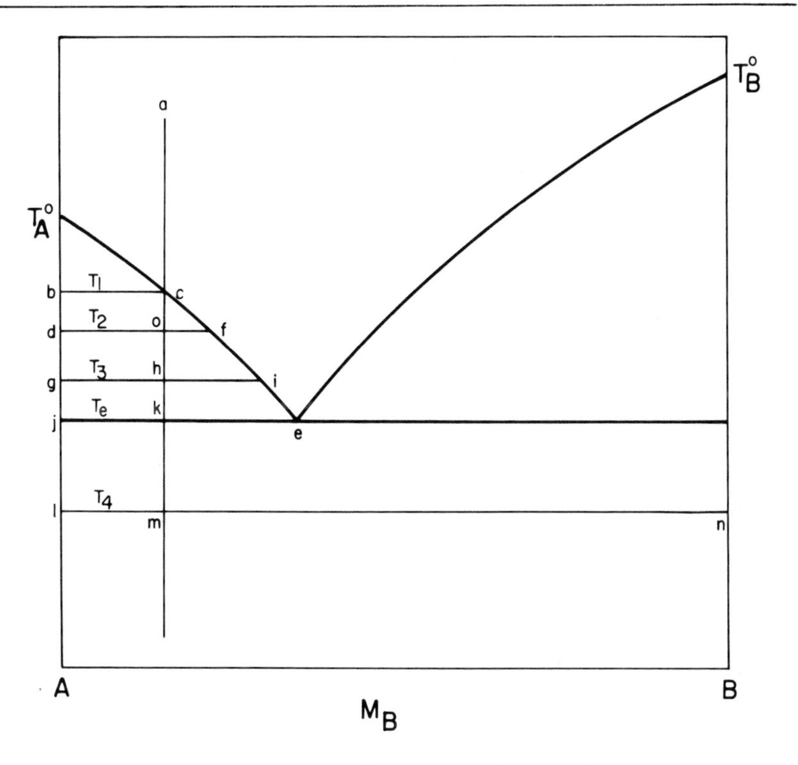

FIG. 10. Tie arms in a simple eutectic binary diagram.

vanishingly small. In the limit, phase diagrams of such systems have the appearance shown in Fig. 9.

The lines T_A°-e and T_B°-e represent the temperature-composition relationships for the two-phase univariant situation when either pure solid A or pure solid B coexists with liquid. These are in fact the solubility curves of A and B as a function of temperature. At the temperature T_e, these solubility, or liquidus, curves as they are called, intersect in a binary triple point which according to the reduced phase rule is isobarically invariant. This point represents the coexistence of "pure" solid A, "pure" solid B, and liquid, the latter always having the composition e. When this liquid freezes, the system becomes univariant again and comprised of a mixture of pure solid A and pure solid B. The triple point is referred to as the eutectic and has associated with it a unique eutectic temperature and composition.

In Fig. 10, a cooling path is defined in order to illustrate the significance of the several regions present.

Starting with a molten composition "a," and cooling to the temperature T_1, the liquid becomes saturated with pure solid A. Since the solid phases are pure, the isothermal tie lines all extend from the A axis to the liquidus curve. Alternatively, for compositions to the right of e, the tie lines connect the curve T_B°-e to the B axis. At T_1 the solid-liquid ratio is essentially zero. If the sample is cooled to T_2 the liquid composition moves along c-f toward f, becoming richer in B while pure solid A continues to precipitate. Such a precipitation is known as the primary crystallization and the pure A phase is termed the primary crystallization phase. At T_2, the mole ratio solid/mole ratio liquid is given by the arm lengths $|f-o|/|d-o|$. The mole ratio, moles solid/initial number of moles is given by the lengths $|f-o|/|d-f|$. Cooling further to T_3 finds the liquid composition becoming richer in B as the liquid composition moves along f-i toward i. The solid-liquid mole ratio increases to i-h/g-h. Finally at T_e the liquid achieves a composition e and becomes saturated with respect to pure B. The solid-liquid mole ratio is e-k/j-k at this temperature. Since three phases coexist isobarically at this eutectic or solidus temperature, the temperature cannot change until one of the phases vanishes. The phase that does vanish is of course the liquid phase. The solids that result must have the combined composition equivalent to the starting composition. Since the solids are pure, the tie line connecting compositions of the coexisting phases below T_e connect the A and B axes. Further, as the phases are pure, the univariant behavior that might be described, i.e., p vs T, is represented by the sum of the univariant properties of each.

Independent of the starting composition of the system, the final liquid remaining at T_e will have the composition e. Since the system is entirely solid below this temperature, T_e is also called the solidus. Different starting compositions will of course yield different solid/liquid mole ratios when T_e is reached, but the compositions of the coexisting phases will be invariant. Further, independent of the starting composition to the left or right of e, the compositions of coexisting phases at any particular temperature will always be the same, even though the solid/liquid mole ratio at these temperatures will vary.

The equation describing the liquidus curve of either A or B is given in an ideal system by

$$\ln M_{A\ or\ B}^{(liquid)} = -\frac{\Delta H^\circ_{A\ or\ B}}{R}\left(\frac{1}{T} - \frac{1}{T^\circ_{A\ or\ B}}\right) , \qquad (16)$$

where $M_{A\ or\ B}$ represents the mole fraction of that component in the liquid which coexists with the pure solid. $\Delta H^\circ_{A\ or\ B}$ is again the molar heat of fusion and $T^\circ_{A\ or\ B}$ is the melting point of the pure end member. It is a simple matter to deduce from Eq. (16) that T°_A or T°_B represents the highest temperature on the solubility curves of either end member in a eutectic interaction.

A significant feature of ideal eutectic interactions is that the analytical curves describing the A and B liquidi are independent of each other, i.e., the behavior of the second end member does not influence the behavior of the first. Consequently, only a single ΔH° appears in each equation.

D. Intermediate Cases

The vast majority of systems exhibit solid solubilities intermediate between those described in Sections II.A and B. The phase diagrams of systems exhibiting this limited solid solubility have the appearance shown in Fig. 11.

In Fig. 11, the solid composition curves, due to the limited solubility, do not coincide with the pure component axes. Consequently, tie lines intersect solid composition curves whose composition varies as a function of T in analogous fashion to the liquidus curves. The liquidus and solidus curves in such cases are described by equations similar to Eqs. (14) and (15), since the heats of fusion of both end

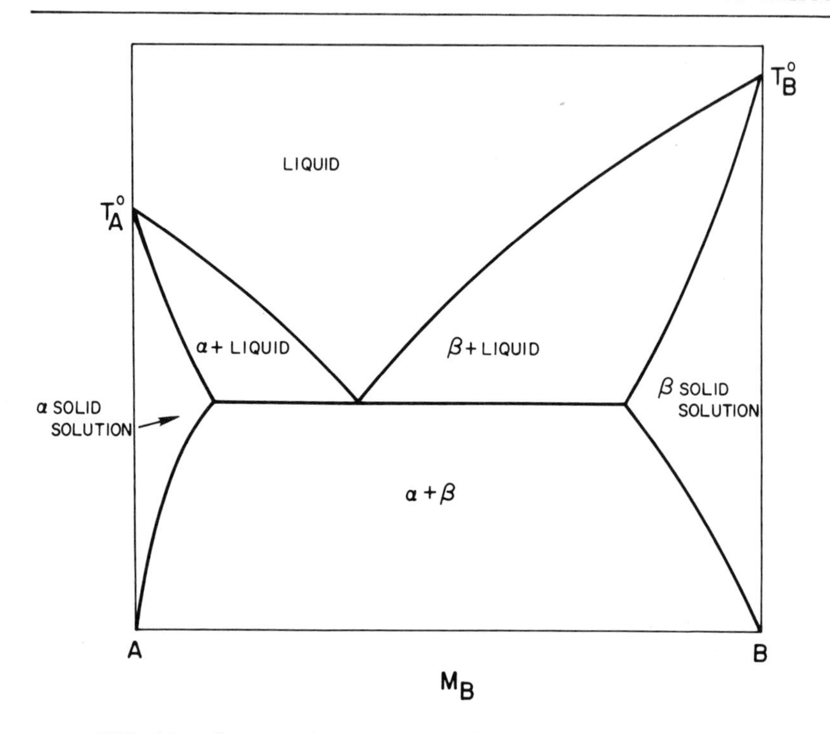

FIG. 11. A eutectic system exhibiting limited solid solubility.

members is involved. In general, the solubility of the end members in each other experiences a maximum at T_e although cases are known where such is not the case (retrograde solubility).

E. Purification

When a system exhibits a behavior of the type shown in Fig. 11, it can be thought of as the case of a pure substance contaminated by one or more impurities, these impurities behaving as a single impurity. Thus, if we consider the component A melting at T_A°, we can visualize the situation where it is contaminated by a host of impurities, each having a lower melting point than A and each forming a phase diagram with A of the type shown in Fig. 7. Together, however, when the solubilities in A are small, the impurities behave, to a first approximation, as if they were a single contaminant. The composite phase diagram,

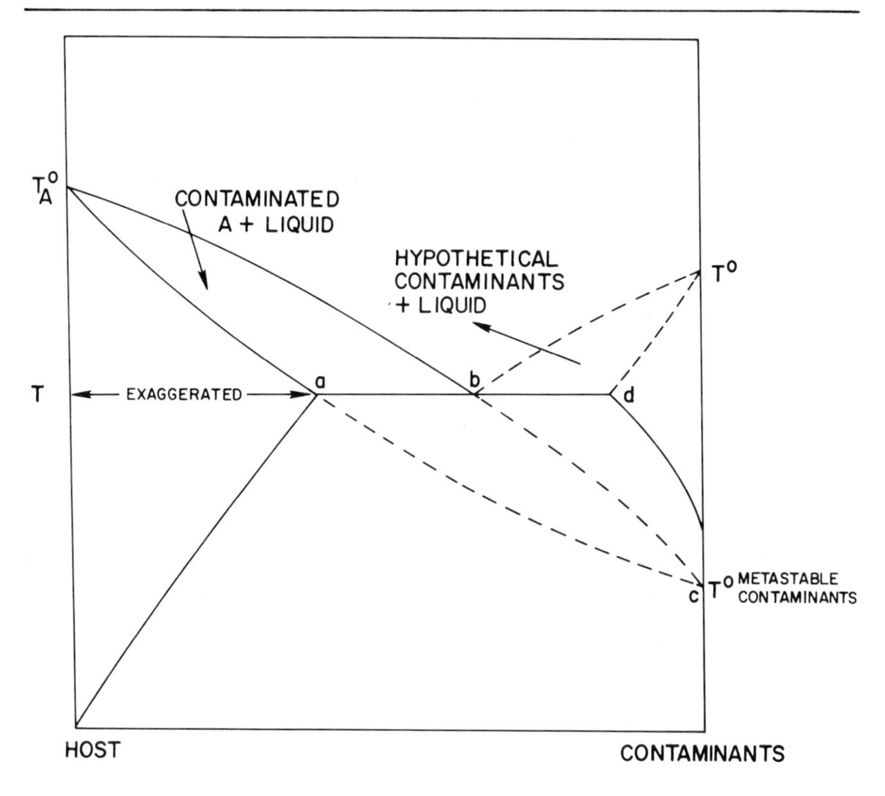

FIG. 12. The phase diagram of a system exhibiting impurity contamination.

A-Contaminants, has the hypothetical appearance shown in Fig. 12. Since the impurities fit into the A lattice, they function as the second component in a continuous solid solution system. Thus as seen in Fig. 12, the curves a-c and b-c extend hypothetically to the same hypothetical metastable melting point $T^\circ_{\text{contaminants}}$. The curves b-T° and d-T° do not really apply, but are dotted in to show what type of system is really being dealt with. Further, in order for our approximation to have greater validity, the solid solubility of the contaminants in pure A should collectively be small. We have exaggerated this solid-solubility region for pedagogical purposes.

The significant feature of the solubility curve for the host lattice, pure A, is that the liquid is richer in the contaminant materials due to the apparent lower melting point of the latter. This concept can be expressed in terms of a so-called segregation coefficient, which for convenience we can describe by the ratio of contaminants in the solid phase to that in the liquid phase:

$$k = \frac{M_{\text{solid}}^{\text{contaminant(s)}}}{M_{\text{liquid}}^{\text{contaminant(s)}}} \quad . \tag{17}$$

This ratio is less than 1 for the case considered. A powerful method for the purification of materials is based on the above considerations. It is known as zone melting or zone refining and functions as follows. If we take a polycrystalline impure ingot of material and pass a narrow heated zone around it such that melting occurs under the heated area, the impurities present will be enriched in the liquid zone as compared to the solid zone immediately in back of it. By passing the zone across the entire ingot, the segregation coefficient will be effective in moving the impurities into the liquid zone. When the process has been completed, it may be repeated to obtain further refinement. Since the process is not an equilibrium one, the effective segregation coefficient is generally larger than the equilibrium value. In other words, the solid is not divested of as much of the impurities as would be predicted from the solubility curves. Further, as the liquid zone is enriched with impurities, the effective composition of the solid-liquid interface moves toward the right in Fig. 12. The purification effect is thus somewhat diminished as the zone sweeps across the ingot. In addition to the nonequilibrium aspects of the zone refining process, a further complication of the simplified model is that some impurities behave as if they had a higher melting point than the host lattice. The effect here is that the impurities are enriched in the solid phase and are not removed as effectively.

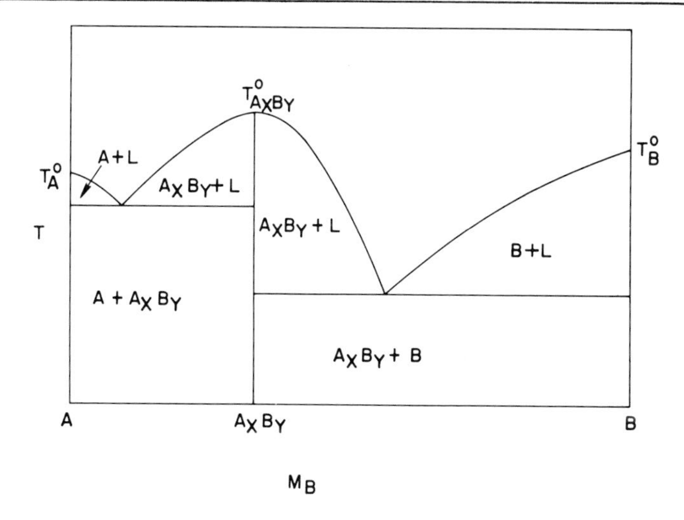

FIG. 13. A binary system exhibiting intermediate compound formation.

F. Intermediate Compound Formation and Nonstoichiometry

The final selected class of two-component systems we shall consider is that in which the end members A and B interact to form an intermediate compound having the generalized formula A_xB_y. We shall restrict our considerations to the special case where the behavior of this intermediate compound with either A or B is of the eutectic type, as shown in Fig. 13. The overall system, if A_xB_y behaves as a pseudo-unary component, may be treated as two independent systems, i.e., $A\text{-}A_xB_y$ and $B\text{-}A_xB_y$. The two systems may be depicted graphically as shown in Fig. 13 or as two separate systems. In the latter instance, the composition axis would of course have to be relabeled. Compounds of electronic interest, such as CdSe, CdS, GaAs, and InSb, are generated in such systems.

A pertinent aspect of intermediate compound systems is that relating to the pseudounary nature of the intermediate compound formed. When the melting point of such a compound exhibits a maximum in the solid-liquid diagram, as in Fig. 13, the compound is said to melt

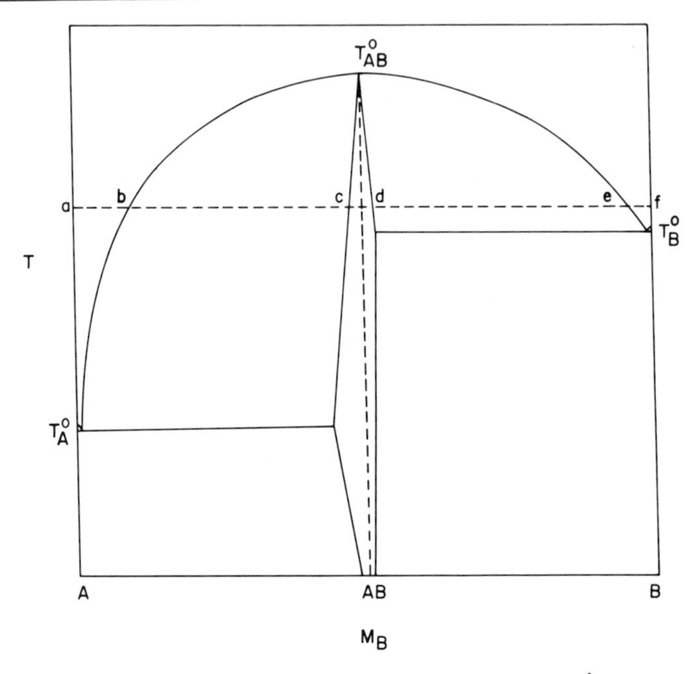

FIG. 14. Nonstoichiometry in a binary compound.

congruently. Theoretically, this implies that the liquid phase of pure A_xB_y has precisely the same composition as the solid phase with which it is in equilibrium. Consequently, the Phase Rule criterion for pseudounary behavior is satisfied. However, if we also consider the vapor phase, it is often the case that the composition of the vapor phase is different from the composition of the pure A_xB_y condensed phases. The compound is then said to vaporize incongruently. In the limit, then, if all coexistent phases are considered, many of these compounds are not pseudounary in behavior. The practical significance of incongruent vaporization behavior relates to the electronic effects caused by the formation of solid lattice vacancies in preparative procedures. These vacancies can function as donor or acceptor sites. The materials purification and preparation difficulties might best be appreciated from a consideration of the following.

As we have pointed out, in eutectic systems there is always some solid solubility of the end members, one in the other. When incongruency of vaporization of the intermediate compound is considered in this context, the coupling of these two facts leads to some rather marked effects. To point up the nature of these, let us assume that one of the components of the intermediate compound is very much more volatile than the other, i. e., in GaAs, the vapor pressure of Ga is very low, that of As quite high. Furthermore, let us assume that the phase diagram has the appearance shown in Fig. 14. Here, the melting point of the intermediate compound is very much greater than the melting points of the unary end members. Such is frequently the case in Group II-VI and Group III-V interactions. This difference in melting points results in a shift of the eutectic compositions to values very close to the end-member axes. Further, the liquidus curves of the intermediate compound lie displaced very close to the end-member axes also. If we examine vapor pressure variations liable to be observed in Fig. 14, assuming a vanishingly small vapor pressure for pure A and a large vapor pressure for pure B, the following qualitative aspects are observed for the isotherm T of Fig. 14. At the point "a," the total vapor pressure tends toward zero. As we move from "a" to "b" in the isothermally univariant two-phase liquid-vapor region (where we do not show the vapor composition but are aware of its presence), the vapor pressure of B increases. This increase for simplicity will be assumed to be in accordance with Raoult's law:

$$p_B = M_B p_B^\dagger \quad . \tag{18}$$

p_B is the vapor pressure of B in the liquid composition interval a-b, M_B is the mole fraction of B in this interval, and $p_B^\dagger$ is the vapor pressure of pure liquid B at the temperature T. Since the composition interval a-b is small, the value M_B is small in this interval, and therefore, when point b is reached, p_B is still a small fraction of $p_B^\dagger$. At point b, a new phase is formed, namely, essentially pure solid AB. The three-phase system, liquid, vapor, and solid AB, is isothermally invariant in the composition range b-c. Consequently, as the composition is varied from b to c, the partial pressure of B does not change. In this entire composition interval, a liquid of composition b is in equilibrium with a solid solution having the composition c. A similar situation applies in the composition interval d-e on the right-hand side of the diagram. At the point e, the mole fraction of B lies close to unity. Consequently, p_B is almost equal to $p_B^\dagger$ as the composition moves to the right of point e, but because of the invariant nature of the region d-e, p_B must have already achieved this value prior to the composition's having acquired the value d. Notice then that whereas at point b, p_B was very low, it approaches the value $p_B^\dagger$ when the composition lies at point e. Yet in the interval b-c no change in p_B can occur, and similarly in the interval d-e. Consequently, it is in the very small isothermally univariant two-phase solid solution-vapor interval c-d that the value of p_B must change from the value it had at point b to the value it has at point e. In fact, changes in the vapor pressure of B of orders of magnitude can occur in minute solid solution intervals, such changes resulting in dramatic changes in the electronic properties of the material in question.

The problems in preparing such incongruently vaporizing compounds can be formidable. Thus, in a very small composition interval, i.e., where the stoichiometry is varying only slightly, the vapor pressure and therefore the lattice defect quantity is varying enormously. The general technique employed is to provide precisely established over-pressures of the volatile component into the apparatus in which such binary compounds are being synthesized, purified, or grown as single crystals. In certain instances, high pressure systems must be employed. In other instances, the material can be prepared only in a deficient state because the solubility on one side of the stoichiometric composition is essentially zero.

The qualitative conclusions concerning the large variation in partial pressure with a small change in concentration in the region of variable concentration can be arrived at analytically by a consideration of simple mass action relationships.

Let us assume that, in the compound AB, the B component is volatile and that it is originally present in the lattice as B^{2-} ions. When this component volatilizes from the lattice, it leaves behind a vacancy containing two electrons. The concentration of such un-ionized vacancies is denoted by $[B^{II}]$. These un-ionized vacancies can lose one or both of the electrons leading to concentrations of singly ionized vacancies $[B^{I}]$ or doubly ionized vacancies $[B^\circ]$. For simplicity let us assume that the double ionization mechanism prevails. The equilibria that lead to the doubly ionized state are given by

$$B^{2-} \rightleftarrows B^{II} + \tfrac{1}{2}B_2\,(\text{vapor}) \quad , \tag{19}$$

$$B^{II} \rightleftarrows B^\circ + 2e^- \quad . \tag{20}$$

The conservation of total vacancy concentration may be expressed by

$$[B^{II}] = [B^\circ] \quad . \tag{21}$$

From Eqs. (20) and (21) we see that the concentrations of doubly ionized vacancies $[B^\circ]$ and electrons $[e^-]$ are related by

$$2[B^\circ] = [e^-] \quad . \tag{22}$$

Further, as the total concentration of vacancies is very small compared to the concentration of B lattice sites, $[B^{2-}]$, the latter concentration is essentially constant. Substituting for B^{II} and $[e^-]$ in Eqs. (19) and (20), the results of Eqs. (21) and (22), and taking into account the constancy of $[B^{2-}]$, we may write the equilibrium constants for Eqs. (19) and (20) as

$$K_1 = p_{B_2}^{\frac{1}{2}}[B^\circ] \quad , \tag{23}$$

$$K_2 = \frac{4[B^\circ][B^\circ]^2}{[B^\circ]} \quad . \tag{24}$$

Solving Eqs. (23) and (24) for p_{B_2} in terms of $[B^\circ]$ and combining constants in a single term, we obtain

$$p_{B_2} = \frac{C}{[B^\circ]^6} \quad . \tag{25}$$

Equation (25) shows an inverse relationship between the 6th power of the vacancy concentration and the partial pressure of B. Thus when the vacancy concentration is largest, the partial pressure of B is smallest, and when the vacancy concentration is smallest, the partial pressure of B is largest. Most significant is the fact that, because of the 6th power dependency, a very minor vacancy concentration change results in a proportionately enormous partial pressure change.

If the ionization mechanism involved primarily the formation of singly ionized vacancies, the dependence would be a 4th power one. In general, the ionization state would probably lie between these two or even between a zero and singly ionized state. The latter would of course yield the smallest p_{B_2} dependency and represent an easier materials stoichiometry control problem. The boundary case we have considered would represent the severest materials control problem, but one which is not difficult to realize in practice.

IV. VAPOR TRANSPORT REACTIONS: MULTI-COMPONENT SOLID-GAS EQUILIBRIA

The selective nature of the treatment of phase equilibria in this chapter precludes consideration of three- and higher-component order systems in conventional fashion. Instead, we will focus specifically on one class of multicomponent equilibria, that in which a single solid phase coexists with a vapor over the entire range of interest. A further restriction is that this solid phase in unary or pseudounary in behavior, all other components being confined to the vapor phase. This type of equilibrium has great practical implications in materials synthesis, purification, and single-crystal growth and is the basis for vapor transport or chemical vapor deposition processes.

Schematically, the kind of equilibrium involved is given by

$$aA\,(\text{solid}) + bB\,(\text{vapor}) \rightleftarrows A_aB_b\,(\text{vapor}) \quad . \tag{26}$$

For the reaction described there will be associated an enthalpy of reaction that is either positive (endothermic) or negative (exothermic). Further, the magnitude of this enthalpy is solely a function of the quantity of reactants present. If, for example, the coefficients a and b of Eq. (26) represent the appropriate numbers to yield one mole of A_aB_b, the enthalpy will represent the molar heat of formation of A_aB_b. There is nothing, however, that demands the formation of one mole of A_aB_b, and we are free to multiply Eq. (26) by any fractional or integral number. Thus, in an actual experiment, the amount of heat liberated or absorbed may vary from almost zero (when very small quantities are employed) to large values (when very large quantities of reactants are employed).

Independent of whether the quantities of reactants employed are small or large, the sign of ΔH accompanying the process will, in the absence of perturbations, remain the same even though the magnitude of ΔH varies.

The variation of the equilibrium constant with temperature is given by the integrated form of the van't Hoff equation:

$$\ln(K_2/K_1) = \exp\left\{-\Delta H_p/R[(1/T_2) - (1/T_1)]\right\} \quad . \tag{27}$$

In Eq. (27), K_2 is the equilibrium constant at some temperature T_2, and K_1 is the equilibrium constant at some other temperature T_1. ΔH_p is the molar enthalpy of the reaction and R is the gas constant. Let us assume that ΔH_p is positive. If $T_2 > T_1$, then $1/T_2 < 1/T_1$ and the right-hand side of the equation is positive in sign. Consequently, $K_2 > K_1$. In other words, with increasing temperature, the equilibrium described by Eq. (26) is driven to the right. The reverse reaction for Eq. (26) has a ΔH_p of identical magnitude, but of opposite sign. If

ΔH_p for Eq. (26) is negative, the reasoning used here leads to the conclusion that with increasing temperature the reaction is driven to the left.

The practical significance of these conclusions may be seen with respect to the reactions of Ge and W with a halogen, Eqs. (28) and (29), respectively:

$$2\,GeI_2(\text{vapor}) \rightleftarrows Ge(\text{solid}) + GeI_4(\text{vapor}), \quad \Delta H \text{ is negative,} \tag{28}$$

$$2\,WCl_3(\text{vapor}) \rightleftarrows W(\text{solid}) + WCl_6(\text{vapor}), \quad \Delta H \text{ is positive.} \tag{29}$$

The reactions shown fall into a class known as disproportionation reactions. Both the Ge and W exist in vapor species exhibiting two oxidation states. The lowest oxidation state species in each case is capable of disproportionating into a higher oxidation state species and a neutral solid.

If GeI_2 present at some temperature is cooled to a lower temperature, the equilibrium constant for the reaction, Eq. (28), as written will increase since ΔH is negative (exothermic) and Ge will be deposited. On the other hand, if WCl_3 present at one temperature is heated to a higher temperature, the equilibrium constant for the reaction described in Eq. (29) will increase and tungsten will deposit. The germanium reaction is termed a hot to cold process since Ge in the vapor phase is deposited by lowering the temperature. The tungsten reaction is termed a cold to hot process since W in the vapor phase is deposited by raising the temperature.

If, in an open tube, shown schematically in Fig. 15, we place an impure supply of Ge and pass I_2 through it, the Ge will react to form both GeI_4 and GeI_2:

$$Ge + I_2 \rightleftarrows GeI_2 \ , \tag{30}$$

$$Ge + 2I_2 \rightleftarrows GeI_4 \ . \tag{31}$$

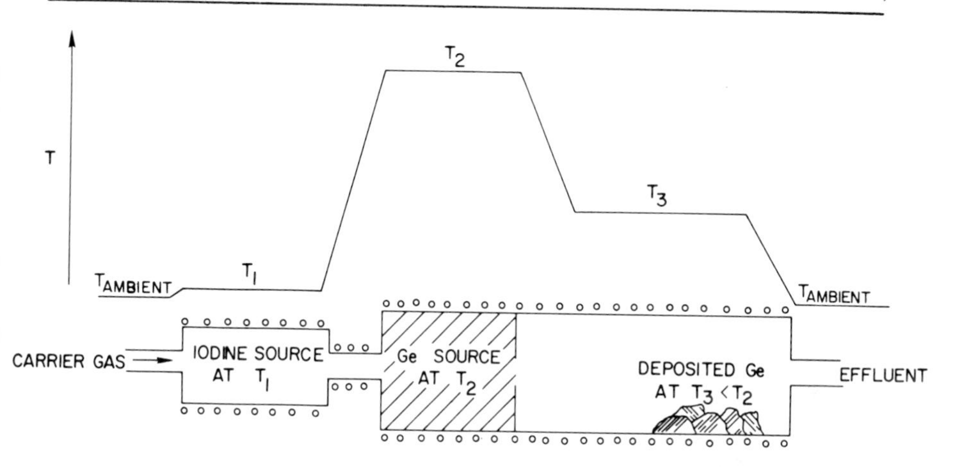

FIG. 15. Schematic representation of a vapor transport system.

Since only GeI_2 is capable of disproportionating, Eq. (28), it is desirable to choose conditions so that Eq. (30) predominates. At temperatures in excess of 600 °C this is the case. Consequently, the impure supply of Ge in the open tube is maintained at this temperature. The iodine is carried in an inert carrier gas such as He at such a rate as to insure equilibration of the halogen with the germanium bed. The effluent from the bed is carried downstream where the temperature maintained is lower. In this decreasing temperature interval, the hot to cold transport process occurs according to Eq. (28), and purified Ge is deposited. The latter may then be collected and zone refined to yield a material of extremely high purity. Precisely the same technique may be employed to grow thin single crystals of Ge on a substrate single crystal located in the cooler zone. If such is the intent, then the high temperature Ge source would be comprised of purified rather than impure material. Such single-crystal growth processes are the basis for all semiconductor transistor device fabrications and are termed epitaxial growth processes. For the fabrication of integrated circuit arrays the deposited epitaxial film would be doped with opposite conductivity type impurities to the wafer single-crystal substrate. The epitaxial structure would then serve as the starting layer for subsequent diffusion processes during which time the active device regions are formed.

It is to be recognized that with the introduction of He as a carrier gas in either purification or epitaxial applications, the system becomes three-component. Since we have three components coexisting in only two phases, the quasiequilibrium established has three degrees of freedom. This unwieldy number may be reduced by fixing the pressure (normally at one atmosphere in an open tube system) and fixing the ratio of He to I_2. The result then in a univariant equilibrium. The characterization of this univariant equilibrium may be handled in a simple manner because all of the I_2 and He are confined to the vapor phase. If a graph of molar Ge/I_2 ratio at fixed molar He/I_2 ratio and fixed total pressure vs temperature is plotted, one can immediately ascertain both qualitatively and quantitatively the effect of temperature on Ge vapor phase concentration. Consequently one can by inspection determine how much Ge is picked up at the source site, how much is deposited at the seed site, the best temperatures for both of these sites, etc. Each different starting He/I_2 ratio would yield a different univariant curve. Figure 16 shows such a series of curves. Since the sum of the partial pressures $p_{He} + p_{I_2} = 760$ Torr at the point where the two materials first mix, it is evident that, if the partial pressure of iodine at its point of introduction into the system is fixed, the ratio p_{He}/p_{I_2} is also fixed. Each curve in Fig. 16 then is characterized by the p_{I_2} value associated with it.

From Fig. 16 it is seen that at a temperature between 600 °C and 700 °C the p_{Ge}/p_{I_2} vapor phase ratio approaches unity. This implies that in this temperature interval the germanium is present in the vapor phase as the desirable species GeI_2. At a temperature of 200-300 °C the p_{Ge}/p_{I_2} vapor phase ratio decreases to approximately 0.5, indicating that the germanium in the vapor phase is present as GeI_4. Thus, if the source bed of Ge is maintained at between 600 and 700 °C and the seed area at 300-400 °C, the hot to cold transport process will

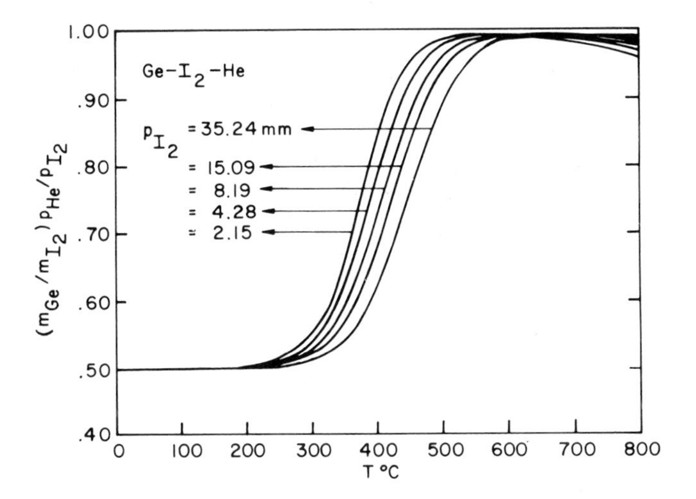

FIG. 16. Vapor phase solubility curves in the three-component solid-gas system Ge-I_2-He.

occur with maximum material transport efficiency. It is seen from Fig. 16 that, starting at 650°C, the efficiency curves (in fact, vapor phase solubility curves) begin to bend downward. This indicates that the reaction changes to a cold to hot process and is related to the fact that the sign of ΔH_p for the reaction depicted in Eq. (30) is negative as written. In other words, the ΔH_p for the decomposition of GeI_2 via the reverse process is positive. Consequently, at higher temperatures the reverse process is favored.

Before proceeding to a discussion of the practical use of the higher temperature decomposition process, it is interesting to consider one further aspect of open tube vapor transport reactions. If instead of employing He as a carrier gas we employed hydrogen, the results obtained in the Ge-He-I_2 system would be greatly perturbed. The perturbation arises because a competitive reaction for iodine is set up between the Ge and H_2:

$$H_2 + I_2 \rightleftarrows 2HI , \qquad \Delta H \text{ negative} \quad . \tag{32}$$

Since ΔH for Eq. (32) is negative, with increasing temperature the

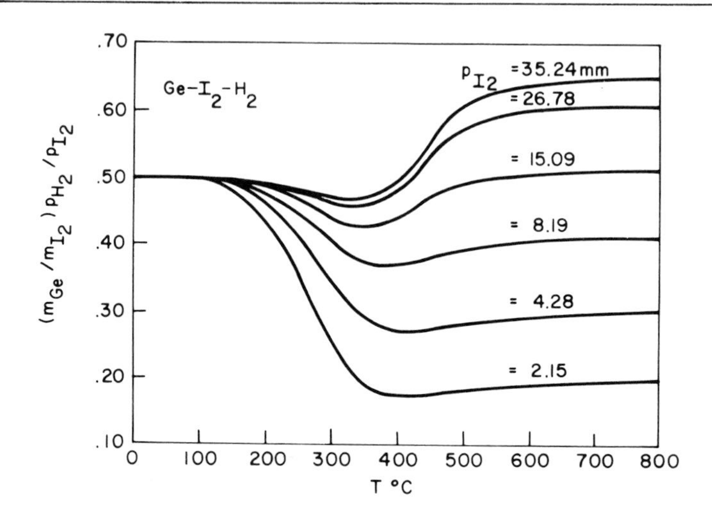

FIG. 17. Vapor phase solubility curves in the three-component solid-gas system Ge-I_2-H_2.

reaction is driven to the left. However, the reactions leading to the formation of GeI_2 and GeI_4, Eqs. (30) and (31), which together yield the reverse of the reaction depicted in Eq. (28), i.e., Eq. (33), has a positive ΔH:

$$Ge(solid) + GeI_4(vapor) \rightleftarrows 2GeI_2(vapor), \quad \Delta H \text{ positive} \quad . \qquad (33)$$

Depending now upon the quantities of H_2 and I_2 reacting in the presence of solid Ge and the temperature, the net overall reaction may exhibit a positive or negative ΔH_p, i.e., the net reaction may be endo or exothermic. By again fixing the total pressure and this time fixing the H_2/I_2 vapor phase ratio, the three-component system in question is again made univariant. A similar efficiency plot to the one shown in Fig. 16 would now have the appearance shown in Fig. 17. It is seen that, with increasing temperature, the curves first show cold to hot transport (net overall ΔH_p positive), then exhibit hot to cold transport as the hydrogen for iodine competition becomes less important.

For more complex systems, i.e., when four components are present, conditions leading to univariance can be established by fixing two component ratios. For example, in an open tube process involving

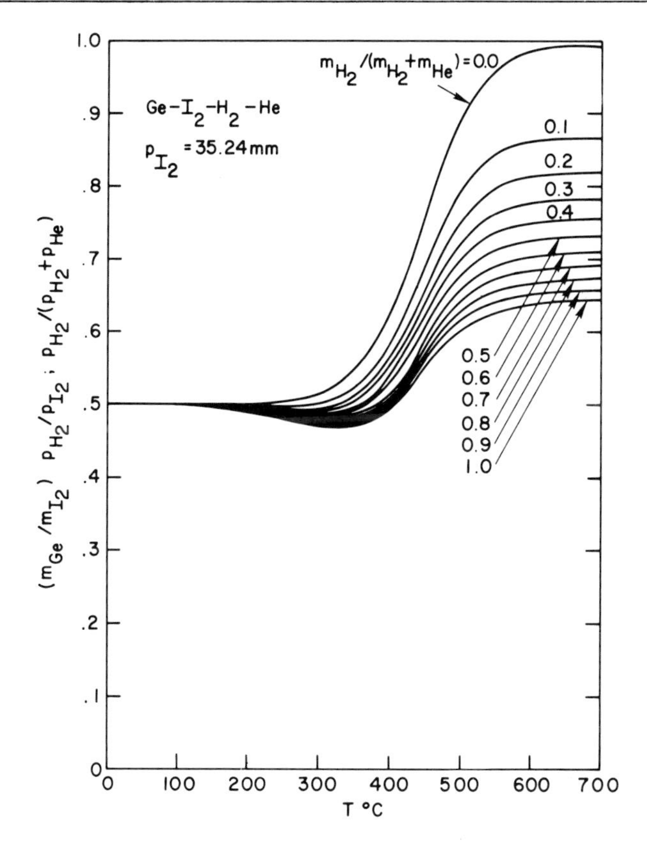

FIG. 18. Vapor phase solubility curves in the four-component solid-gas system Ge-I_2-H_2-He.

H_2, He, I_2, and Ge, we can fix the ratio p_{H_2}/p_{I_2} and $p_{H_2}/(p_{H_2} + p_{He})$. For each fixed value p_{H_2}/p_{I_2} we will generate a series of curves of p_{Ge}/p_{I_2} vs T, each curve in the series representing a different $p_{H_2}/(p_{H_2} + p_{He})$ ratio. One such series is shown in Fig. 18. Alternately, at a fixed ratio $p_{H_2}/(p_{He} + p_{H_2})$ we could generate a series of curves, each curve representing a different fixed vapor phase value p_{H_2}/p_{I_2}. One such series is shown in Fig. 19. The method of representing equilibria of the type being considered is applicable to systems containing many components, and while unconventional it is very informative. It is interesting that although our discussion has been

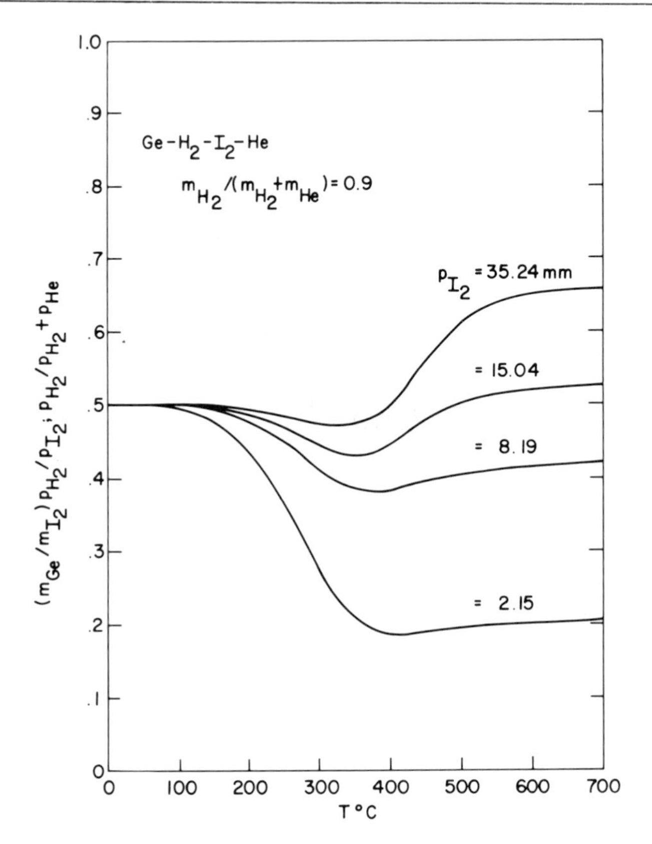

FIG. 19. Vapor phase solubility curves in the four-component solid-gas system Ge-I_2-H_2-He.

confined to solid-gas phenomena it is completely applicable to solid-liquid equilibria in which a pure solid is in equilibrium with a liquid (i.e., eutectic systems exhibiting minuscule solid solubility), since here again all components but one are confined to one of the phases.

An interesting practical application of cold to hot disproportionation reactions is that accruing to the reaction described in Eq. (29). If halogen is added to an electric light bulb, the walls of the bulb and the filament become the cold and hot points of the system, respectively. When the bulb is lit, tungsten vaporizes from the filament to the walls. In so doing, the wire tends to thin unevenly due to selective evaporation, developing hot spots at the thinner portions. At the wall, the tungsten film reacts with the halogen to form WCl_3. Since the disproportionation of WCl_3 occurs via a cold to hot process (positive ΔH_p), the trichloride species transports to the hottest parts of the filament where W is deposited, building up the thin portions until they are at the same temperature as the rest of the filament. This continuous process results in regeneration of the filament and is the basis for fabricating long lived electric light bulbs.

In order for vapor transport properties to be of practical use, three criteria should be satisfied. First, it is necessary that the vapor pressure of the transporting species in some usable temperature range be high enough so that sufficient material is moved in a reasonable time. Second, the equilibrium constant for the reaction depicting the transport should not be too large. Third, the desired solid should exhibit the lowest dew point temperature of all the vapor phase species.

Disproportionation reactions have found great utility in application to a host of materials preparation problems involving the transport of unary, binary, and ternary compounds. In a number of instances transported compounds are reported to deposit in polymorphic varieties not obtainable by other means. The technique is extremely powerful for the growth of single crystals of normally involatile materials or materials which are involatile and/or melt at high temperatures.

Another vapor transport process having considerable practical import is that involving the hydrogen reduction of halide compounds. For example, if impure Si or Ge is reacted with chlorine, the volatile compounds $SiCl_4$ or $GeCl_4$ are formed. These tetrahalides may in turn be distilled in an inert atmosphere to provide an additional purification step. If they are then passed through a heated zone (about 800 °C for Ge and 1150 °C for Si) in the presence of H_2, the tetrahalides are reduced:

$$SiCl_4(\text{vapor}) + 2H_2(\text{vapor}) \rightleftharpoons Si(\text{solid}) + 4HCl \quad . \tag{34}$$

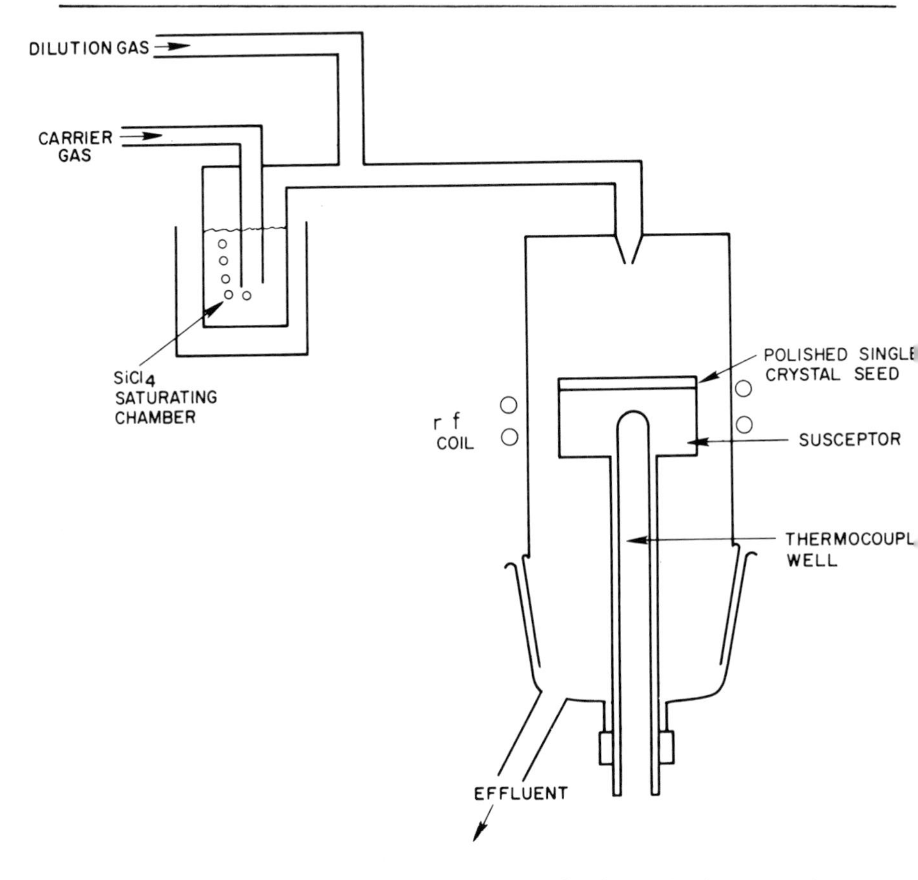

FIG. 20. Schematic representation of a $SiCl_4$ hydrogen reduction epitaxial reactor.

The pure solid Si that deposits may in turn be further purified via a zone refining process if necessary. This same process may be employed to grow epitaxial single-crystal films upon single-crystal wafer substrates in an apparatus depicted schematically in Fig. 20. H_2 is bubbled through the highly purified tetrahalide, and the saturated gas is passed over a radio-frequency heated susceptor upon which the wafer is positioned. Under proper temperature, concentration, and flow conditions, the Si deposited at the hot zone grows epitaxially upon the substrate.

A final vapor transport process of considerable utility is that involving pyrolytic decompositions. In such processes, thermally decomposable materials such as hydrides or metalloorgano oxycompounds are passed through a hot zone in which decomposition is effected. The desired material deposits, while the by-products remain in the vapor phase. An example of such processes is that employed in the formation of dielectric films of SiO_2 or Al_2O_3. These films may be used as insulating layers, diffusion masks, or passivating layers. One method for forming SiO_2 films is

$$Si(OC_2H_5)_4(\text{vapor}) \xrightarrow[\text{heat}]{} SiO_2(\text{solid}) + \text{organic vapor by-products.} \quad (35)$$

Experimental Techniques

The study of phase equilibria can and should involve the use of a number of different experimental techniques. If the gas phase is of interest, mass spectrometric techniques coupled with vapor pressure measurements are frequently employed. If the kinetic aspects of solid-gas phenomena (decompositions, reactions, etc.) are of interest, a technique known as thermogravimetry has considerable applicability. In essence, this method utilizes a balance having a sample support affixed to it, the sample itself being located in a furnace. The change in mass of the sample as a function of time and temperature is monitored continuously. Sophisticated TGA (thermogravimetric analysis) systems utilize balances that graphically record mass changes electronically.

In studies of solid phases, x-ray diffraction techniques are of great value. These enable examination of samples of varying composition either at elevated or room temperature to determine at what compositions new phases appear. In addition, the crystalline properties of the solid phases present may be deduced. For pure materials undergoing phase transformations, x-ray diffractometry is useful in studying polymorphism. The samples are heated or cooled incrementally or even continuously to different temperatures, all the while being monitored by the x-ray beam. Diffraction effects are recorded either on film or

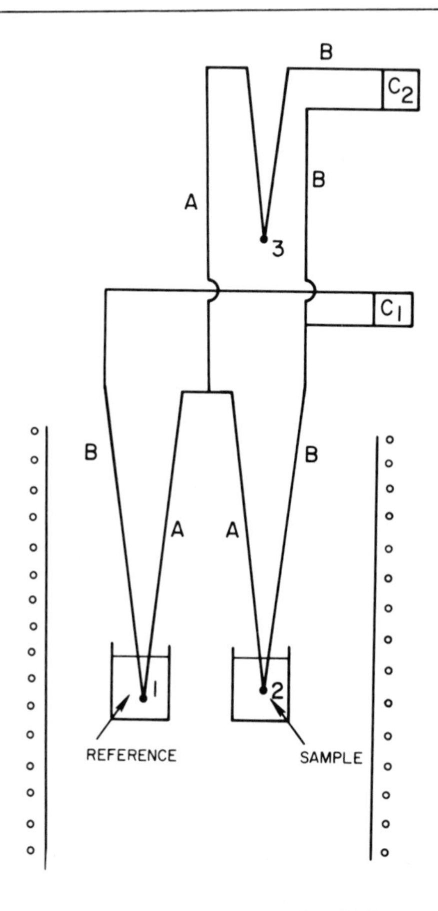

FIG. 21. Schematic representation of a DTA apparatus.

electronically on a strip chart recorder. Other techniques that are frequently used in the study of solid systems are light optical and electron microscopy.

The most powerful technique available for the study of solid-solid or solid-liquid equilibria, particularly in defining composition-temperature phase boundaries, is thermal analysis, specifically a refinement, known as differential thermal analysis (DTA). Its principle of operation is best seen with reference to Fig. 21.

The wires A and B represent dissimilar metals forming the two halves of a temperature measuring device known as a thermocouple. The latter consists of two junctions of these dissimilar metals and develops an emf when the junctions are at different temperatures. The DTA configuration of thermocouples involves two such devices. One of these thermocouples, noted by the junctions 1 and 2 measures the temperatures of both a reference material and the sample, both of which are maintained in a furnace whose temperature is varied. As long as the physical state of the sample does not change with temperature change in the furnace, the temperatures at points 1 and 2 remain the same and the thermocouple does not generate an emf. If, however, a change of state occurs in the sample (melting, freezing, or a solid state phase change), the latter will absorb or liberate heat. As a consequence, the temperature of the sample will lag behind that of the reference and a difference in temperature will develop between points 1 and 2. This difference in temperature manifests itself in the form of an electrical signal which may be measured manually or automatically at the measuring instrument C_1. This differential signal which arises only when the temperatures at points 1 and 2 are different may be amplified electronically enabling the detection of very small latent heat anomalies.

In addition to the thermocouple which measures temperature difference between the sample and some suitable reference material (Al_2O_3 is frequently employed as a reference), a second thermocouple measures the temperature of the sample. The two junctions of this thermocouple are denoted by the points 2 and 3 and the temperature at point 2 is recorded at the measuring station C_2. It is to be pointed out that, in order to measure the actual temperature at point 2, the reference junction at point 3 must itself be kept at constant temperature (generally in an ice bath at 0 °C). Tables are available which for different thermocouple metals list the emf's developed when the reference junction in at 0 °C or 0 °F and the measuring junction is at some other temperature. While

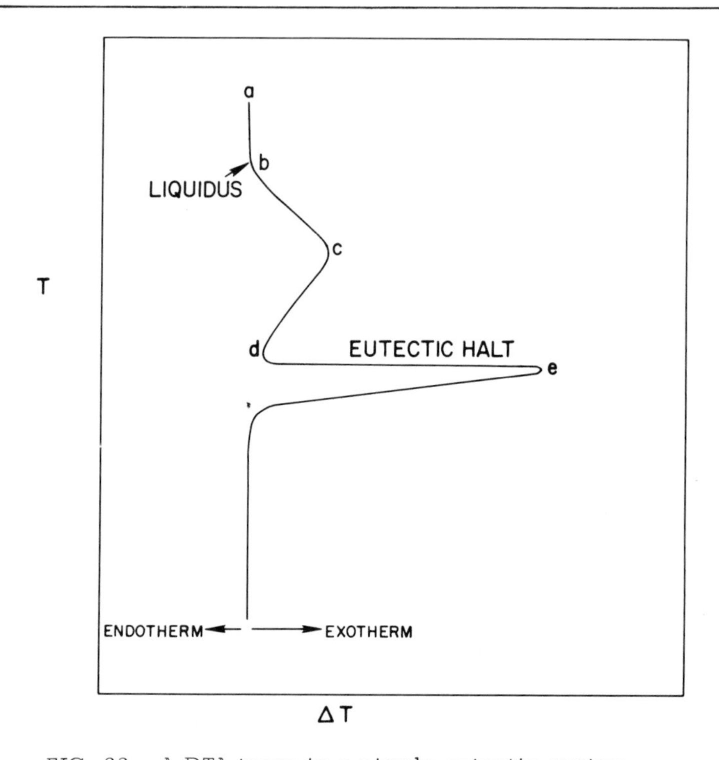

FIG. 22. A DTA trace in a simple eutectic system.

both thermocouples comprising the DTA configuration actually measure temperature differential, one of these (points 2 and 3) measures the temperature differential between a fixed reference point and the sample temperature. The other thermocouple measures the temperature differential between a nonstationary reference point (point 1) and the sample. The latter differential will always be small and thus may be easily amplified.

Figure 22 shows a typical DTA cooling curve in a eutectic system. A binary sample, for example, of defined composition is melted and then subjected to a cooling cycle. The ordinate shows the temperature of the sample while the abscissa shows the differential in temperature between the reference material and the sample. In the initial stages of cooling, both the reference material and sample cool at the same rate.

Consequently, no temperature differential exists between the two, and the line a-b is vertical with respect to the abscissa. When the liquidus temperature is reached, a pure solid begins to settle out, liberating heat in the process. The sample cannot cool as fast as the reference during this liquidus crystallization, since a portion of the energy it is radiating to the cooler furnace walls is in part made up by the latent heat of freezing. As the system is univariant during the liquidus crystallization, the temperature of the sample drops, the curve b-c. At point c, the heat balance between walls and sample is such that heat is lost at a greater rate than it can be made up by the latent heat of freezing, and the sample temperature-reference temperature differential decreases. At point d, eutectic crystallization occurs. The system becomes invariant and the temperature of the sample remains constant up to the point at which the loss of heat to the walls, at point e, becomes greater than the heat evolved by the sample during eutectic crystallization.

If samples of different starting composition are subjected to the same experiment, each will show a different liquidus crystallization temperature, and all samples will show the same temperature of eutectic crystallization. These data can be used to construct the composition-temperature phase diagram.

Of course, we have greatly simplified the preceding discussion; many phenomena serve to confuse the issue. For example, undercooling phenomena (failure of the solid to crystallize at the proper temperature) must be contended with in both the liquidus and eutectic crystallizations. Further, if temperatures are lowered too rapidly, the eutectic arrest may not appear to be isothermal since heat is lost too rapidly to the furnace walls and temperature gradients are established in the sample container (parts of the sample have completely solidified while other portions still contain a liquid phase).

In addition to the methods mentioned, others are also of some use for establishing phase boundaries in solid-solid and solid-liquid equilibria. One such method involves measuring the conductivity of the melt. At the point at which the liquidus is encountered, a change in conductivity frequently occurs. Another method useful when only minute sample quantities are available is visual observation. The micro sample is placed on a strip of resistance heated metal, and while the temperature is raised, the sample is observed microscopically for first and last traces of melting. The sample temperature is monitored with an optical pyrometer.

Chapter 12

CRYSTAL GROWTH

Edward Kostiner*

Department of Chemistry
Cornell University
Ithaca, New York

* Present address: Department of Chemistry and Institute of Materials Science, University of Connecticut, Storrs, Connecticut.

I. INTRODUCTION

Both the solid state chemist and physicist eventually must face the necessity of growing single crystals. The measurement of many physical properties of materials depends on the preparation of well-characterized single crystals. Unfortunately, their growth is more an art than a science. Theories of crystal growth do not seem to be of much help to the experimentalist, since, to quote K. Nassau of the Bell Telephone Laboratories in a recent review article, "the crystal is an entity with a 'will' of its own; the attempt to impose external limitations dictated by some assumed theory of crystal growth may lead to complete frustration." This chapter will attempt to summarize the important techniques of crystal growth from the experimentalist's point of view, outlining the advantages and disadvantages of each.

Broadly speaking, the techniques of crystal growth can be divided into three classifications:

(a) growth from pure materials;
(b) growth from solutions;
(c) growth from the vapor phase.

Each of these classications contains many different types of growth techniques; in some cases the distinction between them is not clear-cut, as will become evident in the discussion.

When choosing a method it is well to keep in mind the reason for growing the particular crystal; for example, submillimeter crystals of high perfection are necessary for single-crystal x-ray crystallographic work, while centimeter-sized crystals are desirable for the investigation of optical properties. Certain growth techniques are suitable for attaining large crystal size and high purity while others are best suited for smaller crystals of high perfection.

Any crystal growth technique must realize certain important criteria to achieve reasonable-sized crystals of sound quality. In order to promote crystal perfection, the temperature of the system should be controlled as closely as possible during the growth process; furthermore, the process should be carried out as slowly as practicable to limit spontaneous nucleation and thereby maximize crystal size. Spontaneous nucleation will give a large number of small crystals which inhibit the growth of a few larger crystals. In the ideal case, growth should take place at a single site in the reaction chamber. An effort should be made to make material available at a constant concentration at the growth surface in order to produce homogeneous crystals, free from concentration gradients and/or deviations from stoichiometry.

The problem of the control of nucleation is a real one. To maximize crystal size or to achieve the growth of one single crystal, spontaneous nucleation must be suppressed. A common technique is the use of a seed crystal to initiate growth (see below). Since large crystals are more stable than small ones (they have a lower surface energy), larger crystals will tend to "outgrow" smaller ones. One can enhance this phenomenon by decreasing the diameter of the growing crystal in a pulling technique of growth ("necking down") to take advantage of the outgrowing process. In the Bridgman-Stockbarger technique the crucible configuration can be such as to provide a necking down of the growing crystal, suppressing polycrystalline growth by allowing the continued growth of only one crystal.

Our approach will be descriptive. As an aid to the illustration of each method, a "recipe" for the growth of a particular material by that method will be given. It is well to realize that every crystal growth method can be modified to a large degree to accommodate the peculiarities of the particular compound being grown, e.g., moisture or air sensitivity, volatility, and decomposition at atmospheric pressure.

II. GROWTH FROM PURE MATERIALS

This classification includes all growth techniques that involve the preparation of single crystals directly from either the pure material or from melts containing only the components of the desired material. In general, growth from pure materials depends on the congruency of the melting point of the compound to be grown (that is, the compound melts without decomposition) and the absence of any solid state phase transition between the melting point and ambient temperature. Since no solvent or other extraneous material is necessarily present, contamination of the grown crystals tends to be relatively low. If the material has a sharp melting point, it is sufficient to melt the polycrystalline solid, then gradually lower the temperature through the melting point. However, if single crystals of good size and perfection are desired, it is well to use one of the following techniques.

A. Czochralski Growth: Crystal Pulling

The material to be grown is melted in a suitable container and a rod, which acts as a nucleating center, is lowered into the melt to induce crystallization. The melt is held slightly above its melting point while the nucleating rod is simultaneously rotated (to promote mixing) and withdrawn from the melt. Under ideal conditions, the nucleated crystal (together with the pulling mechanism) acts as a heat sink, enhancing crystal growth as the rod is withdrawn from the melt. By

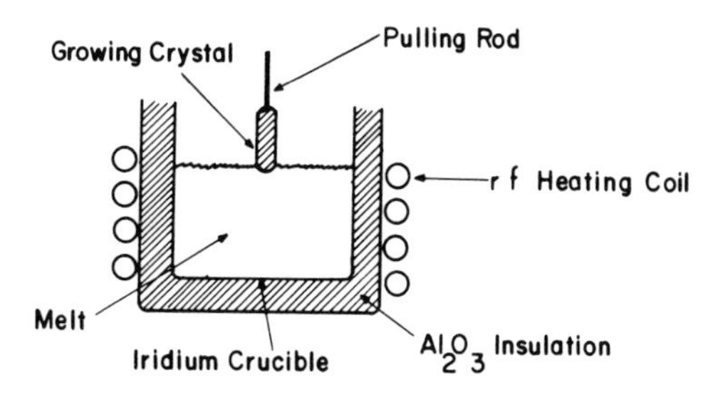

FIG. 1. Czochralski growth.

variation of the pulling and rotation rates, the growing boule can be necked down to assure the growth of one single crystal. Slow pulling rates (~ 0.1-2.0 mm/hr), reasonable rates of rotation (~ 5-100 rpm) to insure homogeneity, and good temperature control of the melt (~ 1-5 °C) to prevent spontaneous nucleation, and concentration zoning are necessary to achieve good growth. Simple binary oxides and fluorides as well as more complex compounds have been successfully grown by this method. An analogous method (and one sometimes used interchangeably with Czochralski growth) is the Kyropoulos technique, in which a seed crystal attached to a "cold finger" (usually a water-cooled seed holder) is lowered into the melt and the crystal grown by slow cooling of the melt.

These two methods have some rather unique advantages. The crystal is visible during the growth process and adjustments can be made to control its growth rate. Although a too rapid growth rate can give crystals of inferior perfection, the growth rate in the Czochralski technique is relatively fast. Additional variations in technique are possible; for example, crystals can be pulled from a flux solution, thereby extending these methods to incongruently melting compounds but also introducing the possibility of flux contamination (see below).

Calcium tungstate [K. Nassau and A. M. Broyer, J. Appl. Phys., 33, 3064 (1962)]. Calcium tungstate powder is packed into a rhodium crucible (1.25 to 2.25 inches in diameter and height, wall thickness

1.5 mm) which is coupled to an induction coil powered by a 450-kcps 10-kW rf generator (see Fig. 1). The melt temperature is held at 1620 ± 2 °C (melting point of $CaWO_4$ - 1575 °C), a seed crystal is introduced into the melt and pulling is initiated at a rate of 0.5 to 3 in./hr with rotation at 25-100 rpm. Alumina powder is used both to support the crucible within the rf coil and as thermal insulation. Crystals have been grown up to 1 in. in diameter and 18 in. long.

B. Bridgman-Stockbarger Growth

Basically, crystal growth by the Bridgman-Stockbarger method consists of passing a crucible containing the melt either horizontally or vertically through a temperature gradient. In the more common vertical arrangement, growth is initiated at the bottom of a cylindrical crucible which has been tapered to a point. The charged crucible is held in the hot zone (temperature > mp) of a two-zone furnace and slowly lowered into the cooler zone (T < mp). A sharp temperature gradient in the furnace is desirable (to reduce the possibility of supercooling) and is usually achieved by placing a baffle between the two zones. This method cannot be used for materials that expand on solidification since the grown crystal is enclosed in the crucible. (In general, this is true for all crucible techniques.) There is the additional problem of the material wetting the crucible; severe wetting by the melt makes

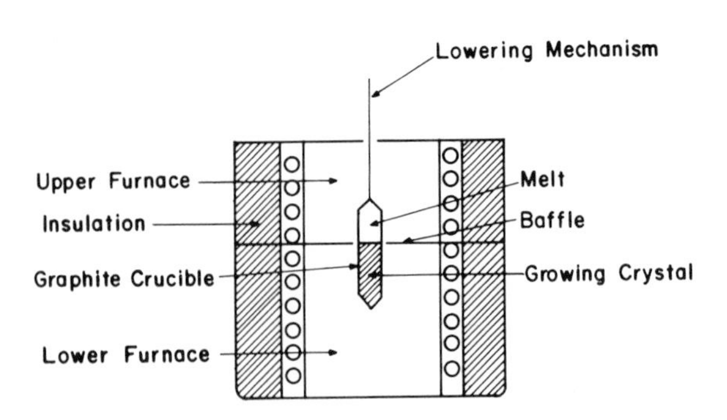

FIG. 2. Bridgman-Stockbarger growth.

it difficult to remove the crystal. The Bridgman-Stockbarger technique is very useful in the preparation of crystals of metals and semiconductors, alkali and alkaline earth halides, and complex ternary fluorides of alkali and transition metals. It is easily adapted for growth at high pressures.

Cadmium fluoride [D.A. Jones and R.V. Jones, Proc. Phys. Soc., 79, 351 (1962)]. High purity cadmium fluoride powder is placed in a cylindrical graphite crucible with a conical bottom. The charged crucible is placed into the hot zone of a two-zone furnace (Fig. 2). The reaction chamber is evacuated and the furnace heated at ~20°/hr until the hot zone temperature reaches 1067°C (mp CdF_2, 1047°C). Pure nitrogen is introduced into the furnace when the temperature reaches 900°C (at a pressure of 2 mm Hg) to inhibit the volatilization of the charge. After heating for 12 hr the crucible is lowered at a rate of 1.2 mm/hr into the cooler zone ($T < mp$). After the lowering is completed the furnace is cooled to room temperature at ~20°/hr. Crystals up to 15 cm in diameter and 12 cm long have been grown.

C. Zone Melting (Floating Zone)

Zone melting is actually a common method for the purification of materials (called zone refining) which is based on the fact that the solubility of impurities in a material is ordinarily different in the solid phase than in the liquid phase with which it is in equilibrium. By successive "passes" of a molten zone through a bar of material, the impurities will segregate in either end of the bar, depending on their distribution coefficients.

The zone melting method of crystal growth is a good example of a "crucible-less" technique, one in which highest purities can be obtained since both the starting charge and the grown crystal are not in contact with any crucible material. A sintered bar of the material to be grown is passed (either horizontally or vertically) through a short

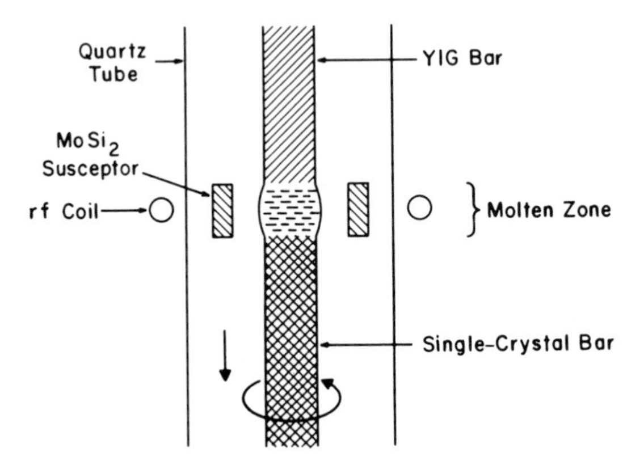

FIG. 3. Floating zone growth.

hot zone (temperature > mp). The resultant molten zone of material passes through the bar, ideally converting the polycrystalline rod into a single-crystal bar. Crystals of refractory metals and oxides can be grown by this technique as well as the more common semiconductor materials.

Yttrium iron garnet [L. L. Abernathy et al., J. Appl. Phys., 32, 376S (1961)]. A sintered polycrystalline bar of yttrium iron garnet ($Y_3Fe_5O_{12}$) is held vertically in an apparatus (Fig. 3) which allows a short hot zone (T > mp) to be passed through it. Since $Y_3Fe_5O_{12}$ has a high volume resistivity, an rf field is coupled to a $MoSi_2$ susceptor to form the hot zone. Oxygen pressure is maintained at 25-100 psi to control the stoichiometry (preventing the reduction of Fe^{3+} to Fe^{2+} at the high operating temperature. Growth is obtained in 0.25-0.40 in. diameter rods at a rate of 0.05-0.35 in./hr by passing the rod through the hot zone.

D. Verneuil (Flame Fusion) Growth

Originally developed in the early 1900's, the flame fusion technique is an effective but difficult method for the growth of certain refractory oxide single crystals. A very finely divided feed material

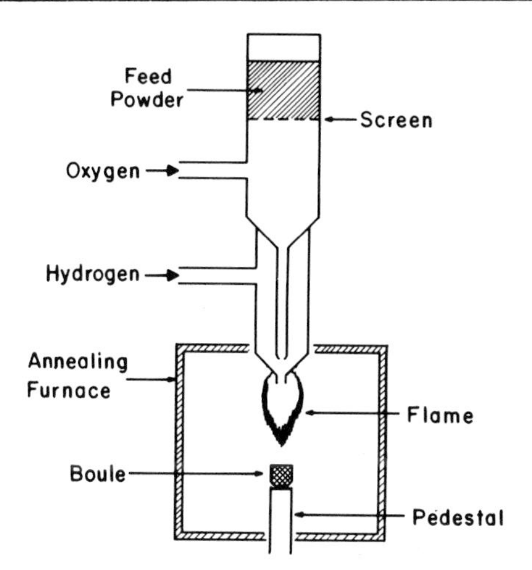

FIG. 4. Verneuil growth.

(~1 μ particle size, essentially free-flowing) is slowly introduced into a specially designed oxygen-hydrogen torch. The flame is directed at the tip of a ceramic pedestal which serves as the crystal growth platform. The fine feed powder is melted as it passes through the flame and deposits on the pedestal, which is slowly lowered as the crystal grows, usually into an after-heater to prevent the crystal from cracking. Higher temperatures (to 4000°K) and an inert environment can be attained by using an argon plasma generated by coupling the flowing gas to an rf induction generator.

Although the Verneuil technique can be a rapid method of growing certain single crystals, it has a great disadvantage in that high temperature gradients occur in the area of the growing crystal. Furthermore, the composition of the flame atmosphere (which can be reducing even in an oxy-hydrogen torch) limits the type of materials that can be grown.

A further critical point is the nature of the feed material; it must be of high purity, low density, and very small particle size. A common method of attaining small particle size is to prepare the feed material by the thermal decomposition (at a temperature below that which would cause sintering) of double salts, e.g., $M_2^{3+}(NH_4)_2(SO_4)_4 \cdot 24H_2O$, or metal oxalates. It is essential to have complete decomposition of these starting materials and to keep the feed powder dry since water or gas evolution during melting or growth will cause bubble formation in the grown crystals.

Calcium titanate [L. Merker, J. Am. Ceram. Soc., 45, 366 (1962)]. The feed material is prepared by mixing solutions of $TiCl_4$, $CaCl_2$, and oxalic acid in the molar proportions 1.0/1.4/4.0. The calcium titanyl oxalate double salt is precipitated, dried, and calcined at 1000 °C to form calcium titanate of stoichiometric composition. This material, screened through a 100-mesh sieve, was determined to have particle size in the range 0.5-2.0μ. Growth was in a burner of the type illustrated in Fig. 4. It was found important to contain the boule in a furnace kept at 1400 °C during growth. After completion of growth, the furnace was raised to 1550 °C, the burner was shut off, and the crystal annealed for 48 hr before cooling at 30°/hr to room temperature. Reduced titanium in the grown crystal gave it a bluish color, which was removed by heating at 1700 °C in air for 6 hr, then cooling at 50°/hr to room temperature. Crystal boules up to 25 mm long and 12 mm in diameter were grown.

III. GROWTH FROM SOLUTION

A. Flux Growth

High temperature flux growth is completely analogous to the growth of water-soluble materials from aqueous solution. Sodium chloride crystals are easily prepared by making a saturated solution of salt in water at a particular temperature (say 80 °C) and then letting the solution cool slowly. As the solution cools, the solubility product of sodium chloride (which decreases with decreasing temperature) is exceeded and crystals precipitate from solution. If care is taken to exclude any foreign matter from the system (which might act as nuclei for precipitation) and the cooling rate is carefully controlled, a few large single crystals of sodium chloride will result. In practice, one suspends a seed crystal in the solution to nucleate crystal growth (suppressing spontaneous nucleation and allowing the growth of one large crystal) and carefully stirs the solution to avoid concentration gradients in the system. Alternately, one could hold the solution at a constant temperature and allow the solvent (water, in this case) to evaporate, thereby exceeding the solubility product of the salt at that particular temperature. The net result is the same, crystal growth from a flux system.

High temperature flux growth utilizes the same principles. A relatively low-melting solvent is saturated with the components of the compound desired, the system held for some hours at a temperature above the liquidus to insure complete dissolution, and then cooled slowly (at about 0.5-10°C/hr). Cooling rates are kept as low as possible because the solubility of the compound being grown usually drops rather rapidly with a decrease in temperature which, of course, would enhance the probability of spontaneous nucleation.

Many variations on this basic method of flux growth suggest themselves. As in the case of growth from aqueous solution, the isothermal flux evaporation method can be used with a flux that has a suitably high vapor pressure well below its boiling point. There are certain advantages to this method due to the fact that the process is isothermal, therefore prompting a high degree of crystal perfection. If a compound is stable only over a narrow temperature range, it may be absolutely necessary to use the isothermal technique. On the other hand, the evaporating flux is usually quite corrosive and can attack the inside of the furnace.

The use of a temperature gradient is another modification of the flux growth method. A cylindrical crucible is filled with a saturated solution (saturated at the soak temperature) of flux and solute. Excess compound is placed in the bottom of the crucible and acts as a nutrient material. The temperature is raised to a point where the solubility of

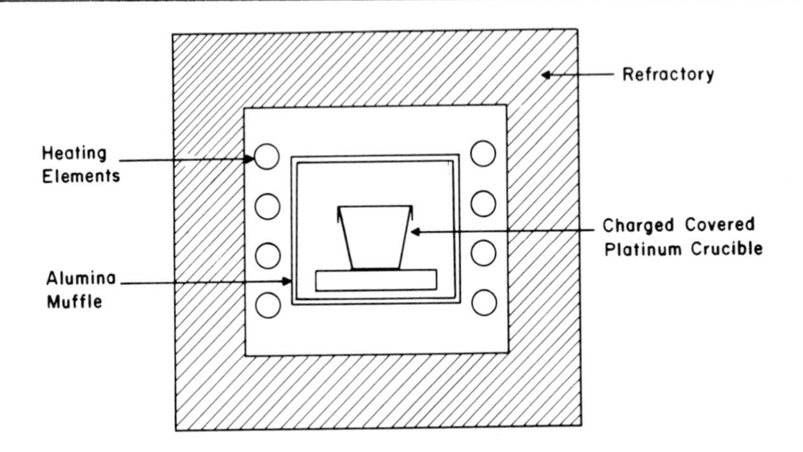

FIG. 5. Flux growth.

the material is such that excess nutrient remains at the bottom of the crucible. A temperature gradient is set up by the use of a two-zone furnace (a baffle is placed within the crucible between the two zones as in the Bridgman-Stockbarger technique), and material is transported from the hot zone to the cooler zone by convection. Since the solubility product of the material is exceeded in the cool zone, crystal growth will occur, again on a seed crystal suspended in that region. As in the case of isothermal flux evaporation, the temperature gradient method has the advantage that crystals are grown under essentially isothermal conditions. In addition, since the solution surrounding the growing crystal is continuously being replenished with solute material, the concentration of nutrient material around the growing crystal remains relatively constant.

As an example of the slow-cooling technique, the flux growth of yttrium aluminum garnet (YAG) will be described. Although YAG melts congruently at 1930°C, crystals can be grown well below this temperature by the flux method.

Yttrium aluminum garnet [R. A. Lefever et al., J. Appl. Phys., 32, 962 (1961)]. 3.4 mole % Y_2O_3 and 7.0 mole % Al_2O_3 are mixed with a flux of 41.5 mole % PbO and 48.1 mole % PbF_2 in a tightly covered 50-ml platinum crucible and placed into a resistance-heated furnace (see Fig. 5). The melt is held at 1150-1160 °C for 24 hr, cooled at 4 °C/hr to about 750 °C, and removed from the furnace. The crystals, ranging in size from 3 to 13 mm in diameter (1-1.5 g maximum weight), are separated by dissolving the flux in hot dilute nitric acid. Average yield is 60-70% based on the original Y_2O_3 content of the melt.

Magnesium aluminate ($MgAl_2O_4$) can be grown by flux evaporation from a PbF_2 flux. An evaporation technique is necessary since MgF_2 and α-Al_2O_3 precipitate from the melt at temperatures below 1150 °C.

Magnesium aluminate [J. D. C. Wood and E. A. D. White, J. Cryst. Growth, 3/4, 480 (1968)]. A 500-ml platinum crucible is filled with 80.6 g MgO (15.7 mole %), 204.0 g Al_2O_3 (15.7 mole %), 2100.0 g PbF_2 (67.4 mole %), and 10.0 g B_2O_3 (1.1 mole %). The crucible is sealed with a platinum cover containing holes to control the rate of flux evaporation and placed in a resistance-heated furnace which is brought to a temperature of 1250 °C over an 8-hr period. The flux is allowed to evaporate over a period of 10-15 days, after which the crucible is removed from the furnace and any remaining flux removed by boiling in dilute nitric acid. Highly perfect crystals up to 1 cm in size are obtained.

The modification of seeding and/or the use of a thermal gradient can be used for the growth of most of those crystals that can be grown by fluxed melt techniques (see above). In summary, crystallization from polycomponent molten salt systems is perhaps the most useful of all crystal growth techniques. A large variety of fluxes have been found useful for crystal growth; among them are lead and bismuth oxides, fluorides and borates, lithium, sodium, and potassium molybdates and tungstates, lead and bismuth vanadates, and sodium and barium borates.

B. Hydrothermal Growth

Hydrothermal crystal growth is the high temperature aqueous analog of temperature gradient flux growth. In order to carry out crystal growth in an aqueous system in the range of 400-800 °C, the entire system must be contained in an autoclave able to withstand internal pressures of 1000-3000 atm, since it is necessary to increase the pressure in order to work with aqueous systems above the normal boiling point of water.

An important advantage of working in an aqueous system at high temperature and pressure is that the ion product of water increases, so that at 600 °C and 2000 atm it is about 10^5 times greater than at room temperature and normal atmospheric pressure. This means that many materials that are not soluble in water under ordinary conditions become sufficiently soluble under hydrothermal conditions for crystal growth to occur. Normally, hydrothermal growth is carried out in a special steel bomb with an alkaline (NaOH or Na_2CO_3) environment to further enhance the solubility of the charge. Large crystals of quartz are grown commercially in this manner, nutrient being dissolved in the hotter part of the system and then transported to (by convection) and deposited on seed crystals mounted in the cooler part of the system. The example below was chosen to illustrate both a simple technique for hydrothermal growth and to show that growth can be carried out in a highly acidic medium.

Gold [H. Rau and A. Rabenau, J. Cryst. Growth, 3/4, 417 (1968)]. 10 g of gold sheet are placed in a quartz tube (15 mm diameter, ~10 ml volume) which is then filled to 65% of capacity with 10 M hydroiodic acid and sealed. The ampoule is placed into an autoclave (Fig. 6) with an amount of dry ice calculated to make the pressure outside greater than that inside the ampoule at the growth temperature to prevent explosion of the quartz tube. (Such tubes can withstand pressure differences if the higher pressure acts from the outside.) The bomb is placed inside a two-zone furnace; a gradient of 480 to 500 °C causes

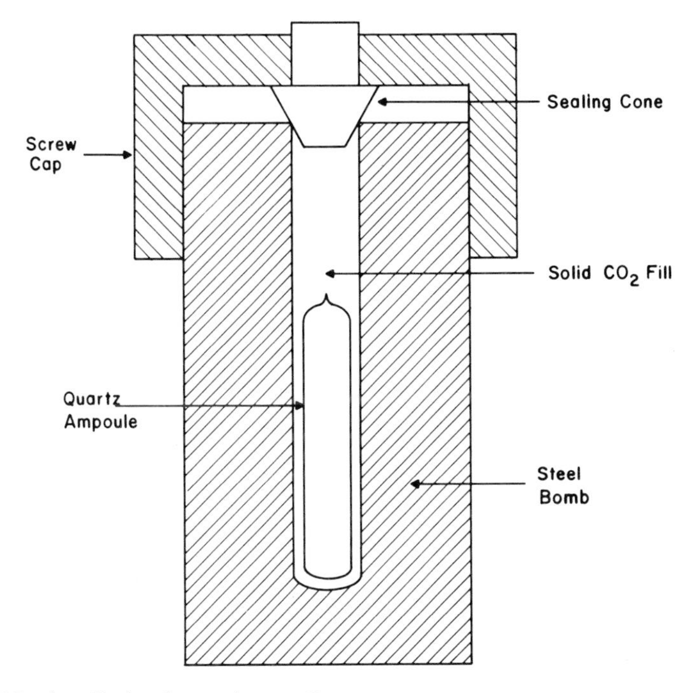

FIG. 6. Hydrothermal growth.

transport and growth of gold crystals in the hotter part of the ampoule, probably through the formation of an intermediate complex which dissociates to form elemental gold at the higher temperature.

C. Electrolytic Growth

An unusual technique for crystal growth is the use of an electric current to modify a flux melt system to cause the controlled crystallization of a particular compound, either by the electrochemical decomposition of the flux itself or of a compound dissolved in the flux solvent. Since the process is carried out isothermally and the rate of crystallization can be controlled by varying the applied current, large sound crystals can be grown. It has been found that low current densities greatly increase the ultimate size and perfection of the crystals since the reaction is carried out more slowly.

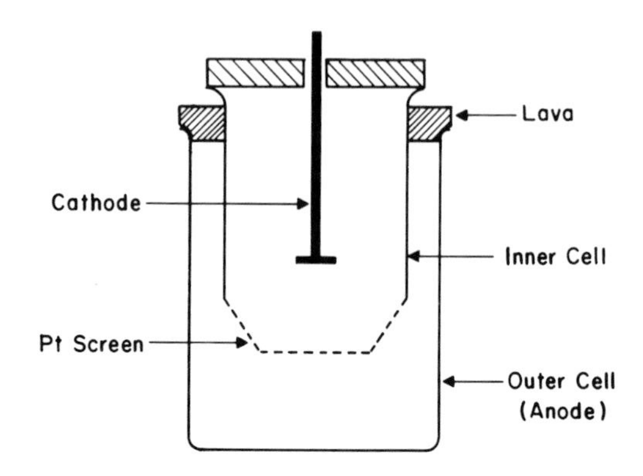

FIG. 7. Electrolytic growth.

Electrolytic growth has been used extensively for the preparation of the so-called tungsten "bronzes" (actually ternary nonstoichiometric tungsten oxides), certain binary and ternary transition metal oxides, and certain transition metal borides, carbides, and phosphides.

An example of the electrolytic decomposition of the flux itself is the growth of MoO_2 crystals by the decomposition of $K_2Mo_2O_7$ at 600°C according to the following reaction:

$$K_2Mo_2O_7(\ell) \rightarrow K_2MoO_4(\ell) + MoO_2(c) + \frac{1}{2}O_2(g) \quad .$$

As the reaction proceeds the composition of the melt is constantly changing; therefore, a high ratio of flux to product yield is used to prevent a great change in chemical potentials. Our detailed example will show the electrolytic decomposition of a compound dissolved in a flux solvent.

Cobalt vanadium spinel [D. B. Rogers, A. Ferretti, and W. Kunnman, J. Phys. Chem. Solids, 27, 1445 (1966)]. Single crystals of composition $Co_{1+\delta}V_{2-\delta}O_4$ are grown from a flux of composition $(1-x)Na_2W_2O_7$-xNa_2WO_4 by the reduction of Co_3O_4 and V_2O_5 dissolved in the flux. By varying the basicity of the flux (the parameter x) the composition of the spinel crystals can be controlled between the values of $0.11 < \delta < 0.56$. Typically, a mixture of 0.055 mole Co_3O_4, 0.165 mole V_2O_5, 0.156 mole WO_3, and

0.624 mole Na_2WO_4 (x = 0.75) is placed in a platinum cell (Fig. 7) and the temperature raised to 900 °C to melt the entire charge. The cell itself is used as the anode while a platinum disk (1 cm^2) is used as the cathode. The initial electrolysis (to reduce the V^{5+}) is carried out at 300 mA/cm^2 for three days, followed by continued electrolysis at 40 mA/cm^2 for 4-7 days to reduce the cobalt and form the spinel crystals. After cooling crystals up to 4 mm (1 cm with seeding) are separated from the flux by washing with hot NaOH solution.

D. Gel Growth

The gel method of crystal growth is an unusually simple technique for the controlled crystallization of certain compounds. An inert gel, such as sodium silicate (waterglass), is used as a matrix through which the components of the desired compound are diffused. Since the rate of diffusion can be controlled through alteration of the gel matrix, single-crystal growth can be attained over a broad range of conditions. Two components can be allowed to diffuse toward each other through a gel-filled U-tube, or one of the reactants can be incorporated into the gel itself and the other diffused into the gel to react. Crystals of compounds which are sparingly soluble in aqueous systems are good candidates for gel growth.

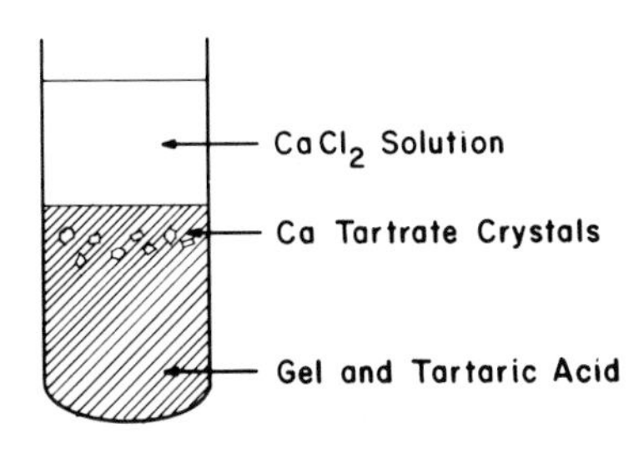

FIG. 8. Gel growth.

Calcium tartrate [H. K. Henisch, J. Dennis, and J. I. Hanoka, J. Phys. Chem. Solids, 26, 493 (1965)]. A solution of sodium silicate (sg 1.06 g/cm^3) is mixed with 0.8-2.0 N tartaric acid and allowed to gel in a test tube immersed in a constant temperature bath (Fig. 8). A solution of 1 N calcium chloride is floated on top of the gel. Crystals of calcium tartrate (up to about 8 mm average size) grow within one week. By controlling the density of the gel, the concentration of reactants, and the temperature of the bath, the size and rate of crystal growth can be optimized.

IV. GROWTH FROM THE VAPOR PHASE

In addition to sublimation, the simplest form of vapor phase crystal growth, one should consider as methods of growth from the vapor phase the techniques of chemical vapor transport, vapor phase reaction (or decomposition) of gaseous species, and the so-called vapor-liquid-solid (VLS) mechanism of growth.

Vapor phase techniques are rather limited in their application to crystal growth, since (aside from direct sublimation) a volatile starting material or intermediate must be available. Except for monocomponent systems, there is invariably the possibility of contaminating the crystal with the transporting material. However, large sound crystals of many types of materials can be grown by vapor phase techniques.

A. Sublimation

Crystal growth by sublimation may be carried out in either a static or a flowing gas system. In a static system, the material to be sublimed is sealed in a suitable tube (usually quartz and normally under vacuum or a slight partial pressure of an inert gas), which is placed in a furnace with a thermal gradient; sublimation takes place in the hotter portion of the furnace and crystal growth in the cooler portion. In a flow system

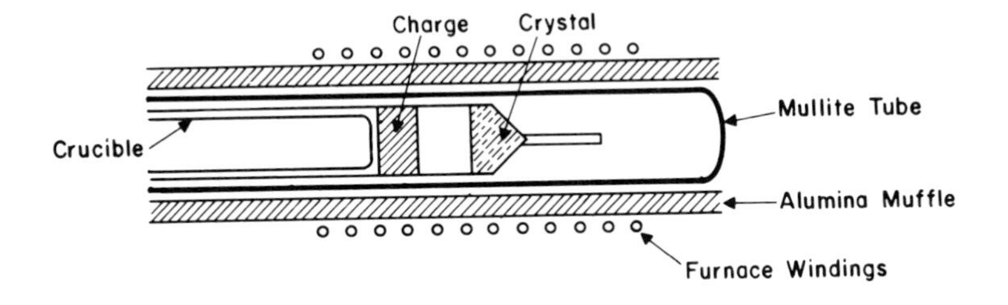

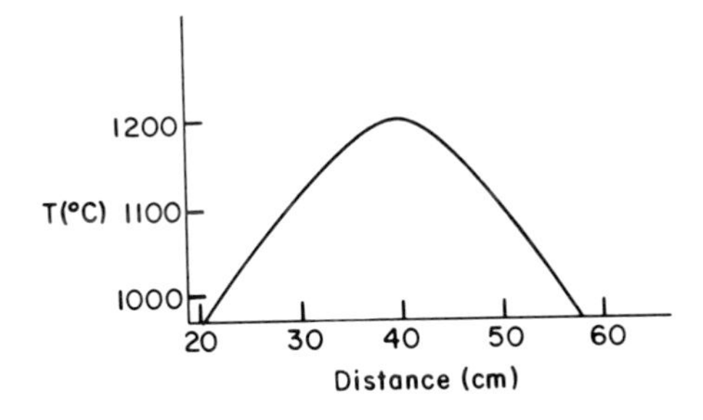

FIG. 9. Sublimation.

an inert gas is passed through the tube over the material in the hot zone, carrying the gaseous species into the cooler zone where it deposits. Since the sublimation process is inherently a purification process, crystals grown in this manner are usually of extremely high purity. Crystal growth by sublimation is obviously applicable to any material that has a reasonably high vapor pressure at temperatures up to about 1000°C. In fact, it is a standard method for purification and crystal growth of many organic compounds.

Cadmium sulfide [W. W. Piper and S. J. Polich, J. Appl. Phys., 32, 1278 (1961)]. The furnace and crucible arrangement are shown in Fig. 9 along with a typical furnace temperature profile. A charge of CdS is placed in the crucible as shown in Fig. 9 and a close-fitting closed quartz tube is inserted. The mullite tube is evacuated and slowly heated to 500°C, baked out for 1 hr, and back-filled with 1 atm of argon. The furnace is heated to the operating temperature (1200°C at the maximum) with the crucible tip initially near the maximum temperature and with the pressure maintained at 1 atm. The entire mechanism is slowly moved (0.3-1.5 mm/hr) so that the tip enters a cooler region. Escaping vapor seals off the crucible by condensing around the closed tube. As nucleation occurs at the conical tip of the crucible, a single crystal up to 13 mm in diameter is grown.

B. Chemical Vapor Transport

The process of chemical vapor transport occurs in chemical reactions in which a solid phase reacts with a gas to form vapor phase products. These vapor phase products then undergo the reverse reaction at a different place in the reaction system, resulting in the reformation of the original solid phase. A substance is said to be chemically transported (not sublimed) if it undergoes a reaction such as

$$aA_{(s)} + bB_{(g)} \rightarrow cC_{(g)}$$

$$T_1 \qquad\qquad T_2 \qquad\qquad \text{where } T_2 > T_1 \quad .$$

Note that the reaction will support transport only if solid material is present on just one side of the chemical reaction.

The sign of the enthalpy change for the reaction as written ($\Delta H°$) determines the direction of transport. If the reaction is exothermic, transport is from T_1 to T_2; if it is endothermic, transport is from T_2 to T_1; if $\Delta H° = 0$, there is no transport. By consulting available tables of thermodynamic data, the optimum temperature for transport should be calculable. Obviously, a necessary criterion for transport is the formation of a gaseous compound.

In addition to metals, single crystals of many binary and ternary chalcogenides (including oxides) and pnictides have been grown by chemical vapor transport. The most useful gases used to form the gaseous intermediate are the halogens and hydrogen halides.

Calcium niobate [F. Emmenegger, J. Cryst. Growth, 2, 109 (1968)]. $CaNb_2O_6$ is prepared by mixing stoichiometric amounts of $CaCO_3$ and

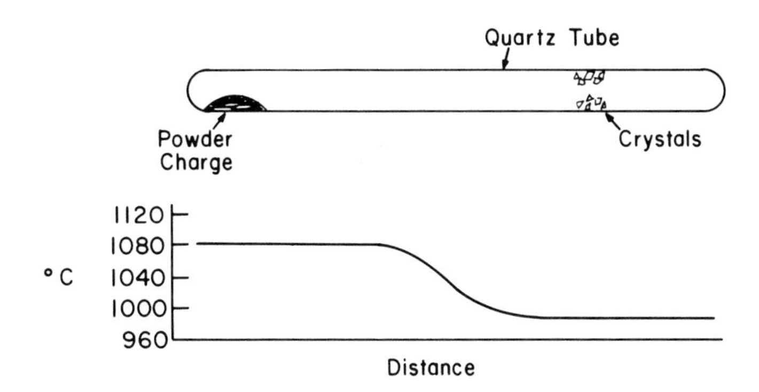

FIG. 10. Chemical vapor transport and a typical temperature profile.

Nb_2O_3 and firing at 1300°C in a platinum crucible. 1 g $CaNb_2O_6$ is placed in one end of a quartz ampoule (110 mm long, 17 mm diameter), evacuated, and back-filled with Cl_2 at 1 atm pressure (room temperature). The sealed ampoule (Fig. 10) is placed in a two-zone furnace with the polycrystalline material at a temperature of 1020°C and the growth zone at 980°C. Within two weeks, crystals of typical size $1 \times 0.5 \times 0.2\ mm^3$ have grown at the cooler end of the ampoule. A simplified equation for the transport reaction can be given as

$$CaNb_2O_6(c) + 4Cl_2(g) \rightleftarrows 2NbOCl_2(g) + CaCl_2(g) + O_2(g) \quad .$$

C. Vapor Decomposition

The method of vapor decomposition utilizes the irreversible decomposition of a gaseous compound by either chemical (e.g., hydrolysis) or thermal means to achieve crystal growth. For example, a gaseous metal halide can be reduced to the metal in a hydrogen atmosphere or on a hot wire (~1500°C) under controlled conditions to give reasonable crystal growth. Crystals grown by this technique are usually of small size but can be of high perfection. Many transition metal oxides can be grown by the hydrolysis of their gaseous halides at high temperatures, and crystals of certain refractory metals can be formed by halide decomposition on a hot wire.

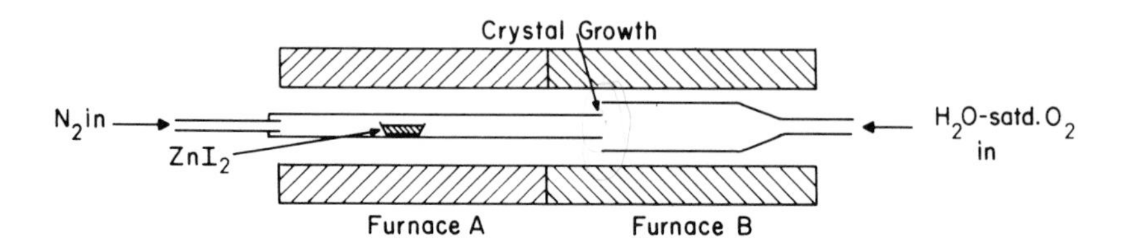

FIG. 11. Vapor decomposition.

Zinc oxide [M. Hirose, Japan. J. Appl. Phys., 10, 401 (1971)]. 20 g ZnI_2 are placed in a quartz tube (24 mm diameter) with a stream of N_2 (flow rate 700 cm^3/min) passing over it. With this tube held in a furnace at 420 °C (Fig. 11), water-saturated oxygen (120 cm^3/min) is passed into an adjoining furnace held at a temperature between 970 and 1200 °C. Crystal growth takes place at the end of the quartz tube holding the ZnI_2 charge. Needles up to 15 mm long or plates up to 8 mm^2 are formed either by hydrolysis (in the temperature range 970-1020 °C) or by oxidation (between 1150 and 1200 °C).

D. Vapor-Liquid-Solid Growth

Vapor-liquid-solid (VLS) crystal growth is a recently discovered mechanism by which filaments (whiskers) of certain materials can be grown from the vapor phase via an impurity present in the liquid state. By this mechanism an impurity, such as gold in the case of the growth of silicon, is placed on a crystalline silicon substrate. Upon heating to 900 °C, the impurity and substrate form a droplet of molten alloy which serves as a sink for the deposition of silicon from the vapor (a flowing stream of $SiCl_4/H_2$). The droplet becomes supersaturated with silicon, which then crystallizes on the substrate. Thus, a crystal filament grows beneath the droplet which is replenished in silicon from the vapor phase. The point is made that the liquid-forming impurity does not have to be in contact with a substrate in order for needles to grow by the VLS mechanism. As long as the liquid droplet is formed and the constituent from the vapor dissolves in it and nucleates, needle crystals

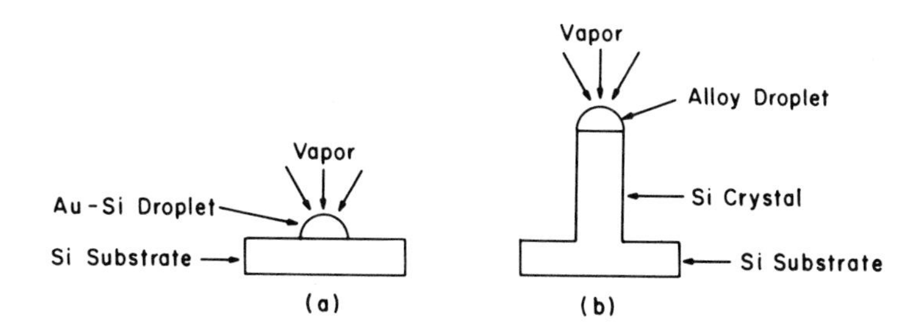

FIG. 12. Vapor-liquid-solid (VLS) mechanism: (a) start of growth, (b) growth of crystal.

will grow. Aside from silicon, only a handful of materials have been grown by this rather specialized technique.

Silicon [R. S. Wagner and W. C. Ellis, Trans. Met. Soc. AIME, 233, 1053 (1965)]. A small piece of gold is placed on the {111} face of a silicon wafer which is then inserted into a quartz tube placed inside a furnace. With the temperature raised to 900°C a liquid gold-silicon alloy forms. Deposition is carried out in a hydrogen flow of 1000 cm^3/min and a $SiCl_4/H_2$ mole ratio of 0.02. Whiskers of millimeter length are grown in the order of 20 hr. Figure 12 schematically illustrates the VLS mechanism.

V. CODA

This short summary of the techniques of crystal growth should serve only as an outline and guide to the methods available to a researcher. It is certainly evident that the permutations of these individual techniques are limitless and that the choice of a particular method must be tailored to the particular compound to be grown. Good luck!

PART IV

SOME IMPORTANT EXAMPLES

Chapter 13

POLYMERIC MATERIALS

Bernhard Wunderlich

Department of Chemistry
Rensselaer Polytechnic Institute
Troy, New York

I. INTRODUCTION TO LINEAR HIGH POLYMERS

The name polymer implies a molecule made of many (πoλυσ) parts (με ρoσ). A small molecule, for example ethylene ($CH_2{=}CH_2$), would accordingly be called a monomer. Dimers, trimers, tetramers would have twice, three times, or four times as many carbon and hydrogen atoms for the molecule. To define the limit above which a polymer can be called a high polymer is the somewhat arbitrary decision as to where "very many" starts. A suitable limit is perhaps a molecule with 1000 atoms, because from this size on properties change only very little with further increase.

In the light of this definition of high polymers, certainly all solids are high polymers. Even a just barely bisible speck of dust, 10^{-3} cm on a side, contains more than a trillion (10^{12}) atoms, and since the atoms in a solid have fixed positions, they form a giant three-dimensional high polymer molecule, which stretches from surface to surface.

The bonds which hold the atoms of a solid in their respective positions are often not the same in all directions. Roughly all bonds can be classified as either strong or weak. The strong bonds are the covalent, ionic, and metallic bonds of about 20- to 150-kcal bond energy. The weak bonds are the van der Waals, dipole, and hydrogen bonds; they cover the range up to about 7 kcal/mole.

As long as strong bonds are formed in all directions of space, a tight network arises. The resulting solid is physically strong and has a high glass-transition or melting temperature. The solid of such a structure could be called a three-dimensional high polymer. Examples are silicate glasses, diamond, quartz crystal, and quartz glass, as well as most salts and metals. At least in the crystalline state, there are examples where strong bonds are limited to two dimensions. The resulting layers of strongly bound atoms are held together by much weaker bonds. The best known examples of such a structure are graphite and mica. All linear high polymers in the crystalline and glassy solid state have strong bonds only in one dimension. Finally it is possible to have only isolated areas of strong bonds, or none at all. Examples of such structures are the solids of noble gases, NH_3, CO_2, and CH_4. In general, these zero dimensionally strong bonded solids are soft and have a low glass transition or melting temperature.

To get an idea of the multitude of linear high polymers, the theories of organic, inorganic, and biological chemistry may be used as guides. Observing valence restrictions and bond angle conditions, large one-dimensional molecules may be constructed with paper and pencil out of many atoms. It is the job of the synthetic polymer scientist to try to actually synthesize these molecules. The physical chemist, ideally, tries to predict before synthesis which would be the most interesting polymer, or, in practice, tries, after the molecule has been made, to understand its properties. Since most high polymers are finally used in the solid state, the study of the solid state is the ultimate result looked for in the science of polymeric materials.

There are many reactions which lead to linear high polymers. An example from the field of inorganic chemistry is the formation of μ-sulfur. On heating molten sulfur, μ-sulfur forms at about 160° C. The liquid sulfur at low temperature is called λ-sulfur and is on a molecular scale made up of 8-membered rings of sulfur atoms. At 160° C these 8-membered rings become unstable, break up, and join with others to form very long chains of μ-sulfur. The physical properties change in a parallel fashion to the change in chemical structure, going from a mobile liquid to a highly viscous melt, despite the fact that the temperature has been raised to accomplish polymerization. Expressed in a chemical formula the ring opening and polymerization are shown below:

$$X \; [S_8 \text{ ring}] \xrightarrow{160^\circ C} -(S-S-S-S-S-S-S-S)-_{X}, \qquad (1)$$

X molecules of λ-sulfur $\longrightarrow$ 1 molecule μ-sulfur with 8X S-atoms.

Two important types of organic reactions leading to linear high polymers are the chain reaction and the step reaction. An example of a chain reaction is the synthesis of polyethylene from ethylene using a peroxide radical catalyst:

$$\mathrm{R:\ddot{\underset{..}{O}}\cdot} \quad + \quad \begin{matrix} \mathrm{H} & & \mathrm{H} \\ \ddot{\underset{..}{\mathrm{C}}} & :: & \ddot{\underset{..}{\mathrm{C}}} \\ \mathrm{H} & & \mathrm{H} \end{matrix} \longrightarrow \mathrm{R}:\ddot{\underset{..}{\mathrm{O}}}:\begin{matrix} \mathrm{H} & & \mathrm{H} \\ \ddot{\underset{..}{\mathrm{C}}} & : & \ddot{\underset{..}{\mathrm{C}}}\cdot \\ \mathrm{H} & & \mathrm{H} \end{matrix} \qquad \text{(initiation)}, \tag{2}$$

peroxide radical — ethylene — initiated polymer chain

$$\mathrm{R}:\ddot{\underset{..}{\mathrm{O}}}:\begin{matrix} \mathrm{H} & & \mathrm{H} \\ \ddot{\underset{..}{\mathrm{C}}} & : & \ddot{\underset{..}{\mathrm{C}}}\cdot \\ \mathrm{H} & & \mathrm{H} \end{matrix} \quad + \quad \begin{matrix} \mathrm{H} & & \mathrm{H} \\ \ddot{\underset{..}{\mathrm{C}}} & :: & \ddot{\underset{..}{\mathrm{C}}} \\ \mathrm{H} & & \mathrm{H} \end{matrix} \longrightarrow \mathrm{R}:\ddot{\underset{..}{\mathrm{O}}}:\begin{matrix} \mathrm{H} & & \mathrm{H} & & \mathrm{H} & & \mathrm{H} \\ \ddot{\underset{..}{\mathrm{C}}} & : & \ddot{\underset{..}{\mathrm{C}}} & : & \ddot{\underset{..}{\mathrm{C}}} & : & \ddot{\underset{..}{\mathrm{C}}}\cdot \\ \mathrm{H} & & \mathrm{H} & & \mathrm{H} & & \mathrm{H} \end{matrix} \quad \text{etc.}$$

initiated polymer chain — ethylene monomer — lengthened polymer chain

(propagation or growth of the chain). (3)

After X monomer atoms have been added, the reaction may come to a halt because, for example, of a radical $\cdot$H abstraction from the solvent:

$$\mathrm{R}:\ddot{\underset{..}{\mathrm{O}}}:\left(\begin{matrix} \mathrm{H} & & \mathrm{H} \\ \ddot{\underset{..}{\mathrm{C}}} & : & \ddot{\underset{..}{\mathrm{C}}}: \\ \mathrm{H} & & \mathrm{H} \end{matrix}\right)_{X-1} \begin{matrix} \mathrm{H} & & \mathrm{H} \\ \ddot{\underset{..}{\mathrm{C}}} & : & \ddot{\underset{..}{\mathrm{C}}}\cdot \\ \mathrm{H} & & \mathrm{H} \end{matrix} \quad + \quad \mathrm{H}\cdot \longrightarrow \mathrm{R}:\ddot{\underset{..}{\mathrm{O}}}:\left(\begin{matrix} \mathrm{H} & & \mathrm{H} \\ \ddot{\underset{..}{\mathrm{C}}} & : & \ddot{\underset{..}{\mathrm{C}}}: \\ \mathrm{H} & & \mathrm{H} \end{matrix}\right)_{X}\mathrm{H}$$

growing polymer chain — radical from solvent — completed polymer chain

(termination step). (4)

The final polyethylene molecule would then have the following chemical formula:

$$\mathrm{R{-}O{-}(CH_2{-}CH_2{-})_X H}\,. \tag{5}$$

The definition of the polymer molecule does not describe the molecule fully in this case, because the two ends of the molecule make it deviate a little from an exact multiple. For large X this deviation is always neglected. X designates the degree of polymerization and the part of the molecule in parentheses is called the repeating unit (CH_2-CH_2-).

An example of a step reaction is the polymerization of an ω-amino acid to a nylon. The ω-amino acids are bifunctional molecules and as such can react on two sides. The reaction between two bifunctional molecules leads to a dimer:

$$2HO-\overset{O}{\overset{\|}{C}}(-CH_2)_a-NH_2 \longrightarrow HO-\overset{O}{\overset{\|}{C}}(-CH_2)_a-NH-\overset{O}{\overset{\|}{C}}(-CH_2)_a-NH_2+H_2O \quad (6)$$

The dimer is still a bifunctional molecule and can thus react with two other monomer, dimer, trimer, or even polymer molecules to form an even bigger molecule which still remains bifunctional and active for further reaction. This continued ability to increase size is characteristic of polymers formed by step reactions. The molecular weight increases long after most of the monomer is used up. The final product, an (a+1) nylon, can be represented as

$$HO-[\overset{O}{\overset{\|}{C}}(-CH_2)_a-NH-]_{X-1}\overset{O}{\overset{\|}{C}}(-CH_2)_a-NH_2\,. \quad (7)$$

A widely used commercial polymer of this series is 6-nylon, which has five CH_2-groups per repeating unit. A polymerization of importance in nature is, for example, the formation of cellulose from glucose:

$$2X\ \text{glucose (ring: } CH_2OH,\ O,\ OH,\ HO,\ OH,\ OH) \longrightarrow H-[O-\text{(ring: } CH_2OH,\ O,\ OH,\ OH)-O-\text{(ring: } OH,\ OH,\ O,\ CH_2OH)]_X-OH + (2X-1)H_2O \quad (8)$$

Consideration of the polymerizations just described indicates that most of these reactions yield what a chemist would call impure

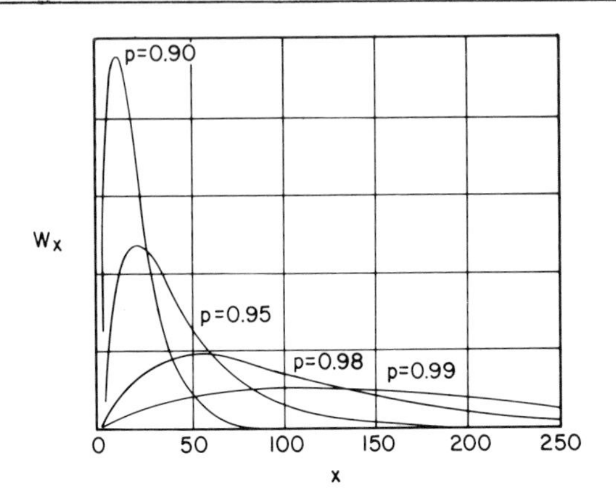

FIG. 1. Weight fraction w_x of molecules with x repeating units. Different curves represent different degrees of completion of the step reaction. For p = 0.90, 90% of all of the functional groups have reacted; analogously, for p = 0.95, 95%, etc. In mathematical terms, $w_x = x(1 - p)^2 p^{x-1}$.

compounds. The length of any particular polymer chain is determined by chance. In any reaction mixture there is a wide distribution of different molecular lengths. Figure 1 represents typical molecular weight distributions for step reactions which go to various degrees of completion. The weight fraction of polymer is plotted vs the number of repeating units in the molecule.

Two things can be learned from Fig. 1. First, in order to get to high molecular weight averages, the reaction must go almost quantitatively to completion; second, there will always be some very short molecules in the reaction mixture. Since many properties of the polymer depend on the molecular weight distribution, polymer samples with identical distributions of molecular weights are necessary to achieve reproducibility in experimentation and application.

Monodisperse polymers consist of chains of only one molecular weight. Examples of monodisperse polymers are known among biological polymers. Some proteins and nucleic acids consist of molecules which are exact replicas of each other. For example, each molecule of ribonuclease, a protein, is $C_{575}H_{901}O_{193}N_{171}S_{12}$ with a molecular weight of 13,682. Monodisperse synthetic polymers can be approached by using certain catalysts which do not allow random termination of the growth of polymer chains. Separation of a broad molecular weight distribution into sharper molecular weight fractions is possible, but quite cumbersome.

Linear polymers may be compared to long strings. One of the characteristics of many such molecules is their flexibility. The flexibility is caused by rotation around some or all of the backbone bonds. Examples of quite flexible chains are polyethylene and sulfur [Eqs. (1) - (5)]. Poly(tetrafluoroethylene) which has a repeating unit of (CF_2CF_2)— is less flexible because of the larger size of the F, bound to the carbon. On rotation the fluorine atoms on neighboring atoms interfere and thus stiffen the molecule. An example of a rigid linear polymer is polysulfone, formed on reaction by the flexible trans-polybutadiene with SO_2:

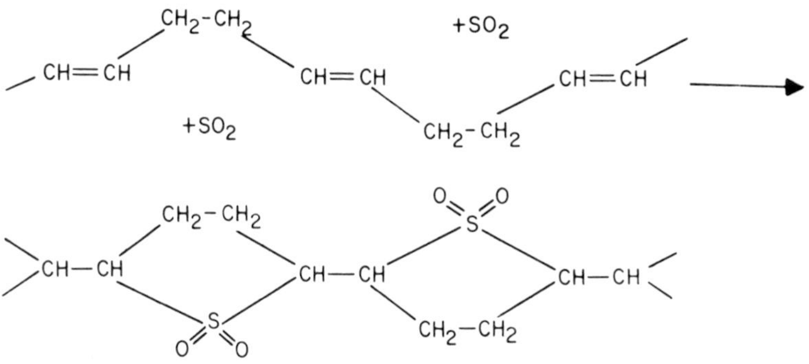

CH=CH—CH_2-CH_2—CH=CH—CH_2-CH_2—CH=CH + SO_2 + SO_2 ⟶

CH—CH(CH_2-CH_2)(S(O_2))CH—CH(CH_2-CH_2)(S(O_2))CH—CH (9)

Polysulfone

Another example of a rigid linear polymer is polyphenylene:

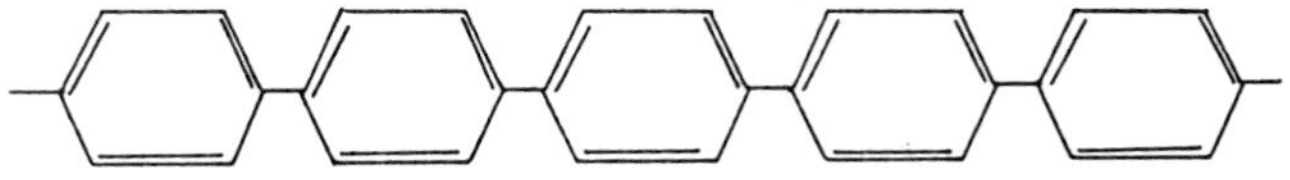

—C_6H_4—C_6H_4—C_6H_4—C_6H_4—C_6H_4— (10)

Polyphenylene

The polysulfone and polyphenylene can be considered rigid because rotation around the remaining backbone bonds does not flex the molecules.

Besides different atoms in a repeating unit of linear high polymers, the same atoms can form different isomers, multiplying the number of possible polymer molecules. Positional and structural isomers on polystyrene and polybutadiene are illustrated below. The most easily formed isomer of polystyrene is called head to tail:

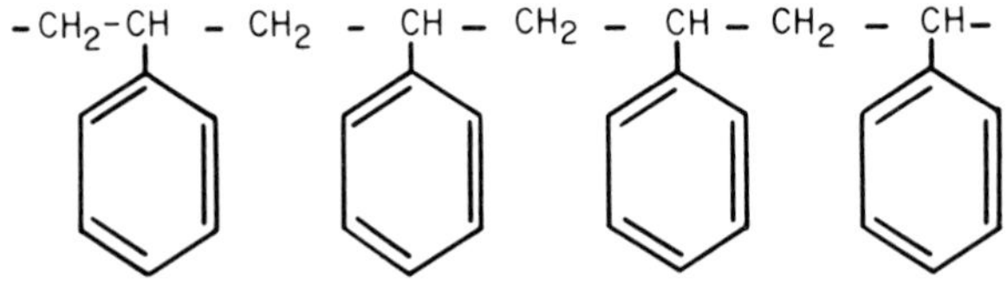

$$-CH_2-\underset{C_6H_5}{CH}-CH_2-\underset{C_6H_5}{CH}-CH_2-\underset{C_6H_5}{CH}-CH_2-\underset{C_6H_5}{CH}- \tag{11}$$

The head-to-head (or tail-to-tail) form, which can only be obtained with great difficulty, is the other positional isomer of the polystyrene molecule:

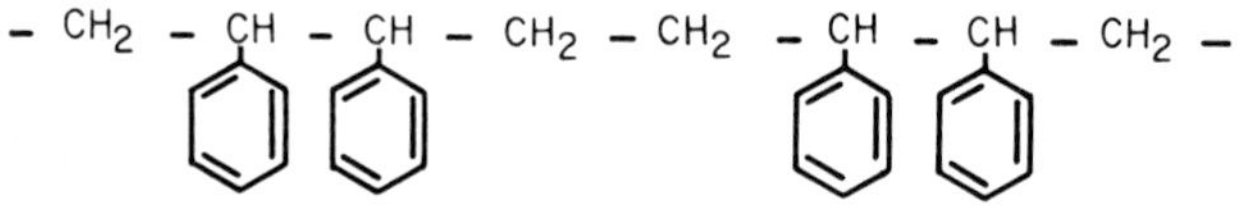

$$-CH_2-\underset{C_6H_5}{CH}-\underset{C_6H_5}{CH}-CH_2-CH_2-\underset{C_6H_5}{CH}-\underset{C_6H_5}{CH}-CH_2- \tag{12}$$

A third isomer would be an irregular mixture of both forms. In ordinary polystyrene the molecules consist of the head-to-tail arrangement with infrequently misfitted repeating units.

The structural isomers of polybutadiene are listed below. In this case the 1,4 isomer is the most common:

$$\text{monomer butadiene: } \overset{1}{C}H_2=\overset{2}{C}H-\overset{3}{C}H=\overset{4}{C}H_2 \tag{13}$$

$$\text{1,4 polybutadiene: } -\overset{1}{C}H_2-\overset{2}{C}H=\overset{3}{C}H-\overset{4}{C}H_2-\overset{1}{C}H_2-\overset{2}{C}H=\overset{3}{C}H-\overset{4}{C}H_2$$

$$\text{1,2 polybutadiene: } -\overset{1}{C}H_2-\underset{\underset{{}^4CH_2}{\|}}{\underset{{}^3CH}{|}}{\overset{2}{C}H}-\overset{1}{C}H_2-\underset{\underset{{}^4CH_2}{\|}}{\underset{{}^3CH}{|}}{\overset{2}{C}H}- \;.$$

Another series of isomers comes about because of the possibility of two sterically different configurations around atoms with tetrahedral bonds. Formulas (14) show the configurations of the stereo isomers of polypropylene:

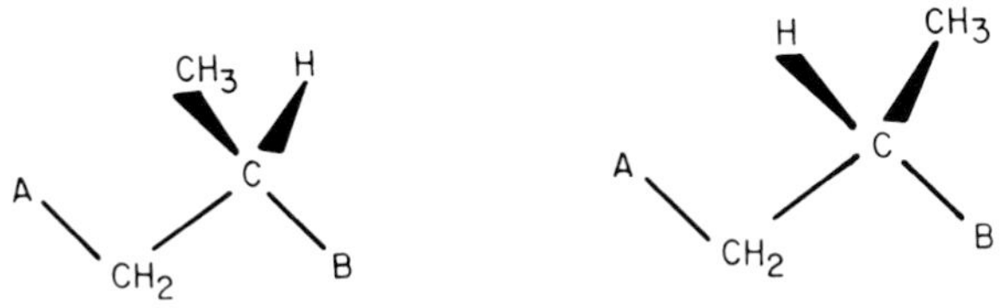

(14)

The backbone chain is stretched out in the plane of the paper; A and B signify the left and right continuation of the backbone, respectively. H and CH_3 are in a plane at right angles to the paper. In the case on the left, CH_3 sticks out of the plane of the paper and H is below. The right isomer has the positions of H and CH_3 reversed; it is the mirror image of the first. If all CH_3 groups of a polypropylene molecule are on one side of the plane of the backbone, the polymer is called *isotactic*. The regularly alternating polymer is called *syndiotactic*, while the random case is called *atactic*. The stereo isomers can have very different physical properties. In polypropylene, for example, the regular isomers, isotactic and syndiotactic, are normally crystalline, while the atactic polypropylene is amorphous. No crystal structure can be found for atactic polypropylene because of the random appearance of the relatively bulky CH_3 on either side of the backbone chain. On cooling, atactic polypropylene forms a glass.

The preceding discussion considers polymers that have only one type of repeating unit, so called *homopolymers*. If more than one repeating unit is used, *copolymers* result, and again many different ways of arrangement of the different copolymers have been found. Besides using more than two components, multiplicity is achieved by *random* or *regular* arrangements of the copolymer units. The regular copolymers can have *alternating* repeating units or have *blocks* of large sequences of identical repeating units.

If copolymers are compared with alloys of metals, the enormously larger number of possible polymeric materials is evident. In the first place there are many more repeating units to make copolymers than there are metals to make alloys. In addition, as was shown above, there are many more ways in which different copolymers can be made out of identical sets of monomers.

II. THE CONFORMATION OF POLYMER CHAINS

To get a feeling for the different shapes or conformations a flexible long chain molecule can assume, it is useful to solve the old mathematical problem of the three-dimensional random walk. Figure 2 shows how such a walk would look in two dimensions. After each step of length ℓ a new direction is chosen at random for the next step. Characteristic of the finished walk would be n, the number of steps traveled, and r, the straight line distance between beginning and end, the end-to-end distance. Mathematically, the end-to-end distance can be expressed as the vector sum over all the steps:

$$\underline{r} = \sum_{i=1}^{n} \underline{\ell}_i \quad . \tag{15}$$

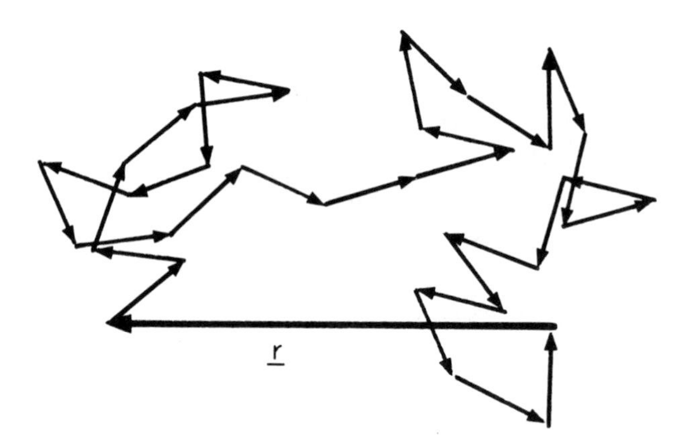

FIG. 2. Two-dimensional random walk. $\underline{r}$ represents the end-to-end distance.

If one is interested only in the absolute magnitude of the end-to-end distance, it simplifies matters to form the square of Eq. (15). The product (dot product or scalar product) of two vectors is mathematically defined such that

$$\underline{\ell}_i \cdot \underline{\ell}_j = |\ell_i|\,|\ell_j| \ \cos(\text{angle between } \underline{\ell}_i \text{ and } \underline{\ell}_j) \ . \tag{16}$$

Equation (15) becomes, on squaring,

$$\underline{r} \cdot \underline{r} = r^2 = \left(\sum_{i=1}^{n} \underline{\ell}_i \right) \left(\sum_{j=1}^{n} \underline{\ell}_j \right) = \sum_{i=1}^{n} \sum_{j=1}^{n} \underline{\ell}_i \cdot \underline{\ell}_j \ . \tag{17}$$

A very large number of different values of r^2 are possible by making different choices of directions for each step of the random walk. For an overall picture it is sufficient to get an average of all possible walks. The average of Eq. (17) is indicated by the angular brackets:

$$< r^2 > = \sum_{i=1}^{n} \sum_{j=1}^{n} < \ell_i \cdot \ell_j > \ . \tag{18}$$

Equation (18) is quite easy to evaluate. For i=j the dot product is simply ℓ^2, since the cos 0 = 1. There are n such terms contributing $n\ell^2$. For i≠j the dot product is equally likely to be positive or negative if the angle between any two vectors is chosen randomly, since the cosine is positive for half the possible angles and negative for the other half. The average over many such random choices must be zero. The evaluated sum of Eq. (18) is, thus,

$$< r^2 > = n\ell^2 \ . \tag{19}$$

This equation shows that the root-mean-square end-to-end distance of a random walk increases only with the square root of the number of steps. A doubling of the root-mean-square end-to-end distance requires a quadrupling of the number of steps.

How far can one apply the simple three-dimensional random walk result to the conformation of an actual polymer chain? The length of each step can easily be identified with the bond length. The bonds are, however, not freely jointed at the atoms. One angle, the bond angle, is practically fixed. Around the other angle the rotation is also not completely free, but may be more or less hindered, depending on the number and size of neighboring groups. To introduce these two restrictions to the random walk means to alter the evaluation of the average of Eq. (18). For the constant bond angle restriction, the cosine for each i-j combination can be evaluated. For the hindered rotation, an average angle is determined by finding the chance of the rotator to be at any one angle around the bond. For large chain length an approximate expression of the mean square end-to-end distance is

$$\langle r^2 \rangle = n\ell^2\left(\frac{1-\cos\alpha}{1+\cos\alpha}\right)\left(\frac{1-\cos\eta}{1+\cos\eta}\right). \tag{20}$$

α and η are the bond angle and the average rotation angle around the bond, respectively. The total effect of these two restrictions to the random walk is to increase the end-to-end distance or to expand the random coil of the polymer. Another restriction to the random walk arises from the fact that a molecule cannot occupy any space more than once. A certain volume is excluded by the polymer molecule. This problem of excluded volume is difficult to treat mathematically. The best results were obtained by simulation of a polymer molecule on an electronic computer. Assuming the elements of a polymer molecule can only take up positions possible on a diamond lattice, it was calculated, for 1123 chains of 800 mobile elements, that the average square end-to-end distance could be represented by

$$\langle r^2 \rangle = 1.40\, n^{1.18}\ell^2 . \tag{21}$$

Equation (21) has lost the simple dependence of $\langle r^2 \rangle$ on the first power of n. Again, the restriction to the random walk resulted in an expansion of the random coil.

To illustrate the results just derived, one may look at the conformations of a polyethylene molecule with molecular weight 280,000.

In such a molecule there are 20,000 CH_2 groups (n = 20,000). The C–C bond length is 1.54 Å, so that the length of a molecule going along the backbone bonds is 30,800 Å, the contour length. Introducing the constant bond angle of 109.5° between two successive bonds, but keeping the molecule fully extended, as in a crystal, a planar zig-zag or all-trans conformation results, which has a length of about 25,300 Å. Table 1 lists these and other root-mean-square end-to-end distances.

Table 1 shows that if we take the end-to-end distance as a measure of the extent of coiling of the molecule, each of the new restrictions expands the molecule somewhat more. A molecule in solution will show further changes in average dimension. There will be interactions between the solvent and the polymer. If the solvent preferentially is in contact with the molecule (good solvent), the chain will be expanded even further because the polymer segments are pushed apart by the solvent molecules. If the solvent molecules do not interact with the polymer or even get repelled from the molecule (poor solvent),

TABLE 1

Root-Mean-Square End-to-End Distances of a 280,000 MW Polyethylene Molecule

Mode of calculation	Equation	$(\langle r^2 \rangle)^{\frac{1}{2}}$ (Å)
Contour length	--	30,800
All-trans conformation	--	25,300
Random walk	(19)	218
Constant bond angle restriction only (η = random)	(20)	308
Constant bond angle restriction and hindered rotation (η = 103°)	(20)	493
Excluded volume calculation	(21)	628

the random polymer coil will contract. Since an increase in temperature often increases solvent power, it is possible to vary the end-to-end distance with temperature. In fact, it is possible to lower the temperature of a solvent so that Eq. (20) describes the dimensions. At this temperature, the θ-temperature, the excluded volume is just compensated by the solvent-caused contraction of the random coil. The θ-temperature of a polymer molecule in solution is similar to the Boyle temperature of a gas, where the attraction due to the intermolecular forces is just compensated by the excluded volume of the gas molecules, and a seemingly ideal gas results. The polymer molecule at the θ-temperature is in the same way an ideal reference state and thermodynamically easily recognizable. Experimentally, the end-to-end distance of polyethylene in solution at the θ-temperature was evaluated to be about 529 Å, in good agreement with the purely theoretical calculations of Table 1, which give 493 Å.

III. GLASSES

The randomly coiled physical structure of a flexible polymer chain in solution was described in Section II. Since the rotations around the bonds go on continuously, the molecule changes its shape continuously. The rotation around the C—C— backbone bond suggests that, for polyethylene, there are three more likely positions (rotational isomers), the trans and the two gauche conformations. To estimate how many different conformations are possible in a polyethylene molecule of 20,000 CH_2 groups, it may be permissible to say that each CH_2 changes only between the three different rotational isomers. For 20,000 CH_2 groups, this leads to the truly astronomical figure of 3^{20000} or 10^{9542} different conformations, between a large number of which the molecule can alternate. In the melt the same situation

FIG. 3. Two-dimensional representation of a polymer melt. Chain ends are marked by a filled circle.

prevails. Here, in addition, the molecules are entangled with each other, as is pictured in Fig. 3. On cooling a melt, the interchange from one conformation to the other becomes increasingly sluggish because of insufficient thermal energy to overcome the barrier to rotation around the bonds and because the number of low energy conformations becomes smaller. Macroscopically, this increasing difficulty to change conformations is recognizable in an increase in viscosity with decreasing temperature. Finally, at the glass transition temperature, the thermal interchange between different conformations stops within a small temperature range, and a rigid solid results. The structure of this solid, the glass, at and below the glass-transition temperature is identical to that of the melt at the glass-transition temperature. The structure can be described only in terms of statistical averages. Outside the general description just given, little detailed information on the structure of polymeric glasses is presently available. A set of parameters that might allow a better understanding are the root-mean-square end-to-end distance distribution, the degree of entanglement between the molecules, and the number of bonds frozen in the different preferred angles of rotation (rotational isomers).

Although the glassy state is often found in linear polymers because of the special difficulties inherent in polymer crystallization, it is by no means restricted to this class of molecules. A large variety of silicate glasses, which form random three-dimensional networks of strong bonds on freezing into the glassy state, are well known. Other materials that form glasses easily can be found among small organic molecules. In fact, most materials can be brought into the glassy state if one uses the proper quenching techniques to avoid crystallization. For example, water and certain alloys have been transformed to the glassy state by shock quenching on liquid nitrogen cooled metal plates. The glass-transition temperature of water was estimated from these experiments to be about −130°C. It should be noted that glass transitions are dependent upon molecular weight and impurity content as well as on measuring time scales (Section V, B), so that data on glass transitions often contain considerable uncertainty, because of poorly defined experiments.

Glasses are known to be strong, quite brittle, and have properties that are nondirectional or isotropic. If one considers the molecular structure, this is easy to understand. The randomly frozen chains have the same average structure in each direction, giving rise to isotropic properties. If one applies a small outside force to such a system of frozen chains one expects only a deformation of bond angles and bond length. No long range cooperative motion of the chain should be possible. As long as a glass conforms exactly to this simple picture, it behaves mechanically like an ideal elastic solid. The fractional deformations for which no change in conformation takes place are, for many glasses, only 0.001 or less. At higher deformation, yielding or fracture may occur. Figure 4 depicts force-deformation data for larger deformations as a function of time, as measured on polystyrene at 25°C. This phenomenon of increasing deformation with time at constant force is called creep and has its origin in the slow and irreversible change in conformations of the chains under the influence of the relatively large external forces.

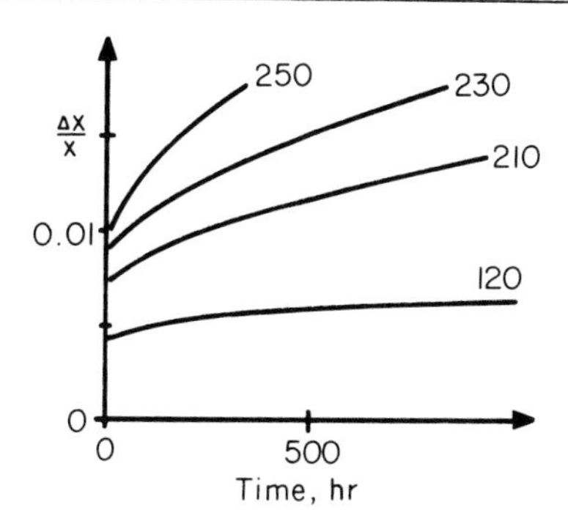

FIG. 4. Creep of a polystyrene bar at 25°C at different applied forces. The numbers in the graph give the force in kg/cm^2.

Glasses under larger strain will develop crazes. These look like small cracks and lie with their plane perpendicular to the stress. Crazes are different from actual cracks in that they still can support a large portion of the force. They represent not a severing of the matter, but a region where extensive flow has taken place. The matter in the cracks is highly oriented and has a density lower than that of the bulk. Under increasing stress these crazes lead to fracture. Glasses usually fracture at an extension of less than 5%.

IV. CRYSTALS

A. General Aspects of Polymer Crystallization

A polymer in the crystalline solid state is distinguished from a glass by the complete or partially ordered arrangement of the molecules. If one looks at the more familiar solids like NaCl, one finds that they crystallize out of the random arrangement in the melt or in solution by adding ion after ion to a growing crystal surface. This process of building the solid phase step after step, if carried out slowly, leads to large and perfect crystals. Trying the same process with a high polymer molecule of dimensions and shape as described above brings out the problem of polymer crystallization. The long, stringlike molecules must be straightened out during the crystallization. The

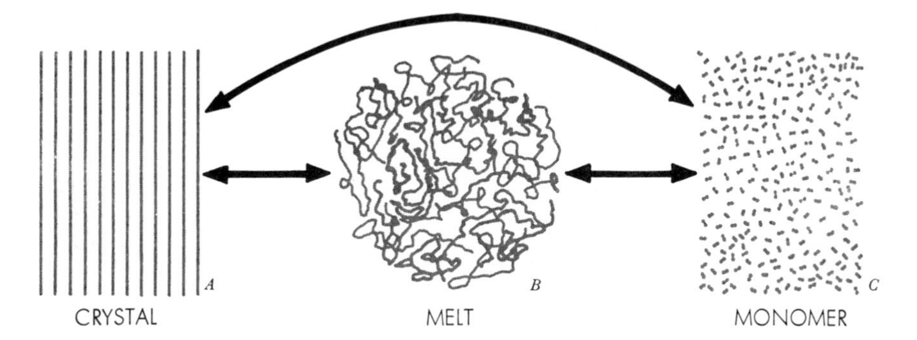

FIG. 5. The possible paths to crystalline linear polymers. A is a schematic representation of the well-ordered crystalline state. B shows the randomly coiled and entangled conformations polymer chains take in solution, in the melt, and also in the glassy state. C represents the structure of the monomer. The double arrow between B and A represents crystallization or melting; the double arrow between C and B, polymerization or decomposition; the double arrow between C and A represents polymerization into the crystalline solid state, the reverse of which would be sublimation.

atoms along the molecular chain axis are already fixed in relative position to each other and must be added to a growing crystal in sequence. Crystallization occurs only in the two dimensions at right angles to the molecular chain. These facts make for a more difficult process and, as a result, usually for much less perfect crystals.

Schematically, the possible paths to crystalline linear high polymers are shown in Fig. 5. From the monomer on the right, polymer is normally produced by polymerization into the random conformation of a melt or solution. The subsequent crystallization step is made difficult by the above mentioned problems of ordering strings. The problem is solved by the polymer system in a way quite similar to the one in which many randomly mixed long strings are disentangled. One string is wound up after another. Similarly, the polymer crystallized from the melt or from solution often crystallizes as a folded chain crystal, shown schematically in Fig. 6. The chains are folded back every few hundred angstroms instead of being completely extended. The reason for folding is the faster transformation of the chain molecules from the

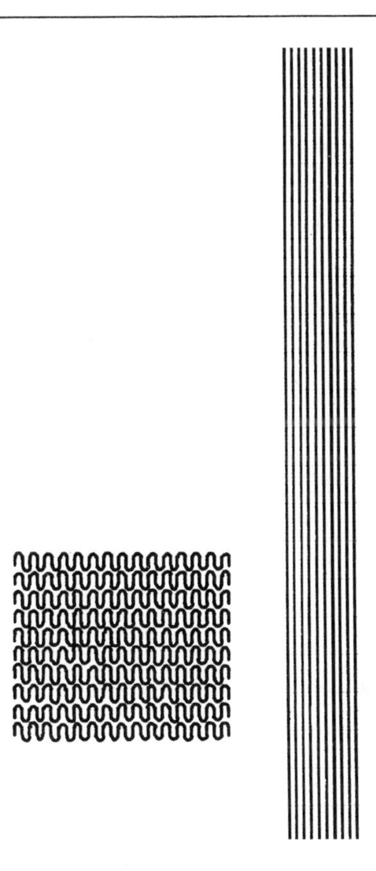

FIG. 6. Two modes of polymer crystallization. The fully extended chains form crystals as thick as molecules are long (typically 10,000 to 100,000 Å), while the folded chain crystals often form stacks of lamellae, each no more than 100 Å thick in the chain direction. (The ratio shown is only 100 to 3600.)

random to the crystalline state. The frequent folding of a polymer chain results in thin platelets (Section IV, C) which have an enormous surface area, causing the folded chain crystals to melt, frequently 10 to 100°C lower than the extended chain crystals. Figure 5 shows that immediate crystallization after polymerization, the polymerization into the solid state, is a simpler mode of crystallization, which avoids chain folding. The polymer molecules never go through the random melt or solution, and thus, no reason exists for chain folding. Indeed, very often long fibrillar crystals with extended chains are found in this case. The process of simultaneous polymerization and crystallization is probably also active in the formation of many crystalline biological polymers.

In addition to the difficulties caused by ordering strings, defects which are built in during the polymerization step cannot be corrected during the crystallization. Copolymer repeating units or chain ends of a different chemical nature cannot be excluded from the crystal as, for example, NaCl can be excluded from ice crystals growing from a NaCl solution in water. Such defects have to be built into the crystal lattice as point defects, or if the size of the foreign chain section is too different, it forms an amorphous defect, a defect which also keeps a number of otherwise crystallizable chain segments from crystallization. Such amorphous defects may concentrate on crystal surfaces or be a major disruption inside the crystals. If these amorphous defects are large and numerous, they may form a continuous "amorphous phase" that surrounds small ordered crystalline areas of the polymer.

B. X-Ray Crystal Structure

Polymer crystals are usually very small and accordingly give a Debye-Scherrer type x-ray diffraction pattern with broad rings of scattered x-rays. If larger portions of the polymer are completely disordered, a very broad ring of scattered x-rays arises from these portions of the solid. Figure 7 gives an example of scattering intensities of x-rays as a function of angle for polypropylene crystallized from the melt. The shaded area indicates the broad scattering from the disordered polymer, while the peaks arise from crystalline polymer. This type of diffraction pattern is found in bulk samples where very many small crystallites are randomly arranged with respect to the incoming x-ray beam, so that all possible diffraction angles are found in one scan.

Comparison of the x-ray intensities scattered into the amorphous broad peak and into the sharper crystalline peaks allows an estimation

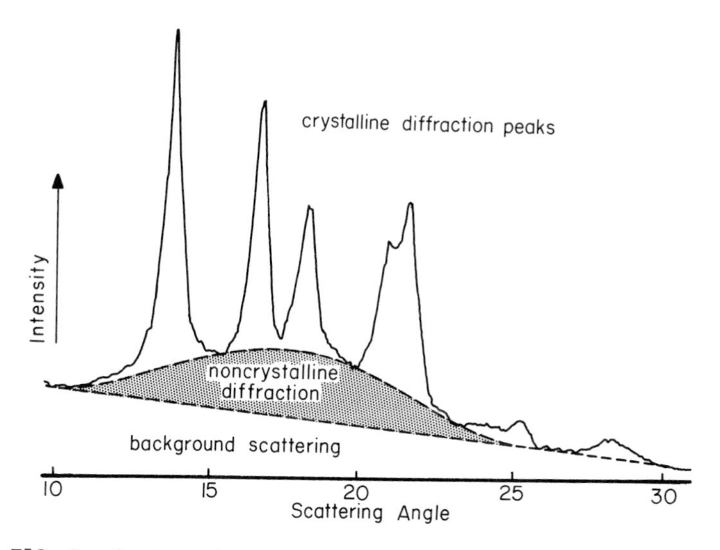

FIG. 7. Scattered x-ray intensity from a partially crystalline sample of polypropylene as a function of the scattering angle. The sharp diffraction lines are superimposed upon a noncrystalline halo.

of the degree of crystallinity. For more exact comparison the ratio of areas of the scan of Fig. 7 must be calibrated by comparison with otherwise characterized samples. The area of the four crystalline peaks of Fig. 7 is 57% of the total scattered intensity between 10° and 25°, while a calibrated comparison gives 65% crystallinity.

The concept of percentage crystallinity is important in the description of semicrystalline polymers. At its root lies the simplifying concept that one can separate crystalline and noncrystalline regions as two building blocks of a polymeric solid. For example, in the calculated x-ray crystallinity it is assumed that only the crystalline regions contribute to the sharp x-ray diffraction pattern and that the noncrystalline regions would scatter like a liquid does. A similar calculation is possible from a volumetric experiment. In this case the weight fraction crystallinity w_c is given by

$$w_c = \frac{v - v_c}{v_a - v_c} . \tag{22}$$

In Eq. (22), v is the measured specific volume, v_c the perfectly crystalline specific volume, and v_a the completely amorphous specific volume. Other methods of evaluating the crystallinity can be based, for example, on heat of fusion or infrared absorption in frequency regions characteristic of the amorphous and crystalline phases. For many samples, mainly when the polymer is crystallized by cooling from the melt, all methods mentioned give results in reasonable agreement, justifying the use of this parameter. In other cases, mainly when samples have been strained, large deviations are found. Clearly the value of crystallinity gives only a limited amount of information.

For the determination of an unknown crystal structure, a trace like Fig. 7 is usually not sufficient. For this purpose the crystals must be oriented to give additional information on the position of the scattering planes in the crystal. One simple way to achieve this for linear high polymers is to draw a polymer fiber. In this process all crystallites are arranged so that the molecular chain axes are largely parallel to the draw direction. A photographic recording of scattered x-rays of drawn polyethylene, a fiber pattern, is reproduced in Fig. 8.

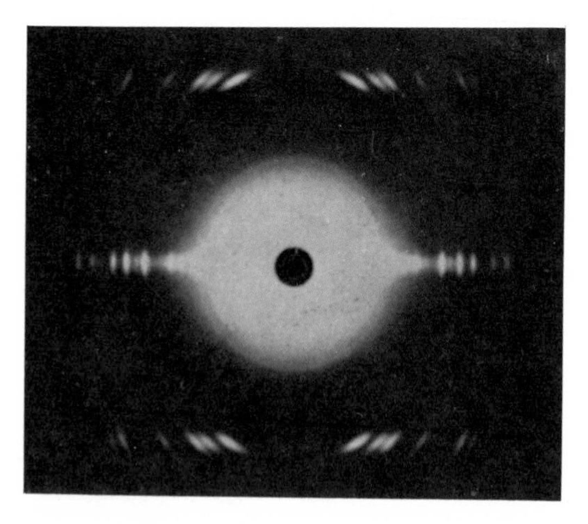

FIG. 8. Photographic recording of a fiber x-ray diffraction pattern of polyethylene.

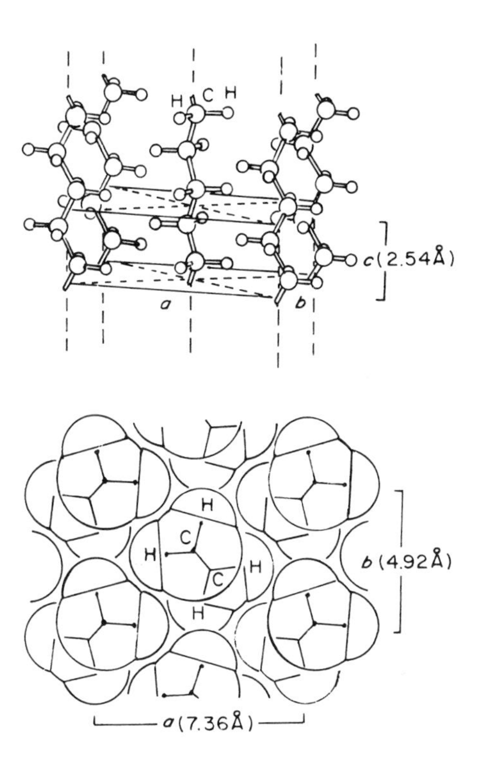

FIG. 9. Crystal structure of polyethylene. The indicated unit cell is orthorhombic with the given dimensions. The polymer chains are fully stretched in the planar all-trans conformation. [Structure after C.W. Bunn, "The Structure of Polyethylene," in Polythene (A. Renfen and P. Morgan, eds.), Iliffe, London, 1960.]

The fiber axis of the sample was vertical with respect to the photograph. The broad inner halo indicates some disordered portions. The reflections found on the equator of the photograph are linked to planes of atoms parallel to the fiber, and in our case also the molecular chain axis. Reflections above and below are caused by planes increasingly tilted with respect to the fiber axis.

The best representation of the crystal structure as derived from x-ray diffraction is offered by the unit cell of the repeating units. Figures 9-12 give representative drawings taken from the literature of known polymer crystal structures. Of particular interest is the

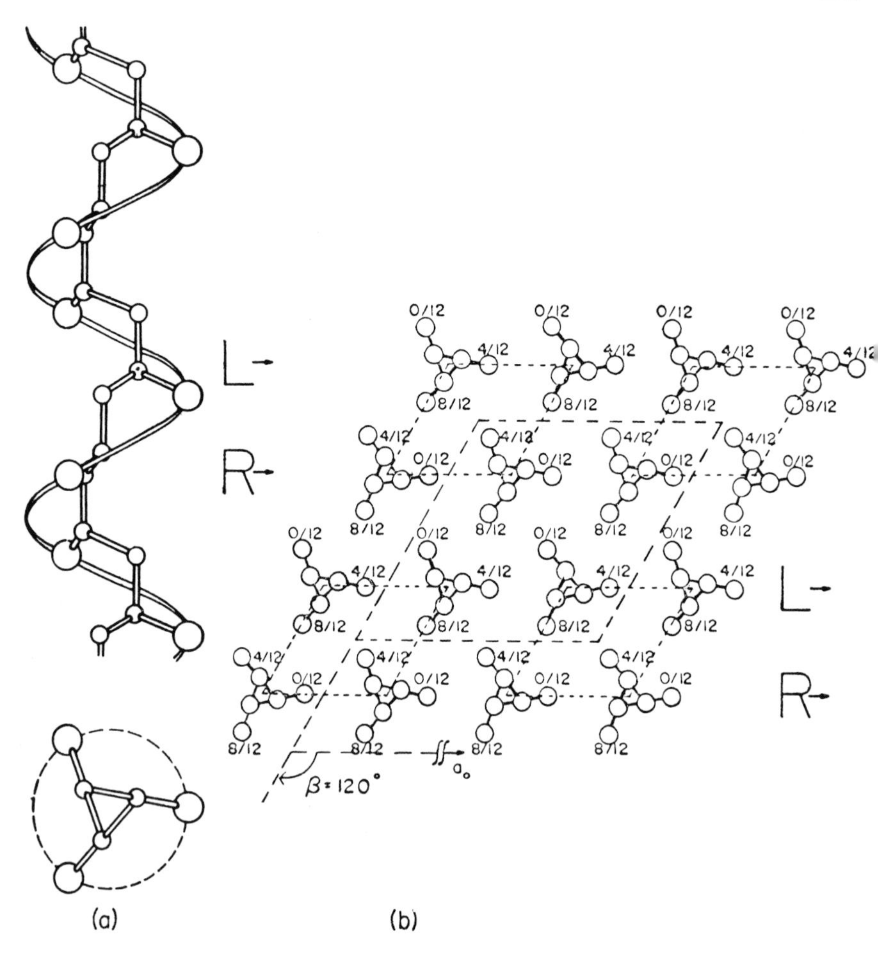

FIG. 10. (a) Helical conformation of polypropylene in the crystalline state as proposed by G. Natta and P. Corradini, Nuovo Cimento, 15, Suppl. 1, 9 (1960). (b) Arrangement of the polypropylene helixes in the "hexagonal" unit cell as suggested by H. D. Keith, F. J. Padden, Jr., N. M. Walter, and H. W. Wyckoff, J. Appl. Phys., 30, 1485 (1959); unit cell dimensions a = b = 12.74 Å, c = 6.35 Å.

polyethylene unit cell where all carbon atoms are in the trans conformation, which has the lowest energy. A sizable portion of the heat of crystallization comes from this change on crystallization of many bonds in the gauche conformation to the trans conformation. In the polypropylene crystal structure, the helical twist of the main chain is of interest. This twist leads to a periodicity of three repeating units along

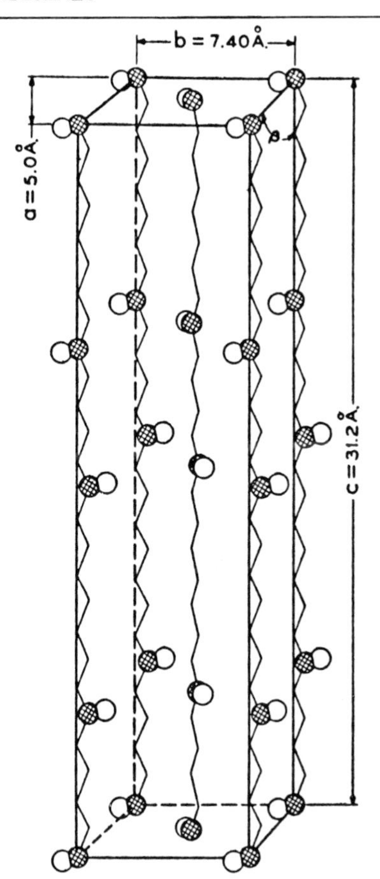

FIG. 11. Orthorhombic unit cell of poly(ethylene azelate) repeating units. [After C. S. Fuller, Chem. Rev., 26, 143 (1940).]

the chain and offers, in a dense crystal structure, the best possibility of accommodating the bulky CH_3 group, which sticks out of the chain every second carbon. The polyester has almost the same crystal structure as polyethylene except that the ester groups are lined up to give maximum interaction. Nylon has strong hydrogen bonds between chains; certainly in the ideal crystal structure all hydrogen bonds are formed.

C. Morphology

The unit-cell structures described in the previous section illuminate the ideal arrangement of the repeating units of crystalline polymers.

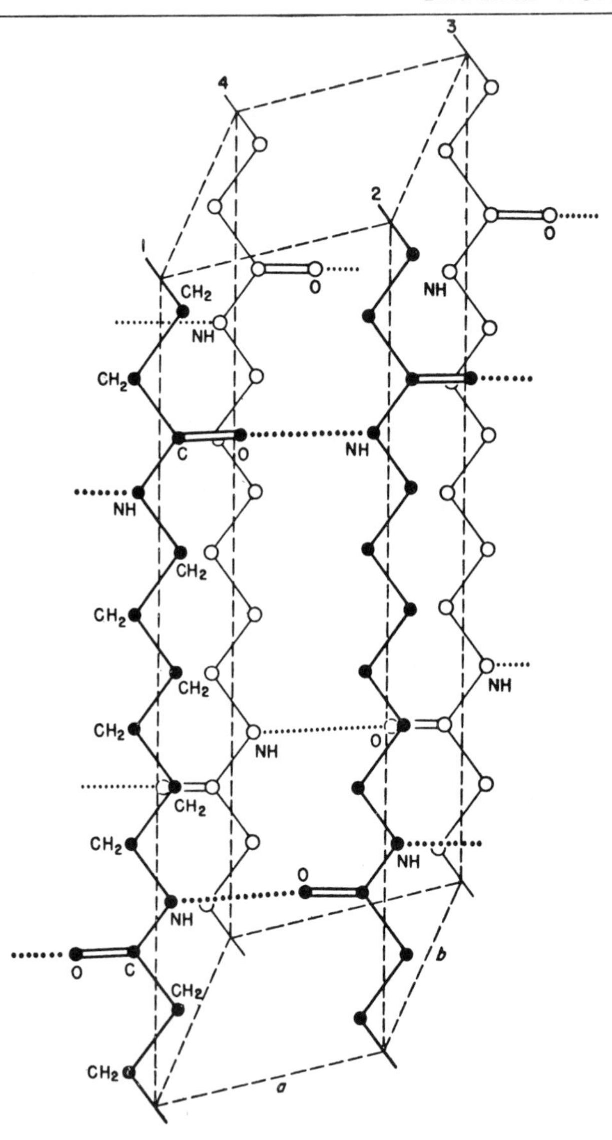

FIG. 12. Arrangement of nylon 6,6 repeating units in the triclinic unit cell. Note that all H bonds are formed in the (010) plane. The unit-cell dimensions: a = 4.9 Å, b = 5.4 Å, c = 17.2 Å, α = 48.5°, β = 77°, γ = 63.5°. [C. W. Bunn and E. V. Garner, Proc. Roy. Soc. (London), 189A, 39 (1947).]

In this section the morphology, the external shape of polymer crystals, will be discussed.

FIG. 13. Micrograph of a growth spiral of folded chain crystals of polyethylene from dilute solution. The lamellae are of the order of millimeters with a thickness of about 135 Å. The specific surface area is about 10^6 cm^2/g; the weight of the largest lamella about 10^{-8} g. The micrograph was taken with interference contrast.

In dilute solution, polymer molecules are farthest apart from one another and least entangled, which leads to more perfect crystals upon crystallization. A dilute solution (~ 0.05% by weight) of polyethylene in toluene starts crystallizing isothermally at about 90°C. The resulting crystals are folded chain crystal lamellae of about 135 Å thickness. Figure 13 is a micrograph of such polyethylene crystals, in the form of a growth spiral. Crystals with perfect growth surfaces may be found on isothermal crystallization down to 75°C. Their lateral size seems to be limited only by the patience and skill of the researcher. Below a 75°C crystallization temperature, folded chain dendrites appear. These arise from skeletonlike crystal growth similar to that in snow flakes. Figure 14 shows a micrograph of a dendrite flattened out on a microscope slide. The larger the supercooling, the finer the structure of the dendrites. Still, the basic structure is thin, folded chain lamellae, shaped more and more like ribbons.

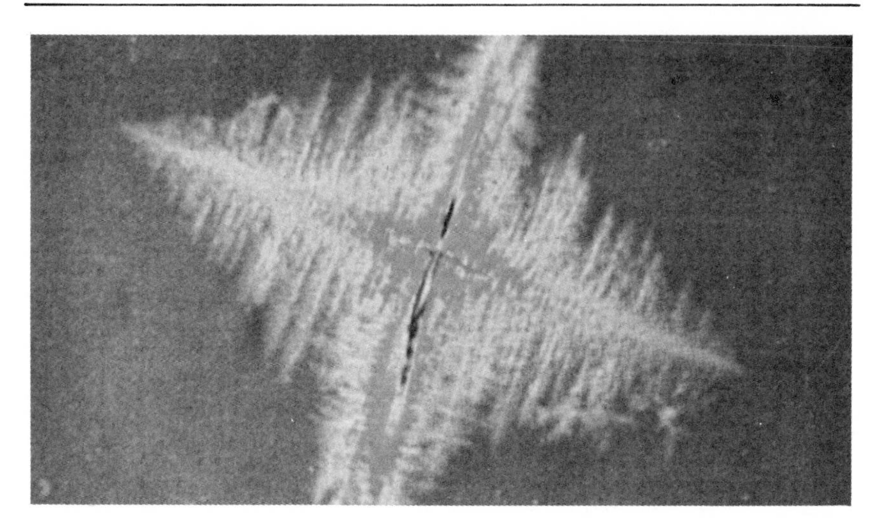

FIG. 14. Dendrite of polyethylene flattened on a microscope slide. The size is about 0.05 mm. The dendrite is made up of many stacked lamellae of 100 Å thickness with folded chains. The micrograph was taken with interference contrast.

A more detailed investigation of the polyethylene single-crystal platelets, as shown in Fig. 14, reveals that they are not completely flat, but have a tentlike structure. Upon collapse on a microscope slide, the tentlike crystal either slides to a flat plate by internal chain slippage or settles and develops a pleat. This tentlike structure comes about since successive chain folds direct each other into a staggered array. The dendrites are multilayered. Many platelets of about 100 Å thickness are piled on top of each other. These layers are interconnected only by screw dislocations in the center and accordingly are not flat when floating in solution. Since the separate branches are quite flexible, an almost spherical shape is taken on. Figure 15 shows an interference micrograph of a dendrite similar to that shown in Fig. 14, but floating in solution. Hardly any of the regularity of the collapsed dendrite is visible in this structure.

For many linear polymers, crystallization from the melt is similar to crystallization from solution. The study of melt crystallized polymers is, however, much more difficult because of the intimate intermeshing of the growing crystals. Study of fracture surfaces show the lamellar structure for slowly cooled melt crystallizations. Cooling a melt more rapidly leads, as in the case of solutions, to dendritic crystal aggregates. As for the solution-grown dendrites shown in Fig. 15, those from the melt take on a spherical shape. Small ribbons of folded crystals radiate out from the crystal center. When observed under the polarizing microscope, the optical anisotropy of the birefringent crystalline ribbons gives rise to the picture shown in Fig. 16. The ribbons in Fig. 16 are so small that the optical microscope does not resolve the detailed structure. This crystal morphology is called spherulitic, the name being derived from the appearance of the sample under the polarizing microscope.

Finally, the equilibrium morphology has also been obtained for several polymers. Extended chain crystal lamellae, with a thickness in the chain direction equal to the molecular weight, have been produced from a polyethylene melt by crystallization under elevated pressure. Figure 17 shows an electron micrograph of a replica of the fracture surface of such an extended chain crystal. The striations visible are produced during fracture. The molecular chains are parallel to the striations. The thickness of the lamella shown is about 2 μ, the length of the average molecule in the mixture.

All morphologies here discussed, folded chain crystal lamellae, dendrites, spherulites, and extended chain single crystals, are easily recognizable and are the better understood examples of well-crystallized polymers. Very little is known about polymers of a lower degree of crystallinity.

D. Defects

The study of defects in polymer crystals has long been hampered by an overabundance of imperfect crystals. Only recently has the study of defects, necessary to understand the properties of polymer crystals, been possible. It has been found that the defects in polymer crystals are similar to the defects found in the crystals of simpler molecules.

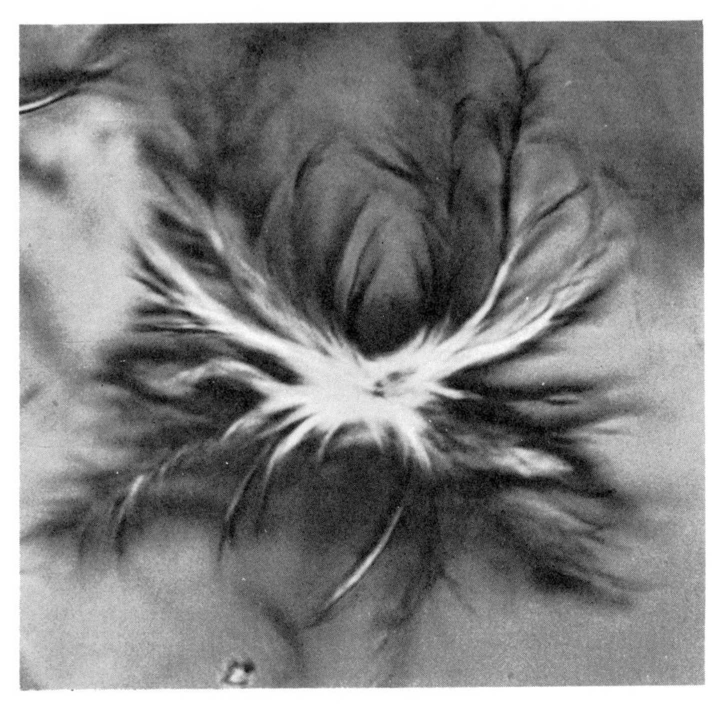

FIG. 15. Interference micrograph of a dendrite floating in solution. Overall size of the dendrite ~0.1 mm.

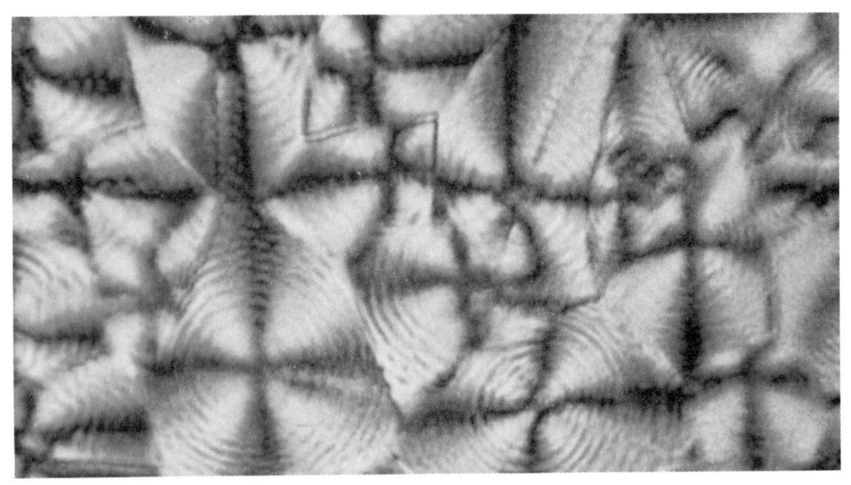

FIG. 16. Spherulites of polyethylene as observed under the polarizing microscope. Magnification 1000x.

FIG. 17. Electron micrograph of a replica of a fracture surface of an extended chain polyethylene single crystal. The molecular chains are parallel to the striations and are of the same length as the crystal.

(1) Vacant lattice sites and interstitial atoms can be classified as point defects; they behave differently in linear polymer crystals and in monomeric crystals. The vacancies arise mainly in connection with other defects such as the ends of chains, which must be classified as foreign atoms because of their different chemical nature. Single interstitial atoms are possible only as monomeric impurities (foreign atoms). Side chains of the proper character may also be called interstitial but show the characteristic of being immobile. Often an interstitial element in a polymer crystal gives rise to chain disorder or an amorphous defect.

(2) Dislocations are one-dimensional or line imperfections. Two main types are of interest, namely, the screw dislocation and edge

dislocation. The screw dislocation of polymer crystals is illustrated by Fig. 13. The main characteristic of screw dislocations of this type is the enormous size of the Burgers vector, more than 100 Å.

Edge dislocations play an important role in the understanding of mechanical properties of metals. In polymers, their existence has been proven by observation of moiré lines between two layers of folded chain lamellae.

(3) Two-dimensional imperfections include all types of external and internal surfaces on polymer crystals. The surface of particular interest in folded chain polymer crystals is that which contains the aligned folds. A particularly high surface free energy is expected for this surface because of the extra free energy expended on folding and packing the more bulky folds.

(4) Chain disorder defects are defects particular to polymers. They include folds as well as simple switches from one lattice alignment to another and kinks. Only when aligned in a two-dimensional defect in form of the fold surface is anything known about these defects. Much speculation and semiquantitative calculation of chain disorder in the crystal has been carried out recently.

(5) The amorphous defect is caused by somewhat larger disturbance of the lattice, such as a side group which cannot crystallize within the lattice or a major chain disorder. Its characteristic is a decrease of the crystallinity by an amount larger than its own weight. This indicates that an amorphous defect causes several otherwise crystallizable chain units to remain amorphous. For example, inclusion of one $-CH(CH_3)-$ in polyethylene was shown to decrease the crystallinity by several additional $-CH_2-$ units. The exact structure of such a defect is not known, and it is not expected that there will be a sharp dividing line between amorphous defects and chain disorder. If the amorphous defect gets larger, it may be called a three-dimensional imperfection.

E. Mechanical Properties

Polymer crystals have as a main characteristic parallel molecular chains. In the chain direction, they have strong primary bonds, while at right angles much weaker secondary bonds hold the crystal together. This anisotropy of bond strength governs the mechanical properties of the crystals. In the chain direction the crystals are quite strong, while at right angles to the chain axis they may be weak. All crystalline samples investigated to date, however, have been polycrystalline. In these samples the overall properties are determined, to a large degree, by the way the crystals are held together.

Information on the mechanical properties of polymers is available for polyethylene. For spherulitic bulk samples it was shown that, on stretching, many fibers are pulled out between adjacent spherulites. This can happen only if there are tie molecules between the two separating parts, which are pulled out in the form of tightly packed bunches. A large part of the mechanical properties, in particular the ultimate strength of the material, depends on the structure of these tie molecules. The mechanical properties as a whole depend on the following hierarchy of structure information:

1. the chemical structure of the polymer chain;
2. the unit cell structure of the crystal;
3. the interior defects of the crystal;
4. the morphology;
5. the way the morphological entities are "glued" together.

Presently, there are no samples for which all of this required information has been accumulated along with the necessary mechanical property measurements.

As is true for all crystals, polymer crystals cannot tolerate a large amount of elastic deformation. On applying a sufficient tensile force, the sample becomes permanently deformed. Small deformations

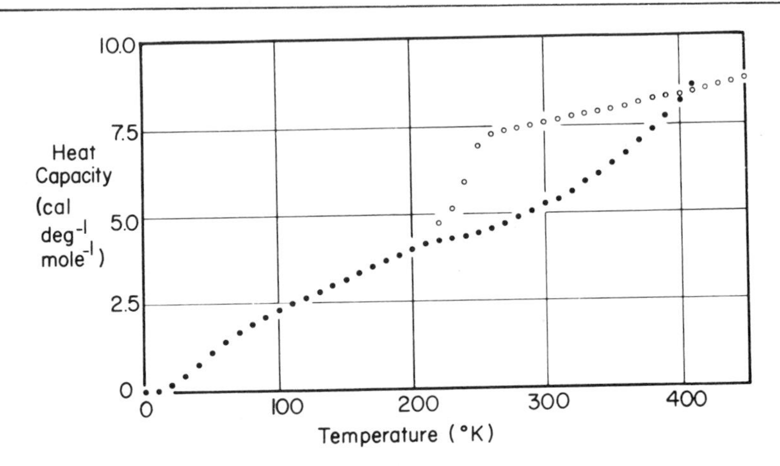

FIG. 18. Heat capacity of polyethylene as a function of temperature. Open circles represent amorphous polyethylene, filled circles crystalline polyethylene. The latter also represent heat capacity calculated from the frequency spectrum of Fig. 19.

may be caused by phase transformation or twin formation, small rearrangements in the unit cell structure which relieve the applied stress. Larger deformations, however, occur by chain tilting and slip. Finally, fibrils are drawn out. This "fibrillar" material does not contain completely stretched out molecular chains but is still largely folded. The orientation of the fibrils is parallel to the applied stress, and the molecular chains lie parallel to the fibrils. On a large scale this process is used in drawing fibers of polymers for commercial application.

F. Molecular Motion

With the closer packing in crystal polymers, molecular motion is more restricted to vibrations than in the glass form. To elucidate the vibrational characteristics of completely crystalline polymers, heat capacity, infrared and Raman spectra, and neutron diffraction results have been used. In Fig. 18, data on the heat capacity of completely crystalline polyethylene up to the vicinity of the melting point are reproduced.

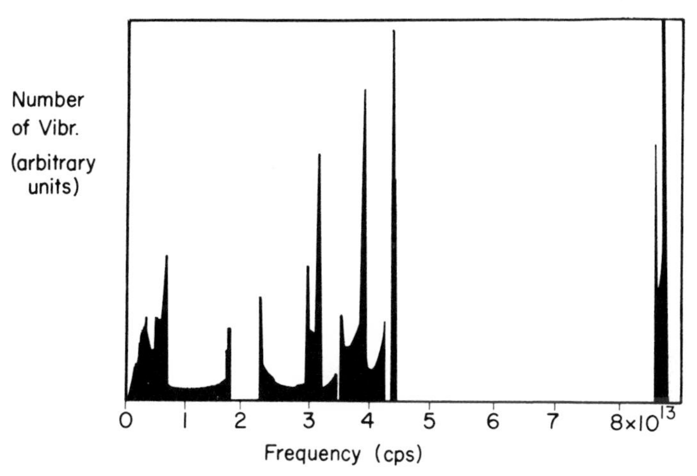

FIG. 19. Frequency spectrum of crystalline polyethylene. The number of vibrations in a certain unspecified amount of crystal are plotted against the frequency in cycles per second.

Since the heat capacity-temperature curves of polyethylene and other polymers deviate from those of, for example, metals, the vibrational frequency spectra of polyethylene and metals should also be significantly different. Only below 10°K is a Debye T^3 law followed for polyethylene. Then a linear portion exists in the heat capacity-temperature curve between 110 and 210°K which is followed by a level section and then a new rise in heat capacity starting at about 250°K. The heat capacity at 400°K is only 3.5 R/mole of CH_2 units. One would expect, on full excitation, a heat capacity of 9 R.

Besides heat capacity measurements, Raman and infrared spectroscopy give information on vibrations in solids. Every ir and Raman active frequency corresponds to a possible vibration in the solid. In this way some direct information is gained. Since the theory of vibrational spectra is well developed, it is possible to calculate, with the help of the experimental ir and Raman data, a frequency spectrum for all of the higher frequencies (above 2×10^{13} cps). Combining this data with heat capacity information permits the development of a complete spectrum down to the lowest frequencies. Figure 19 shows the frequency spectrum of polyethylene. It holds the key to all the features of the heat capacity shown in Fig. 18. In fact, the filled circles in Fig. 18 represent the spectrum of Fig. 19. The initial increase in heat capacity, following a T^3 law, comes from a quadratic increase in the number of vibrations with frequency. Several peaks in the frequency distribution curve interrupt the T^3 law above 10-20°K. The linear portion of the heat capacity curve is the result of the almost flat portion in the frequency spectrum. The slight leveling off of heat capacity above 210°K is caused by the gap in frequencies at about 2×10^{13} cps. The new rise in heat capacity comes from the series of vibrations between 2.2 and 4.4×10^{13} cps. The very high frequencies do not contribute to the heat capacities at temperature below 400°K. Even the intermediate group is excited only to a limited degree.

Since the frequency spectrum was evaluated by theoretical calculation as well as experimentation, it is also possible to say more about the type of motion the molecules carry out. The lowest frequency series of vibrations is mainly a skeletal vibration. The chain is contracting and expanding by bending C—C—C bonds. The higher portion, up to 1.5×10^{13} cps, includes skeletal twisting vibrations. These two modes of vibration can contribute 2R to the heat capacity, a value which is reached at about 250°K. Above the gap, the so called optical vibrations start. Their higher frequency is caused by a motion stretching the C—C bond, which takes a much higher energy than bending or twisting. The main portion of this frequency group, however, involves bending of C—H bonds. A total of 5R would be contributed to the heat capacity if all these vibrations were excited. Since at 400°K only 3.5R is measured and 2R is contributed by the skeletal motion, these low optical vibrations are excited at this temperature to about 30%. The last series of vibrations are C—H stretching vibrations. Their frequency is very high because of a strong C—H bond and a small H mass. These vibrations could contribute 2R to the heat capacity. Calculations show that at 400°K they are no more than 2% excited.

V. TRANSITIONS

Two transitions are of major importance to delineate the solid state. For glassy materials it is the glass transition which separates the solid region from the liquid. For crystalline materials the melting transition separates the solid region from the liquid. For semicrystalline polymers it is even possible to find both transitions in the same material. If a substance can be obtained in both the crystalline and glassy states, or is found as a semicrystalline material, the glass transition is always the lower of the transition temperatures. Usually the ratio of melting to glass-transition temperatures (Kelvin) is 1.3 to 2.0.

A. Glass Transition

A glass was described in Section III as a solid with a frozen-in, more or less random conformation of the polymer chain. Its molecular motion is largely restricted to vibrations. Heating a glass through the transition temperature region changes its properties from a brittle material to a rubbery substance which can, because of the possible extension of molecular chains, reach an extension as great as 1000% at the break point, compared to a normal 5% extension.

The nature of the glass transition can best be understood by considering the different contributions to the heat capacity of a polymeric melt. Figure 20 represents data on polyethylene. The right-hand side represents the melt, the left-hand side the glass. Motion within the backbone chain of the polymer (skeletal motion) shows a contribution to the heat capacity which decreases with increasing temperature from 2R to 1.5R. This is true since one of the two modes of skeletal motion, the torsional mode, becomes more and more a rotator at higher temperature with a correspondingly lower heat capacity. The optical vibrations which involve C—H bonds (asymmetrical stretching, symmetrical stretching, bending, rocking, twisting, and wagging modes)

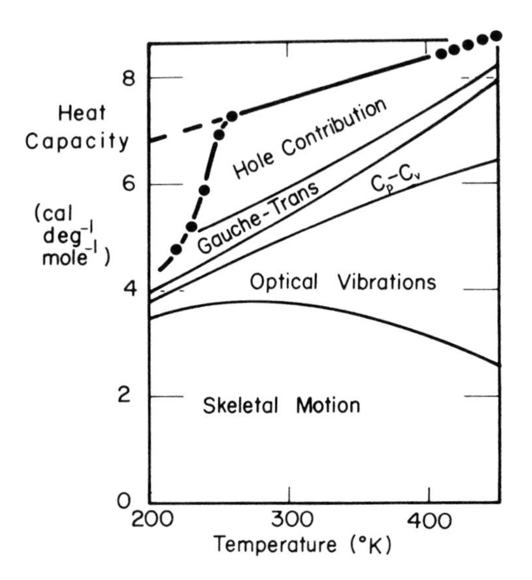

FIG. 20. Heat capacity of amorphous polyethylene as a function of temperature. The separate contributions of skeletal motion and optical vibrations, high energy conformation contribution, and hole contribution are indicated. The glass transition temperature is about 237°K. (The C_p-C_V contribution corrects for the difference of heat capacities measured at constant pressure and constant volume.)

and the C—C stretching mode are similar in the melt and the crystal; they are vibrations that are quite localized and change only little with changes of state. Another contribution to the heat capacity in the melt arises from an increasing number of high energy conformations of the chain at elevated temperatures. For polyethylene this high energy conformation is the gauche form, which is about 800 cal/mole higher in potential energy than the trans form. The final contribution comes from the larger expansion coefficient of the melt and has been labeled the "hole contribution." This contribution can be considered to be the result of the necessary introduction of holes to account for the larger macroscopic volume increase of the melt. But the holes also permit easier changes from one conformation to the other, macroscopically detectable by a lower viscosity. It is shown in Fig. 20 that at the glass transition this "hole contribution" and also the gauche-trans contribution freeze-in.

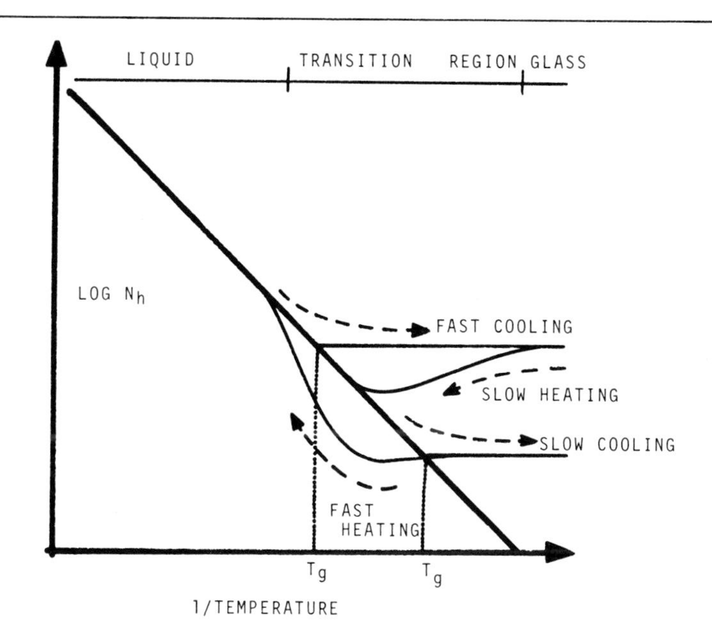

FIG. 21. Logarithm of number of holes plotted vs the reciprocal of temperature. The heavy solid line represents the equilibrium hole number.

The "hole model" of the liquid state is a simple model developed in the 1930's by Frenkel and Eyring. It explains, in agreement with the data of Fig. 20, the glass transition as the freezing of the "hole equilibrium." Below the glass-transition temperature the number of holes remains constant. Parallel with a freezing of the number of holes goes the freezing of the large scale conformation changes of the polymer chain. Holes as well as molecules are immobilized.

There are more detailed theories of the glass transition which try to evaluate the different possible chain conformations statistically. However, the simple hole model can already explain the most salient features. Figure 21 shows, in a plot of the logarithm of the hole concentration as a function of inverse temperature, a dependence we would expect for a hole equilibrium. On cooling, progressing from the left to right in Fig. 21, a point is reached where conformation changes are so slow that the system freezes. For fast cooling this point will be reached at a higher temperature than for slower cooling. As a result

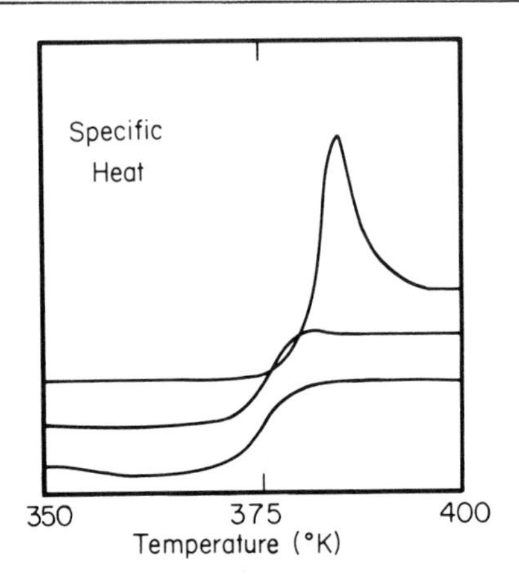

FIG. 22. Hole contribution to the heat capacity of polystyrene in arbitrary units with successive curves replaced upward. Measurements were made with a 5°C/min heating rate. The measurement for the lowest curve was made after cooling at a rate of 30°C/min through the glass-transition region. The second and third curves were measured after cooling at rates of 5°C/min and 0.2°C/min, respectively.

there is a large variety of glasses that are produced by changing cooling rates. These glasses are not only distinct in their different frozen chain conformation and hole concentration, but also mechanically and in their macroscopic functions of state, volume, and enthalpy, for example.

On reheating glasses cooled at different rates, hysteresis phenomena can appear. For example, if a slowly cooled glass with a low freezing temperature is rapidly reheated, it will not attain melt equilibrium at the low glass-transition temperature. Instead, it will continue to act as a glass. It will superheat and at a somewhat higher temperature achieve equilibrium by a faster increase in hole concentration. This causes a hysteresis maximum in the time-dependent apparent heat capacity, as is shown in Fig. 22. Similarly, a fast cooled sample reheated slowly gives rise to a minimum in the heat capacity curve, measured during slow reheating. Via time-dependent heat capacity measurements, it is thus possible to detect and distinguish differently cooled glasses. Measurements of this sort are possible by quantitative differential thermal analysis.

B. Melting Transition

A separate discussion of melting of polymeric materials is necessary because in the past much confusion arose from wrong interpretation of experimental data. During the last 10 years much has been done to show that the melting behavior of polymers can be understood when the multicomponent nature of most polymers and their metastable crystals are recognized.

We can define melting, broadly, as the transformation of a fully or partially crystallized substance into the liquid state, the melt. Thermodynamically, melting can be discussed using Eq. (23):

$$T_m = \Delta H/\Delta S \quad . \tag{23}$$

The melting temperature is T_m; ΔH and ΔS are the enthalpy and entropy of fusion, respectively. Equation (23) applies to systems in equilibrium as well as to metastable systems. For monatomic elements, ΔS is nearly constant, as illustrated in Table 2. This is known as Richard's

TABLE 2

Molar Heats of Fusion, Melting Points, and Entropies of Fusion of Elements

Element	ΔH_f (cal/mole)	T_m (°K)	ΔS_f (cal/deg-mole)
Ne	80	24.5	3.26
Ar	281	83.5	3.35
Kr	390	116	3.36
Rb	520	312	1.70
Hg	560	234	2.40
Na	610	371	1.64
Cd	1500	594	2.53
Zn	1800	692	2.60
Al	2500	930	2.70
Ag	2700	1234	2.18
Cu	2700	1356	1.99

TABLE 3

Heats of Fusion, Melting Points, and Entropies of Fusion of Polymers per Mole of Chain Atom

Polymer	ΔH_f (cal/mole)	T_m (°K)	ΔS_f (cal/deg-mole)
Polyethylene	980	414.6	2.36
Polypropylene	737	449	1.64
Polystyrene[a]	1200	512	(2.34)
Poly(vinyl fluoride)	900	470	1.91
Polytetrafluoroethylene	730	600	1.22
Poly(ethylene oxide)	660	339	1.94
Poly(methylene oxide)	795	453	1.76
Polyacrylonitrile	600	590	1.02
Poly(ethylene terephthalate)	785	540	1.45
Poly(ethylene sebacate)	571	349	1.64
Nylon 6, 6	736	540	1.36
Nylon 6, 10	666	499	1.33
Nylon 6	728	498	1.46

[a] In the case of polystyrene, the ΔH_f and ΔS_f should not be compared per chain atom. The C_6H_5 side group is so big that it should be taken as a separate unit, reducing ΔS_f to 1.56.

rule, which places ΔS between 1.8 and 3.4 cal deg^{-1} mole^{-1}. If ΔS changes only slightly for a group of crystals, the melting temperatures are fixed by the enthalpy changes. Furthermore, since the gain in disorder is similar for a series of substances, the number of "broken bonds" in the liquid at any given time should also be similar, so that the melting points are nearly proportional to the bond strength in the crystal (or the cohesive energy density). Table 2 shows that, for example, argon, which is held together by van der Waals bonds, melts at a low temperature, while gold with strong metallic bonds melts at a high temperature. Data for linear high polymers are collected in Table 3.

The ΔH and ΔS values are calculated for polymers per mole of backbone atoms. The very flexible polymers, for example, polyethylene, have ΔS similar to that of the monatomic elements. None of these polymers melts at a very high temperature, since the heat of fusion is relatively low. If, on the other hand, the polymer chains become less flexible, as in polytetrafluoroethylene, ΔS decreases, and despite a small ΔH, the melting point increases. High melting points in polymers can be caused in linear high polymers by stiff backbone chains.

The simple description of melting offered by Eq. (23) is often complicated by one or more of the following: (1) The transition from crystal to melt is not instantaneous but needs a certain amount of time. If one heats a crystal faster than it can melt, the interior superheats temporarily. Experimentally, a melting range which broadens to higher temperatures on faster heating is observed. (2) Small crystals have a large surface free energy, so that

$$G_c = G_c'^{\circ} + G_s \quad (24)$$

G_s represents the surface energy per mole of substance. The point of equal free energy of melt and small crystal is at a lower temperature than for large crystals. In addition, small crystals are, because of higher free energy, metastable with respect to larger crystals. Whenever the time scale allows, small crystals will reorganize to bigger crystals with smaller free energy. Defect crystals behave in a fashion similar to small crystals. (3) Impure melts have a lower free energy due to the entropy of mixing of the impurities with the crystallizable component. The change in free energy is represented by the change in chemical potential:

$$\mu_p = \partial G_\ell / \partial n \quad (25)$$

The temperature of equal chemical potentials of crystal and crystallizing component in the melt is changed according to the following equation:

$$(1/T_m) - (1/T_m^{\circ}) = (R/\Delta H)[-\ln(v_p) + (x-1)v_i - x^2 v_i^2 \chi_i], \quad (26)$$

TABLE 4

Melting Points of Extended Chain Polymer Crystals

Polymer	Melting Point (°C)
Polyethylene	141.4
Polytetrafluoroethylene	327.0
Polyoxymethylene	182.5
Polycaprolactam	228

where T_m is the observed equilibrium melting point of the polymer dissolving into a melt of volume fraction v_p, T_m° is the equilibrium melting point of the same polymer into its pure melt, ΔH is the heat of fusion of one mole of polymer, x is the ratio of molar volume of polymer to that of the second component i, and χ is an appropriate interaction parameter defined per mole of the second component, which may also be polymeric.

It turns out that polymer crystals are often slow melting, form notoriously small crystals, and are mostly impure. Experiments on melting must then be interpreted by taking into account the time scale of the experiment, the crystal sizes, and the purity.

Extended chain crystals (see Fig. 17) are close to thermodynamic equilibrium, so that on very slow heating a sharp thermodynamic equilibrium melting point can be achieved. This melting point will be lowered and broadened as soon as a molecular weight distribution is present (impurity effect). Table 4 lists results of melting point determinations on extended chain crystals. The first three polymer melting points are probably close to the equilibrium melting point for infinite molecular weight, while the last may be somewhat too low because of insufficiently high molecular weight.

These melting points are distinguished from many monomer melting points by the fact that they can be measured only in the heating mode. A polymer chain, once removed from the crystal, will not recrystallize

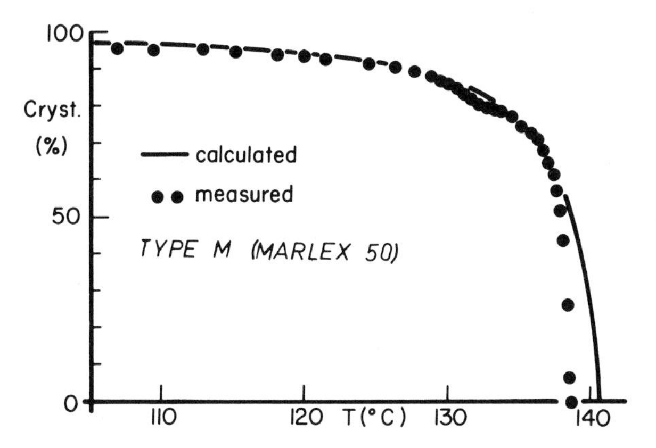

FIG. 23. Equilibrium crystallinity of a broad molecular weight polyethylene as a function of temperature. The solid line is calculated for eutectic separation of all different molecular weights using Eq. (23). Filled circles are experimental points.

into the same crystal morphology at the melting temperature. Considerable supercooling and often special crystallization conditions (as for example high hydrostatic pressure or crystallization from the monomer) may be necessary to reverse the melting process.

A broad molecular weight distribution of polyethylene has also been analyzed. The polyethylene was crystallized into an extended chain conformation by the use of about 4000 atm pressure. The resulting extended chain lamellae showed a broad distribution in length parallel to the molecular chains. The dilatometric melting curve is shown in Fig. 23. The lower portion of the dilatometric melting curves (crystallinity vs temperature), up to about 135°C, could be perfectly matched using Eq. (26). This match suggests that separation into different molecular weights occurred according to a eutectic multicomponent phase diagram. At the higher temperature the sample melted more sharply than calculated. This can be taken as an indication that the polymer chains in the remaining crystals were cocrystallizing in a common lamella. In agreement with the idea of cocrystallization, no lamellar thicknesses corresponding to the largest molecular weight

TABLE 5

Estimated Melting Temperature in °C of the Last Trace of Extended Chain Crystal Samples for Different Heating Rates

Polymer	Heating rate			
	1°C/min	5°C/min	10°C/min	50°C/min
Polyethylene	151	154	157	165
Polyoxymethylene	191	196	201	206
Polytetrafluoroethylene	332	337	340	350

fractions were observed. The lowering of the maximum experimental melting point from the high molecular weight value in Table 4 can be attributed to mixed-crystal formation of the different molecular weights.

Such a simple melting behavior is found only on slow heating. All extended chain polymer crystals analyzed superheated as soon as the heating rates were increased. The crystals could be heated faster than the melt-crystal boundary could progress toward the interior. The remaining inner portion of the crystal thus superheats temporarily and melts at a much higher temperature. Table 5 lists some typical values of superheating for melting of the last crystal trace. All values were obtained by differential thermal analysis.

Spherulites grown from the melt (see Fig. 16) are metastable with respect to the extended chain crystals. The temperature at which the free energy of these crystals equals that of the melt is usually 5 to 10°C below that of the extended chain crystals. Frequently these crystals also contain large portions of ill-crystallized polymer (low crystallinity), which extends the melting range to very low temperatures. For cooling rates from the melt between 1°C/hour and 1°C/sec, which encompass most actual situations, the melting peak temperature of the resulting spherulitic crystals, as measured by differential thermal analysis, varies only little with heating rate, as is shown in Fig. 24. Many of the resulting crystals have large enough fold length to be metastable

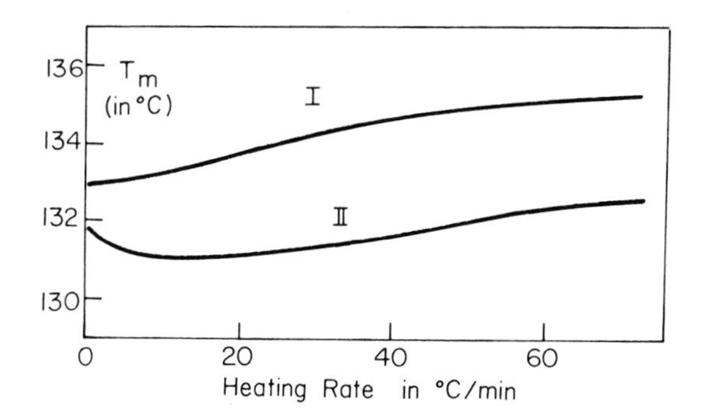

FIG. 24. Melting of spherulitic polyethylene crystals. The melting peaks of differential thermal analysis experiments at variable heating rates are plotted. Curve I shows a crystalline polymer obtained by slow cooling, and Curve II a fast cooled polymer.

up to temperatures close to the melting point. If it is possible to transform a metastable crystal without reorganization on heating directly into the supercooled melt, a path of zero entropy production melting is followed. No irreversible process has occurred during this experiment. In such a case melting occurs where the free energy of the metastable crystal (which can be analyzed by x-ray diffraction, microscopy, etc., at low temperatures) is equal to the free energy of the supercooled melt (whose properties can be extrapolated from the stable melt). Zero entropy production melting points are important for a thermodynamic characterization of defect or small crystals. Only a small amount of superheating is discovered in these crystals, which are 1/10 or less the length in the molecular chain direction found in the extended chain crystals.

It must not be overlooked, however, that by careful crystallization or annealing all intermediate stages to the extended crystals can be produced. Also, on excessive quenching very unstable spherulites may be produced as, for example, on cold crystallization from the glassy state. These crystallite distributions will behave similarly to the dendrites described below. Besides experimental time scale variation, polymers of different inherent crystallization, melting, and reorganization rate will produce varying effects. The region of zero entropy production is limited at low heating rates by reorganization or recrystallization, at high heating rates by superheating. The region where at least approximate zero entropy production melting can be achieved is dictated by the chemical nature of the polymer and the detailed crystal structure on hand. If the thermal behavior of a polymer is quite well known, observation of melting as a function of heating rate can be used as an analytical tool.

The morphology of folded chain crystal lamellae (Fig. 13) can be easily analyzed. Interference microscopy, electron microscopy, and x-ray diffraction enable a complete description of unit cell, crystal shape and dimensions, chain length between folds, and sometimes internal defects. Because of the usually small fold length, even for laterally large lamellae, reorganization, mainly by fold length increase, occurs on heating. The actual process is different for different polymers. Polyethylene, for example, can be caused to increase smoothly in fold thickness via a solid state thickening mechanism over a range of heating rates. As a result, the microscopically observable melting point of solution grown, folded chain lamellae decreased smoothly from 130°C for 0.5°C/min heating rate to a constant 121°C for heating rates from 20 to 3000°C/min. Zero entropy production is reached only above 20°C/min. Polyoxymethylene and polycaprolactam, in contrast, showed thickening, mainly by recrystallization. In the latter case, differential thermal analysis shows two or more melting peaks on slow heating, often separated by a clearly visible recrystallization exotherm. In all cases, a zero entropy production melting region could be established with reasonable certainty. Superheating is of no concern in these extremely thin crystals. Equation (24) permits calculation of a surface free energy if we assume that no other defects contribute to the equilibrium melting point lowering. The melting point lowering calculation for thin lamellae is based on the Thomson equation (1871) and was first applied to lamellar crystals by Tammann (1920):

$$\Delta T = (2\sigma T_0)/(\Delta H_f)\ell. \tag{27}$$

ΔH_f is the heat of fusion per cm^3, σ the specific surface free energy, and ℓ the fold length in cm. For polyethylene, polyoxymethylene, and polycaprolactam, specific surface free energies of 80, 50, and 30 $ergs/cm^2$ have been calculated from fold length determination and melting point data, on the assumption that all melting point lowering is caused by the surface free energy.

BIBLIOGRAPHY

1. Textbooks:

   P. J. Flory, Principles of Polymer Chemistry, Cornell Univ. Press, Ithaca, New York, 1953.

   F. W. Billmeyer, Jr., Textbook of Polymer Science, Wiley-Interscience, New York, 1962.

   P. H. Geil, Polymer Single Crystals, Wiley-Interscience, New York, 1962.

2. Reference books:

   B. Wunderlich, Macromolecular Physics. Vol. 1: Crystal Structure, Morphology, Defects, 1973; Vol. 2: Crystallization, Annealing, Melting, in press, Academic Press, New York.

   J. Brandrup and E. H. Immergut, eds., Polymer Handbook, Wiley-Interscience, New York, 1966.

   H. F. Mark, N. G. Gaylord, N.M. Bikales, eds., Encyclopaedia of Polymer Science and Technology, Wiley-Interscience, New York, 1964—.

3. Journals

   Journal of Polymer Science, Polymer Physics Ed.

   Advances in Polymer Science

   Journal of Macromolecular Science, Physics

Chapter 14

BIOLOGY AND SEMICONDUCTION

H. Ti Tien

Department of Biophysics
Michigan State University
East Lansing, Michigan

I. INTRODUCTION

Until quite recently the possibility of electronic conduction in living systems was not seriously considered, with perhaps a few notable exceptions to be mentioned below. Traditionally, the origin of electrical potentials observed in biological systems, for instance, have been attributed almost exclusively to ionic permeability. The measured current was a result of the movement of ions in an aqueous environment. This traditional point of view was apparently influenced by the fact that electronic conduction could only take place in metallic conductors. The application of modern concepts of electronic conduction and solid state physics to biology is generally accredited to A. Szent-Györgyi [1] who, in a brief note published in 1941 entitled "Towards a New Biochemistry," drew attention to the fact that energy in living systems might be transferred by conduction bands. Szent-Györgyi's short note thus ushered in a new era of research and understanding in life sciences. The possibility that mechanisms analogous to electronic semiconduction might play a critical role in living systems was indeed appealing to physical-minded investigators, who have since initiated extensive studies covering practically every major aspect of biological processes in terms of semiconduction mechanisms. To name a few, these include photosynthesis, respiration, vision, nerve excitation, and carcinogenesis.

Prior to the development of solid state physics, it should be mentioned that several investigators had considered the possibilities of electronic conduction in certain biological systems. Perhaps the most comprehensive but least known was the redox theory. In 1928, Lund [2] suggested that transmembrane potentials were caused by redox potentials across the membrane. Electron flow in certain brain cells was proposed by Stiehler and Flexner [3] in 1938. Similar ideas concerning active transport across a plant membrane by a redox mechanism were proposed by Lundegårdh [4]. The suggestions of these earlier authors should be considered as forerunners of applying the concepts of electronic conduction to biology before the development of solid state physics.

To a casual reader the connection between semiconduction and biology might seem far-fetched. If, however, one considers that living systems are composed of biological materials which are also organic in nature, the existence of semiconductive properties in living systems should not be too surprising, in view of the fact that many organic compounds exhibit some of the properties of the well-understood inorganic semiconductors. Therefore, in this chapter we will begin our discussion by considering briefly the properties of organic semiconductors. Next, we will describe living systems in terms of the biological membranes, since without due consideration to the structural organization and functions of living systems, the extension of concepts of semiconduction to biology would not be very meaningful. Finally, the remainder of this chapter will be devoted to a description of various biocompounds and biological systems in which the relevance of semiconduction has been demonstrated.

The aims of this chapter are therefore twofold: (i) to bring together the various investigations of semiconductivity of important biocompounds, and (ii) to present a coherent picture concerning the role of semiconduction in biology in terms of biological membranes.

II. ORGANIC SEMICONDUCTORS AND BIOCOMPOUNDS

A. Basic Aspects of Organic Semiconductors

In this section a very brief summary of some of the salient features of organic semiconductors is given. This serves as a prelude to the discussion that follows on semiconductivity of biocompounds.

Organic semiconductors in general differ from inorganic semiconductors in that the molecules are held together by very weak forces. The hydrocarbons with conjugated systems have been the most extensively studied organic semiconductors. Two typical examples of conjugated hydrocarbons are shown in Fig. 1. In each molecule the carbon atoms are joined together alternately by single and double bonds. Actually, the bond distance between the carbon atoms is about the same regardless of whether they are joined by single or double bonds in conjugated compounds. Thus, the delocalized electrons (π-electrons) can move throughout the whole molecule. It is generally accepted that it is the excitation of these π-electrons that leads to the generation of mobile charged carriers.

Experimentally, it has been found that organic compounds reported to be semiconductors obey the relation

$$\sigma = \sigma_0 \exp\left(-\frac{E}{kT}\right), \tag{1}$$

where σ is the "dark" conductivity (Ω^{-1} cm^{-1}), σ_0 is a constant, and E is the energy associated with the process [5]. In general, organic semiconductors appear to obey Ohm's law up to field strengths of the order of 10^3 V/cm. The quantity E, evaluated from a plot of $\log \sigma$ vs

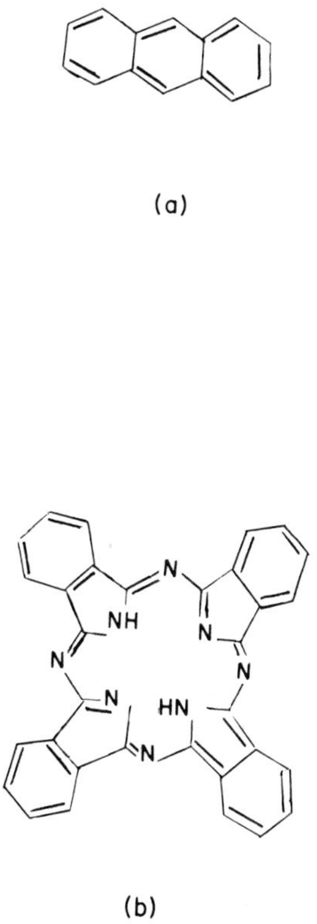

(a)

(b)

FIG. 1. Structural formulas for two typical organic semiconductors. (a) anthracene, (b) phthalocyanine. The carbon atoms (not shown) in each molecule are joined by alternate single and double bonds. Actually the electrons involved in bond formation are delocalized π-electrons which can take part in conduction.

$1/T$, has been interpreted as follows: (i) E is identified with the energy gap between the ground state and the lowest excited singlet or triplet state of the molecule [5, 6]. (ii) E is associated with the conduction band in terms of the ionization potential (IP) and the electron affinity (EA) of the molecule [7]. (iii) The magnitude of E is a measure of an energy barrier inherent in the transport process of carrier migration [8].

It should be pointed out that a number of authors prefer to use 2kT in the exponential term of Eq. (1). The reported value is, therefore, twice the value of E. In the second case (ii), it has been suggested that

$$E = \frac{1}{2}(IP - EA) - P \quad , \qquad (2)$$

where P is the polarization energy produced by the interaction of charge carriers with the surrounding molecules [7]. Apart from a measurable conductivity in the dark, most organic semiconductors, including the two examples in Fig. 1, are also found to be photoconductors. Since in organic semiconductors the molecules are held together by very weak forces owing to a lack of overlap between the orbitals of neighboring molecules in the compound, the spectral excitation curve for photoconductivity and the absorption spectrum of the compound are generally found to be very similar.

For organic semiconductors, the mechanism of conduction has been explained in terms of exciton theory. For illustration, we will again use the anthracene crystal as a typical organic semiconductor (see Fig. 1). As is the case with inorganic semiconductors, the basic equation for conductivity is given by

$$\sigma = ne\mu \quad , \qquad (3)$$

where n is the charge carrier density, e is the electronic charge, and μ is the mobility of the carrier. The charge carrier density in an organic crystal such as anthracene is determined by the energy introduced into the crystal. This energy can be introduced thermally, optically, or electrically. Thus, the first stage in conduction is the production of excitons, for instance, by light. It is assumed that during its lifetime an exciton can travel through the whole molecular aggregate. Depending on the nature of the optically excited state, either a singlet exciton or a triplet exciton may be formed. Since an exciton is generally depicted as consisting of an electron-hole pair, an organic semiconductor will be conductive only when dissociation has taken place. One way this dissociation can come about is by the interaction of pairs of excitons [9]. However, a more likely mechanism of dissociation is for an exciton to interact with imperfections in the crystal. These imperfections in the crystal in the case of an organic semiconductor are usually due to impurities (or donor and acceptor molecules) situated at the electrode-crystal interface [6].

In the absence of impurities, the dark conductivity of organic semiconductors is often identified with conventional intrinsic semiconductivity. In general, the conduction mechanism of organic semiconductors is explained in terms of either the electronic band model or the hopping model developed for inorganic semiconductors. The only novelty is that aggregates of organic compounds instead of covalent-bonded inorganic crystals are under consideration.

B. Biocompounds

Extending the ideas developed for organic semiconductors, we can imagine that a conduction band also exists in the compounds of biological origin (i. e., biocompounds). Excitation of an electron in a biocompound, by light, thermal energy, or externally applied field, promotes it to a higher energy level, the so-called first "singlet" state above the ground state. The minimum energy necessary for this process may be similarly evaluated from measurements of conductivity as a function of temperature. The excitation energy may exist in the form of excitons capable of moving around in the biocompound. Presumably an exciton thus formed can dissociate into an electron and a positive hole. Although the exact nature of this step is not known, two possible mechanisms have been proposed: (i) by interacting with an "imperfection" in the biocompound, and (ii) by colliding with another exciton. In either case, conductivity may take place in a biocompound as a result of movement of both electrons and holes, in analogy with inorganic semiconductors.

The basic problems to be solved in biological semiconductors are the mechanism of charge transport and the amount of energy necessary for charge generation. As we will see presently, biocompounds in living systems are usually organized in the form of membranes separating two aqueous phases. Thus, if sufficient energy is absorbed by a membrane, electrons and holes may be generated. Under the influence of an applied electrical field or a redox potential across the membrane, positive holes and electrons should move toward opposite sides of the membrane.

It should be mentioned that purification of biocompounds is exceedingly difficult, if not impossible, owing to heat instability and other labile factors. Therefore, semiconduction in the case of biological materials or biocompounds is not "intrinsic" but due to extrinsic "impurities." Since single crystals of biocompounds of sufficient size are practically nonexistent, the experimenters interested in studying biocompounds are forced to use polycrystalline powders, which are usually studied between electrodes under pressure. To exclude impurities such as air or water vapor, most of the measurements have been carried out either under high vacuum or in a dry state [6].

III. MEMBRANE BIOPHYSICS

The study of living systems or biology by modern techniques has revealed the ever present and highly ordered structures known collectively as biological membranes. Before describing some of the basic types of biological membranes relevant to our later discussion, it would be helpful to give a brief consideration of important constituent molecules that have been found in the cellular membrane. These compounds of biological interest will be designated as biocompounds. A list of representative biocompounds is given in Table 1.

As far as is known, lipids form an integral part of all biological membranes. The ubiquitous presence of protein is also indicated. Pigments, quinones, and sterols (especially cholesterol), on the other

TABLE 1

A List of Important Biocompounds

Classification	Example	Remarks
Lipids	Fatty acids (oleic acid) Phospholipids (lecithin)	Found in all biological membranes
Proteins	Various structural proteins Cytochromes Ferredoxins ATPase	Present in specialized membranes
Pigments	Chlorophylls (chlorophyll a, b, etc.) Retinenes (vitamin A, retinal) Carotenoids (β-carotene)	Involved in primary energy transducing steps
Quinones	Ubiquinones Plastoquinone Vitamin K	Involved in electron transport
Nucleic acids	DNA RNA	Carrier of genetic information
Sterols	Cholesterol	In most membranes except bacteria

hand, appear to play a more specialized structural and functional role. For instance, pigments are found only in certain types of membranes. Nucleic acids, performing the well-known function as carriers of genetic information, are included in the list, since they have been found to exhibit some semiconductivity properties [6].

A. Interfacial Chemistry of Lipids

The contribution of lipids to biological membrane structure has been recognized by many investigators. Gorter and Grendel, using the Langmuir trough technique [10], deduced that the plasma membrane of various cells was about two molecules in thickness [11]. This concept proposed by Gorter and Grendel has dominated our thinking concerning the structure and function of biological membranes and the

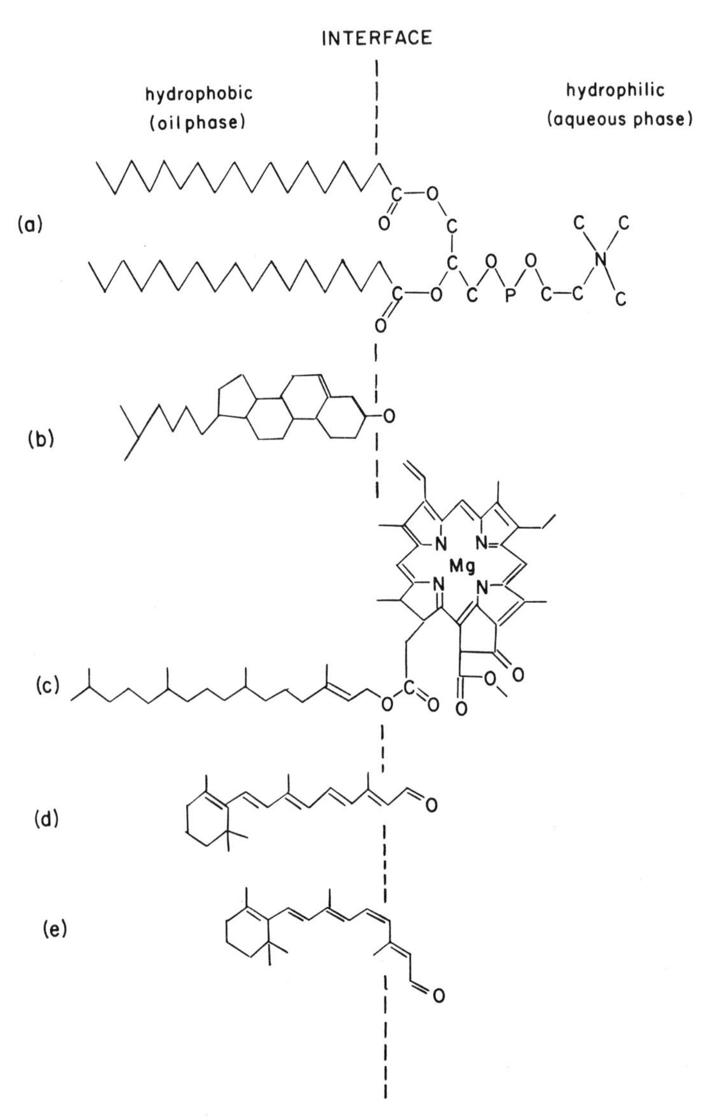

FIG. 2. Orientation of some of the most important biocompounds at an oil-water interface: (a) phosphatidyl-choline (lecithin), (b) cholesterol, (c) chlorophyll a, (d) all-trans retinal, (e) 11-cis retinal. They are shown in an energetically favorable configuration. The dielectric constant of the oil and aqueous phase are about, respectively, 2 and 80.

subsequent model membrane studies which will be further discussed. Unlike all other biocompounds given in Table 1, the structure of lipids is most unique in that one part of the molecule is water soluble (hydrophilic) and the other oil soluble (hydrophobic). For example, a lecithin molecule (phosphatidyl choline) when introduced at an oil-water interface will spontaneously orient itself with the long-chain hydrocarbon portion immersed in the oil phase (see Fig. 2). The consequences of orientation of lipid molecules at interfaces, leading to a reduction of interfacial free energy, are of fundamental importance in the formation of molecular aggregates and biological membranes. In Fig. 2, the structural orientations of some most important biocompounds at an oil/water interface are illustrated. For further details on the effects of adsorption at interfaces and molecular aggregation in solution, review articles and monographs are available [10, 12-14].

B. Types of Biological Membranes

Our knowledge concerning biological membranes comes from electron microscopy and biochemical and physicochemical studies. Biochemical analysis reveals the composition of the membrane, whereas the physicochemical approach provides us with possible structural arrangements in terms of known biocompounds. Electron microscopy gives us the visual images of biological membranes. At present, the resolving power and experimental techniques of electron microscopy are such that skillful interpretations are necessary. Nevertheless, through electron microscopy the most important biological membranes thus far investigated give an appearance of a triple-layered image. This triple-layered image consists of two denser regions sandwiched in a lighter region; the overall thickness of this limiting structure is about 50-100 Å [15]. From a functional point of view, biological membranes may be divided into five basic types, as summarized in Table 2 (see also Fig. 5).

TABLE 2

Some Basic Types of Biological Membranes[a]

Membrane type	Composition	Structure
Plasma	Lipids: 40-50% phospholipids cholesterol Proteins: 60-50%	Two opaque layers separated by a lighter layer, overall thickness ~75 Å (i.e., bilayer type)
Thylakoid (chloroplasts)	Lipids: 40% Pigments: 8% Protein: 35-55%	Bilayer type, 50-100Å
Mitochondrial (inner or cristae)	Lipids: 20-40% phospholipids Protein: 80-60%	Triple-layered arrangement, thickness about 75Å, bilayer type
Rod outer segment	Lipids: 40%, of which 80% are phospholipids Retinene: 4-10% Opsin	Layered disks or sacs (bilayer type?)
Nerve	Lipids: 80% Proteins: 20%	Bilayer type

[a] Data taken from the following sources:

Symposium on the Plasma Membrane, American Heart Association, Inc., New York, 1962.

Biochemistry of Chloroplasts (T. W. Goodwin, ed.), Academic, New York, 1966.

J. S. O'Brien, J. Theoret. Biol., 15, 307 (1967).

S. K. Malhotra and A. van Harreveld, in Biological Basis of Medicine (E. E. Bittar and N. Bittar, eds.), Vol. 1, Academic, New York, 1968.

C. Theoretical and Experimental Models for Biological Membranes

From the sketches of the various biological membranes presented in Fig. 3, it seems probable that all cellular membranes are constructed according to a basic design: an ultrathin layer of lipid of about bimolecular thickness (~50Å) with a layer of protein molecules and/or other materials adsorbed on the lipid surface. In other words, the structure of all basic biological membranes is depicted as a protein-lipid-protein sandwich separating two aqueous phases (see Fig. 3).

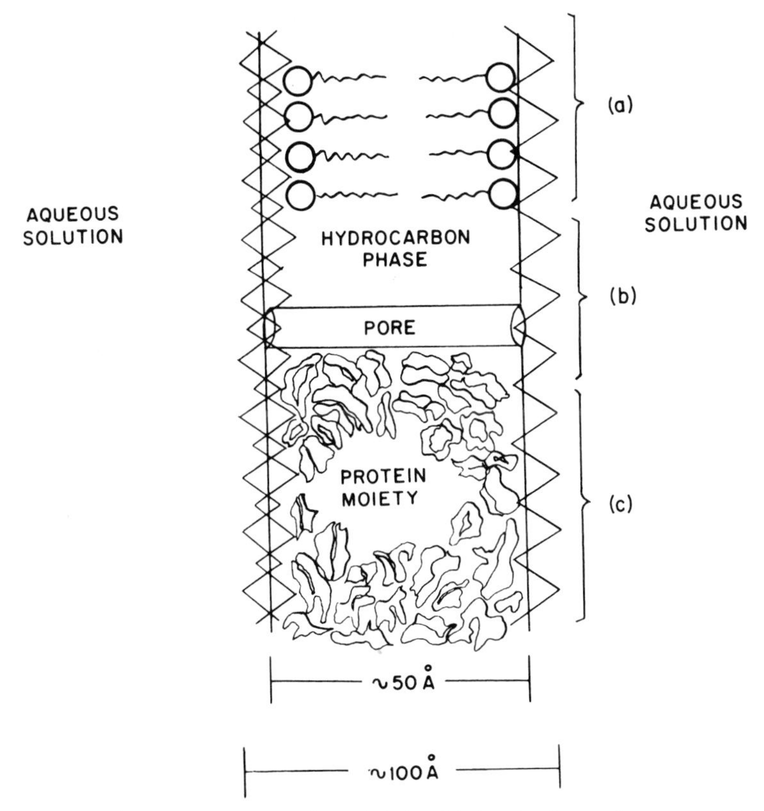

FIG. 3. A suggested model for biological membranes. This is the popular bimolecular leaflet model consisting of a Gorter-Grendel lipid bilayer sandwiched between two thin layers of nonlipid materials. (a) The "unit" membrane; (b) modified BLM containing pores; (c) modified BLM containing protein. See text and Ref. [12] for further details.

The interfacially adsorbed layers on each side of the lipid need not be identical. In fact, according to the "unit membrane" hypothesis of Robertson [16], all biological membranes consist of a single lipid bilayer of unspecified composition coated on one side by a monolayer of nonlipid and on the other by a monolayer of a different kind of nonlipid material (see Fig. 3). As far as theoretical models of biological membranes are concerned, they appear to be all based on a variation of the lipid bilayer concept (for a review, see Ref. [17]).

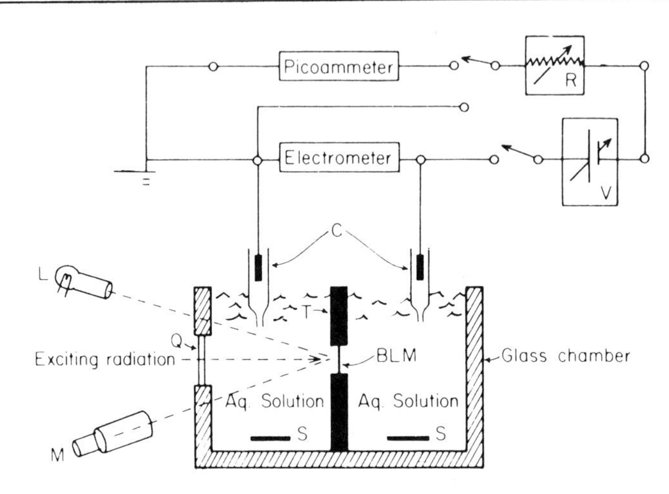

FIG. 4. Apparatus for studying isolated bimolecular (black) lipid membranes (BLM). R, resistor bank (0-10^{10} Ω); V, polarizing voltage source; L, light for observing membrane formation; M, viewing tube or microscope (20-40 x); Q, quartz window for uv excitation; C, calomel electrodes with salt bridge; T, teflon chamber or partition; S, stirring bar (for details see Ref. [19]).

The development of an experimental model, incorporating the findings of electron microscopy and other experimental and theoretical investigations, has been accomplished recently [18, 19]. An ultrathin lipid film less than 100 Å thick can be formed in aqueous solution. This is done as follows:

A Teflon cup with a small hole in the side sits in a glass container. The cup and the outer container are filled with an aqueous solution just above the hole. A minute quantity of a lipid solution is then spread over the hole to form a thin lipid layer. Under favorable circumstances, the thin lipid layer exhibits interference colors followed by the appearance of "black" spots, leading eventually to the formation of a "black" lipid membrane separating two aqueous phases (Fig. 4). Since the thickness of this type of membrane does not differ greatly from the length of two lipid molecules used, and owing to the optically "black" appearance, membranes of this type have been referred to as bimolecular, bilayer, or black lipid membranes (or BLM for short). As

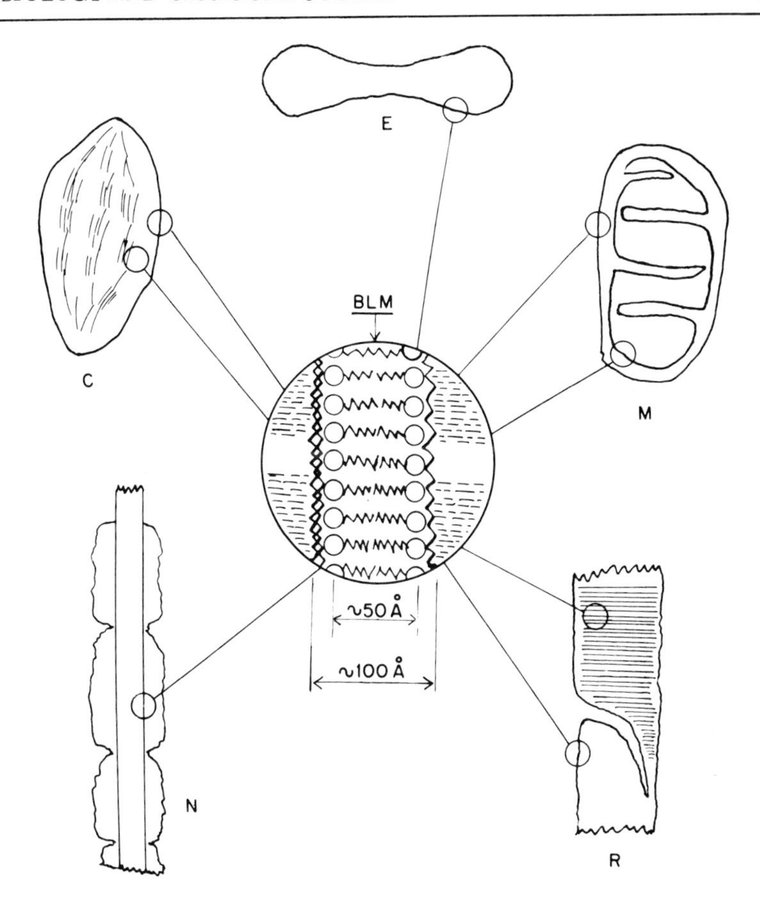

FIG. 5. Schematic illustration of the five basic types of biological membranes as they are generally visualized in electron micrographs and their molecular interpretation according to the bimolecular leaflet model (shown in the center-BLM). E, erythrocyte (plasma membrane); C, chloroplast (thylakoid membrane); M, mitochondria (cristae membrane); N, nerve axon (myelin); R, rod outer segment membrane.

is shown in Fig. 4, by placing electrodes across the BLM its transverse electrical properties can be measured, and controlled chemical studies can be undertaken. At present, it is generally recognized that BLM are the closest approach to models for biological membranes. As we will discuss in later sections, BLM have in fact been used as models for all the five basic types of biological membranes summarized in Table 2 (see also Fig. 5).

TABLE 3

Biological Transducers

Functional unit	Basic structure involved	Transformation
Chloroplast	Thylakoid membrane	Light to chemical energy
Mitochondrion	Cristae membrane	Food to chemical energy
Retina	Rod and cone outer segment membranes	Light to electrical energy
Nervous system	Nerve membrane	Chemical to electrical energy
Inner ear	Tectorial membrane	Sound to electrical energy

D. Biological Transducers

As has been suggested by Green [20], certain biological units can be described as transduction devices of modern technology. A list of such biological transducers in which semiconduction mechanisms may be involved is given in Table 3. Although the molecular constructs of these transducers are not known, there is little doubt that such "devices" must exist in the living system in order to carry out highly complex and integrated vital processes of transforming energy from one form into another. In view of the fact that the lamellar structure is characteristic of most if not all biological membranes, it seems probable that there is a basic design common to all biological transducers. Therefore, in our discussion of semiconduction and biology, the crucial role played by membranous structures will be stressed. In addition, experiments on model membranes will be described, since they provide unique opportunities to test our hypothesis about the operation of living systems at the molecular level.

E. Electronic Conduction as Primary Driving Force of Life

According to a suggestion by Szent-Györgyi [21], the life cycle on this planet can be considered basically as a movement of electrons.

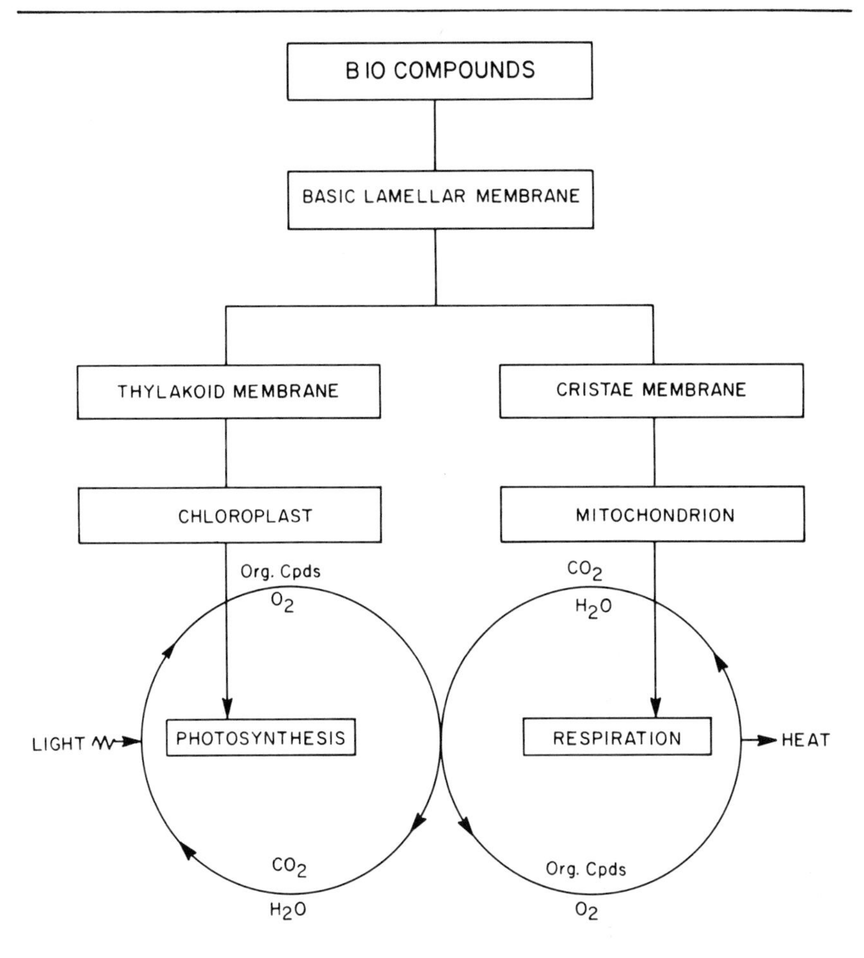

FIG. 6. The life cycle and its relationship to biocompounds via biological transducers (chloroplast and mitochondrion) and membranes. See text for a full explanation.

Before elaborating on this theme, we will depict the overall view of the life cycle. This is represented by an infinity sign shown at the bottom of Fig. 6. These coupled circles represent a steady state transducing system of life. The two basic processes of life are photosynthesis and respiration, which are carried out, respectively, in the chloroplasts and mitochondria. The overall life cycle in terms of biocompounds and biological membranes is summarized in Fig. 6. In photosynthesis, the sun's energy in the form of photons promotes electrons in the ground

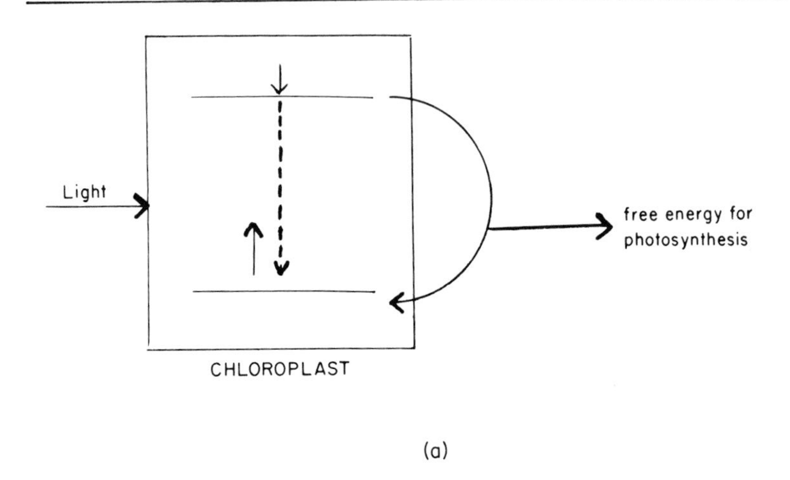

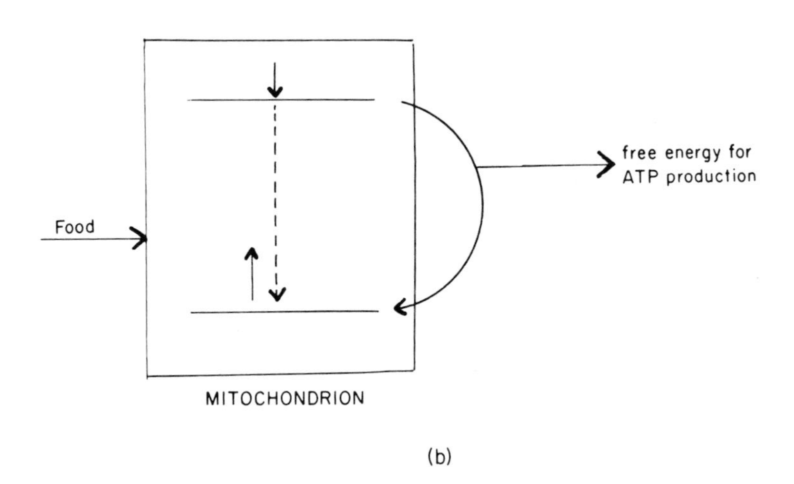

FIG. 7. (a) Symbol of photosynthesis. (b) Symbol of oxidative phosphorylation. Electrons are represented by arrows (after Szent-Györgyi [21]).

state (valence band) to a higher energy state (conduction band). Before falling back to the ground state, the free energy of the electron is converted into chemical bond energy. The process of photosynthesis, taking place in the chloroplast, is explained by Szent-Györgyi in terms of symbols. This is shown in Fig. 7(a). The organic compounds and oxygen produced by photosynthesis are utilized by other living systems (i.e., animals) that are incapable of making direct use of photon energy of the sun. The conversion of chemical bond energy by animals, termed respiration, takes place in a membranous structure called mitochondrion. In a mitochondrion, electrons are again promoted to higher states, the driving force being provided by combustion processes of organic compounds (i.e., food stuff) and oxygen. The chemical energy is stored in the pyrophosphate bonds of adenosine triphosphate (ATP), which is the universal currency of life. The process of respiration in terms of electron excitation is illustrated in Fig. 7(b). Thus, we see that life, as suggested by Szent-Györgyi, is driven only by electrons. The movement of electrons from higher potentials to which they have been promoted by the sun's photons to their original lower potentials via biological transducers constitutes an electric current. This electric current or electronic conduction, viewed as the primary driving force of life, will be the main theme of our discussion in the following sections.

IV. SEMICONDUCTION AND PHOTOSYNTHESIS

A. General Considerations

In simplest terms, photosynthesis can be considered as a process by which the green plant reduces CO_2 to carbohydrates and oxidizes water to oxygen with the aid of sunlight. In terms of its magnitude and importance, it can be said without any hesitation that photosynthesis is by far the most important chemical reaction on this planet. In regard to the topics to be covered in this chapter, the description of photosynthesis from the viewpoints of solid state physics has been most extensive. The main evidence that semiconduction is relevant to photosynthesis may be summarized as follows:

(a) In dried plant chloroplasts and chromatophores of bacteria, the dark conductivity varies with the absolute temperature according to Eq. (1).

(b) The dried samples are photoconductive.

(c) The photoconductivity excitation spectra of dried materials are very similar to the absorption spectra of the chloroplasts.

(d) The photocurrent is dependent on the light intensity.

(e) A reconstituted membrane using chloroplast extract exhibits both the photovoltaic effect and photoconductivity.

We will consider these findings in more detail after a brief description of the physical and chemical aspects of photosynthesis accepted at present.

Basically photosynthesis is composed of two distinct types of reactions: a photophysical process and a series of dark reactions. The dark reactions have been elucidated in detail by Calvin and Bassham [22]; these reactions appear to be mainly a biochemical problem. The photophysical process or the primary step of photosynthesis has not been completely understood and constitutes an area of much current research. Perhaps it is because of this lack of precise understanding at the present time that a large number of schemes have been proposed. We will consider only the most popular "Z" scheme, in which the electronic conduction is central to the argument [23].

To begin with, the "Z" scheme postulates two photosystems, I and II. As depicted in Fig. 8, the sequence of reactions starts with the absorption of light (< 680 nm) by system II which promotes an electron (originated from H_2O ?) to a higher energetic state. The excited electron then cascades down the so-called "electron-transport chain" (linked circles, Fig. 8) from about 0 V potential (relative to the standard hydrogen electrode potential) to about +400 mV, whereupon it receives another boost from light energy (far-red) absorbed by system I. Systems I and II are coupled by a system of electron carriers including

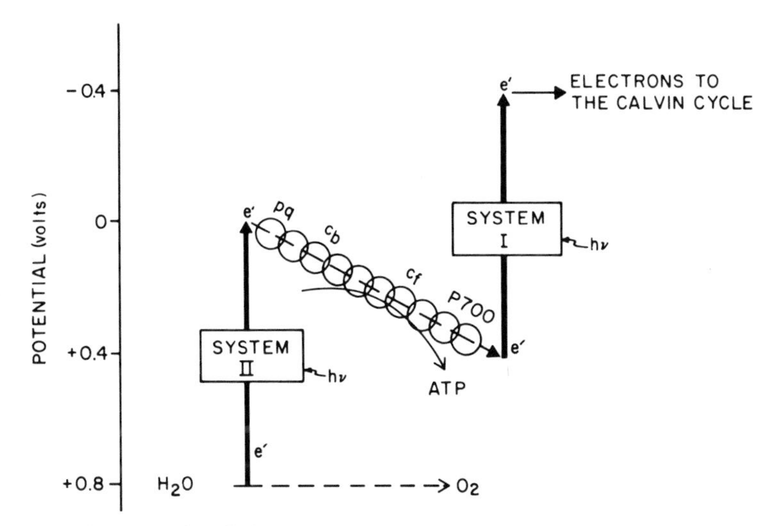

FIG. 8. The "Z" scheme of photosynthesis (see text).

plastoquinone (pq), cytochromes b and f (C_b and C_f), and pigment-700 (absorption band at 700 nm). The free energy of the electron, as it travels downhill, is trapped in the chemical bond energy of ATP. The second light quantum excites the electron from about +400 mV to about -400 mV. The energetic electron, a powerful reducing agent, thus produced is believed to be associated with ferredoxin (an iron-containing protein) to initiate the dark reactions. It should be stressed that the above scheme involving the movement of electrons is merely a working hypothesis. There are shortcomings and omissions. For instance, the mechanism by which oxygen is evolved is not considered at all. Alternate schemes, equally plausible, have also been proposed [23].

In the above description, we have left the so-called photosystem I and photosystem II completely unspecified. The reason is our lack of knowledge at the present time. Nevertheless, these "systems," appropriately called "photosynthetic apparatus," must reside in the chloroplast. More specifically, the photosynthetic apparatus and thylakoid membranes (also known as grana) of the green plant, as revealed by the electron microscope, must be synonymous [24].

The chemical composition of thylakoid membranes is given in Table 2. The basic structure of the membranes is believed to be made of a lipid bilayer together with the pigment molecules, the structural and enzymatic proteins, and the electron transport chain. Under the electron microscope, an individual thylakoid membrane is about 50-100 Å thick. The molecular arrangement of the constituent biocompounds in the thylakoid membrane is unknown, but from the principles of interfacial chemistry briefly described in Section III, the hydrophilic groups of the various components must be located at solution-membrane interfaces. For example, the magnesium-containing porphorin group is oriented toward the aqueous phase, whereas the phytol chain (hydrophobic) extends itself into the interior of the membrane (see Figs. 2 and 5).

B. Mechanisms of Light Conversion

The primary process of light conversion in the photosynthetic apparatus may be considered in terms of thylakoid membranes. A schematic picture showing a possible molecular arrangement in a thylakoid membrane is given in Fig. 5. The classical experiments of Emerson and Arnold [25], who were studying photosynthesis with flashing light, demonstrated that a photosynthetic apparatus consisted of several hundred energetically coupled chlorophyll pigments. It is now generally believed that two types of mechanisms are possible in the primary light conversion process: (i) energy migration and (ii) electron-hole separation. In the first case, light energy absorbed by any one of a number of different pigments is rapidly transferred without significant loss to a so-called "reaction center." By a process still unknown, the migrated excitation energy upon arrival at the reaction center is separated into two parts; one part is of reducing nature (as electrons) and the other part is of oxidizing character (as positive holes). The electrons and holes thus separated are prevented from recombining, possibly by the hydrophobic (lipid) layer present in the interior of the membrane. The second mechanism, suggested by Calvin, postulates that electrons and holes may be transferred between different reaction centers. The origin of this idea was first conceived by van Niel, who formulated an overall view of photosynthesis in terms of the oxidant (OH) and reductant (H). The translation of these chemical terms into "holes" and "electrons" was made by Katz [26] on the basis of many combined experiments of photosynthesis and fluorescence of various monocellular organisms (e.g., purple sulfur bacteria and chlorella). Katz suggested that on theoretical grounds two types of processes may occur in the chloroplast when light energy is absorbed. The first possibility is the formation of an exciton. Quantum mechanically the exciton is not necessarily localized at a given chlorophyll molecule but is spread out over the whole photosynthetic unit. The second possibility, according to Katz, is to assume that chlorophyll molecules are in two-dimensional "crystals" which are photoconductive. The absorption of light energy excites an electron to the conduction band and leaves a positive hole in the valence band. Thus, electron and hole are free to move around. Katz further suggested that the electrons can be transferred to an electron acceptor, which is thereby effectively reduced, and can then react with CO_2 or its derivatives to produce carbohydrate. Similarly, the holes can be combined with an electron donor, which is thereby oxidized. The advantage of the second scheme is that the primary light-generated products (electrons and holes) are mobile charge carriers and will permit oxidation and reduction to be carried out at different places. As was pointed out by Katz, the exciton or energy transfer scheme on the other hand might initiate almost any type of chemical reaction and therefore requires rather detailed assumptions about the chemical nature of the energy acceptor.

The scheme of Calvin involves both possibilities suggested by Katz but operates in a sequential manner. Figure 9 illustrates the scheme envisioned by Calvin [27]. It is postulated that chlorophyll in the ground state absorbs light which excites it to its lowest singlet

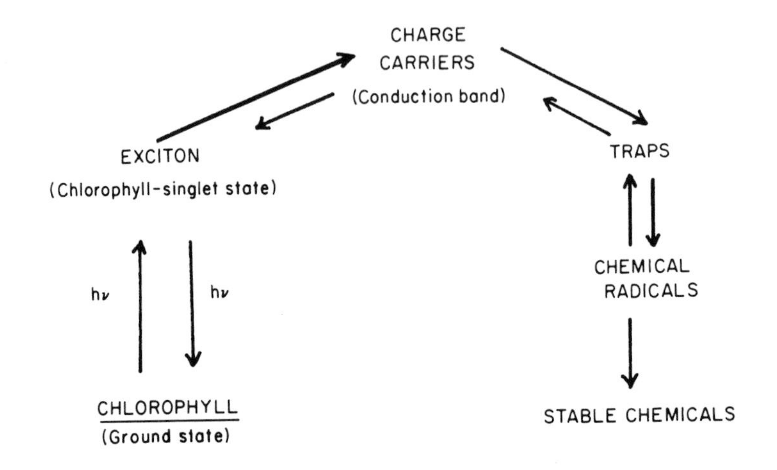

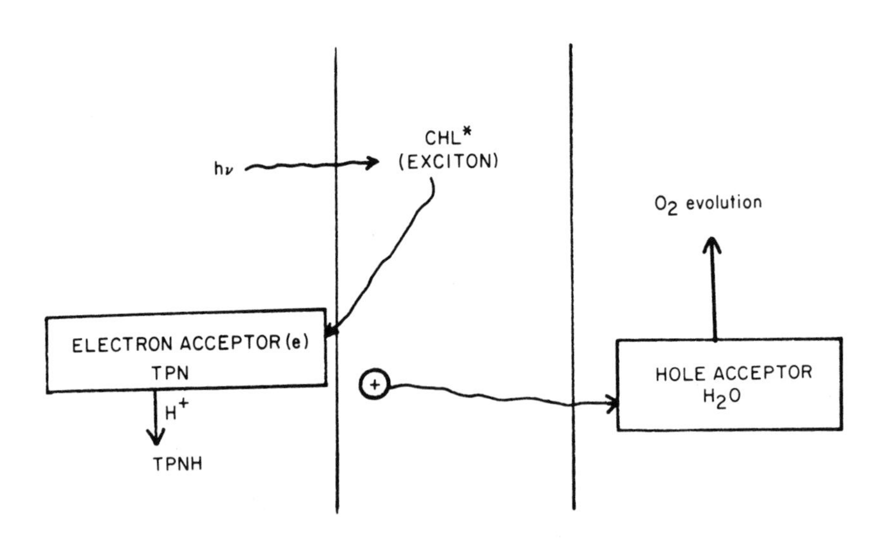

FIG. 9. Scheme for light transduction by chlorophylls in the chloroplast (after Calvin [27]).

state. The excited state (exciton) can move around among the chlorophyll molecules by resonance transfer until a point is reached where dissociation takes place. In the conduction and valence bands, electrons and holes can move in different directions. The electrons and holes will move around until they find suitable sites of lower potential energy into which they become trapped. They remain in separate electron and hole traps until suitable chemicals can interact with them. Finally, stable chemicals are produced from trapped electrons and holes via a series of steps. An essential feature of Calvin's theory, not unlike the scheme proposed by Katz, is that it makes possible the location of the primary photoproducts on opposite sides of a lamellar structure. The importance of photoconduction is implied in this scheme. Conceptually, the Katz-Calvin scheme is very attractive. The high energy electrons and holes generated by light are physically separate, thus effectively preventing recombination.

The theory of electron transfer between donor and acceptor molecules in photosynthesis leading to the production of ATP and O_2 evolution can be viewed as, basically, a redox reaction [28, 29]. The overall scheme, in its most basic form, can be written in the following manner:

$$\mathrm{Chl} + h\nu \rightarrow \mathrm{Chl}^{*} , \tag{4}$$

$$\mathrm{Chl}^{*} + \mathrm{Chl} \xrightarrow[\text{Ionization}]{\text{exciton migration}} \mathrm{Chl}^{\ominus} + \mathrm{Chl}^{\oplus} , \tag{5}$$

$$\mathrm{Chl}^{\ominus} + \mathrm{A} \rightarrow \mathrm{Chl} + \mathrm{A}^{\ominus} , \tag{6}$$

$$\mathrm{ADP} + \mathrm{P_i} + \mathrm{A}^{\ominus} + \mathrm{H}^{+} \rightarrow \mathrm{ATP} + \mathrm{AH} , \tag{7}$$

$$\mathrm{Chl}^{\oplus} + \mathrm{D} \rightarrow \mathrm{Chl} + \mathrm{D}^{\oplus} , \tag{8}$$

$$\tfrac{1}{2}\mathrm{H_2O} + \mathrm{D}^{\oplus} \rightarrow \tfrac{1}{4}\mathrm{O_2} + \mathrm{H}^{+} + \mathrm{D} . \tag{9}$$

Overall,

$$\mathrm{ADP} + \mathrm{P_i} + \mathrm{A} + \tfrac{1}{2}\mathrm{H_2O} \xrightarrow[\text{Chl}]{h\nu} \mathrm{ATP} + \tfrac{1}{4}\mathrm{O_2} + \mathrm{AH} . \tag{10}$$

In the equations above, a light quantum ($h\nu$) is absorbed by the pigment Chl intimately associated with the thylakoid membrane; the excited molecule Chl^* after moving to a suitable site (possibly via exciton migration) is ionized by an unknown mechanism, thus forming an electron and a positive hole. The electron and hole interact, respectively, with an acceptor (A) and donor (D), which are used in turn to generate the primary products, ATP from ADP and inorganic phosphate (P_i) and oxygen from water. The overall reaction as indicated in Eq. (10) represents the redox scheme of photosynthesis in terms of the electron donor and acceptor concept. The light absorbed by the phototransducer provides the necessary free energy to initiate electron and hole formation mediated through the pigments located in the lamellar structure of the thylakoid membrane.

In summary, it may be said that, among nature's most vital processes insofar as living organisms are concerned, photosynthesis is the most likely process in which the ideas developed by solid state physicists and chemists can be readily applied in the understanding of the primary photophysical and photochemical steps of energy transduction. The structure of thylakoid membrane in the chloroplast, as revealed by the electron microscope together with other chemical and physical evidence, appears to be organized in a liquid crystalline array. The presence of conjugated biocompounds in the membrane phase provides favorable pathways for electronic conduction. Although model experiments to date have lent support to semiconduction mechanisms, it is fair to state that much work needs to be done before a complete understanding of the mechanisms involved in photosynthesis is achieved.

C. Experimental Studies

The experimental demonstration of the existence of semiconductivity and its relevance in photosynthesis and other biological processes has met with great difficulties from both practical and theoretical standpoints. If semiconduction existed in the plant chloroplast, as suggested by Calvin, the most direct kind of experiment would be to isolate a thylakoid membrane, put a pair of electrodes across it, and measure the potential generated upon illumination by light [27]. The thylakoid membranes revealed under the electron microscope [30] are too small for this to be carried out with techniques available at present. Furthermore, the complexity of ill-defined biological membranes is not amenable to an understanding in physical chemical terms. Therefore, many investigators interested in demonstrating and understanding biological structure and function have been compelled to study model systems. In the following paragraphs we will give a description of various systems studied in support of the semiconduction theory of photosynthesis.

1. Semiconductive Properties of Dried Chloroplasts

The theory of Katz mentioned above, that photoconductivity might play a role in photosynthesis, was put to an experimental test in 1957. Arnold and Sherwood [31] studied the optical and electrical properties of dried chloroplasts and obtained glow "curves" (light emission as a function of temperature) similar to those made with inorganic semiconductors. Further, the electrical conductivity of these preparations measured as a function of temperature followed the equation for an intrinsic semiconductor [i.e., Eq. (1)].

2. Photoconductivity on Chlorophyll Derivatives

The work of Arnold was soon followed by other investigations. Nelson [32], employing the techniques he developed earlier, measured the photoelectric properties of deposited films made from chlorophyll derivatives. The compounds he studied were all photoconductive. There were two time constants for rise and decay of photocurrent, one less than a second and the other of the order of several minutes. Nelson further observed that the action spectrum of the fast photoconductivity was similar to the absorption spectrum of the pigment used. The experiments of Nelson and Arnold provide clear evidence that dried chloroplasts and chlorophyll derivatives, in the solid state at least, behave like organic semiconductors. It should be mentioned that as early as 1941 many workers in Russia had observed that various organic dyes in a condensed state also behave as typical semiconductors [33, 34].

Other closely related types of experiments have been carried out by Tollin et al. [35] and by Rosenberg and Camiscoli [36]. The former group investigated a number of compounds such as metal-free phthalocyanine and chloranil as a "quantum conversion model." The latter group evaluated activation energies according to Eq. (1). In the experiments of Tollin et al., both the semiconductivity and the photoconductivity of the phthalocyanine were found to increase greatly upon addition of the electron acceptor (chloranil). The phthalocyanine-chloranil model was thought to be analogous to the naturally occurring system in which the primary quantum-conversion process in photosynthesis involves formation of oxidizing and reducing entities by light. The rise time of the photocurrent in their light-pulse experiment was about 1 μsec or greater, which was interpreted to mean that the first excited singlet state was probably involved.

In Rosenberg's experiments, activation energies of chlorophyll a and b in the form of compressed pellets were determined according to Eq. (1). For chlorophyll a and b, E's were found to be 1.12 eV and 1.44 eV, respectively. Rosenberg and Camiscoli further measured the activation energy for photoconduction in chlorophyll b to be 0.36 eV. A much lower value of 0.08 eV was obtained by Eley and Snart [37]. Contrary to the photoconduction interpretation, the results of Rosenberg and Camiscoli were said to lend support to the view of Rabinowitch [38], that photoconduction in chlorophyll was not involved in the primary process of photosynthesis.

3. The Nature of Charge Carriers and Energy Transfer in Model Systems

In addition to various organic dyes investigated, Terenin and his associates [33] have ascertained the nature of charge carriers in solid layers of chlorophyll and its derivatives. They have found that the mobile charge carriers for most biocompounds examined are positive holes. Further, they have also reported that excitation spectra of the photoconduction for these compounds are similar to absorption spectra, consistent with the view of weak interaction forces among the molecules. Nevertheless, energy transfer, according to the finding of Terenin et al., seems likely because in a close-packed layer of chlorophyll molecules a conduction band may exist, which clearly is not possible in a group of isolated molecules.

While the above findings are both stimulating and interesting, the legitimacy of extrapolating the data obtained in systems in a dry state or in vacuo to the thylakoid membrane in the chloroplast must be questioned. For example, these model systems are deficient in at least two aspects: (i) in the chloroplast, membranous structures are involved, and (ii) the aqueous environment that surrounds these membranes is always present. Hence, there is an obvious need for new approaches to these problems before the application of semiconduction theory to photosynthesis can be fully realized. One such new approach, accomplished recently using black lipid membranes, is described in the following paragraphs.

4. Electronic Processes in Black Lipid Membranes (BLM)

As we have stressed in Section III, in living systems, membranous structures are central for both structural organization and function. Therefore recent experiments using black lipid membranes (BLM) formed from chloroplast extracts and photosynthetic pigments are of interest [39]. These BLM were formed on a Teflon support separating two aqueous phases (see Fig. 4).

There are at least two fundamental questions one would like to ask:

(i) Is there any experimental evidence to support the idea of electron conduction in a thin membrane immersed in an aqueous environment?

(ii) If so, what is the source of electrons in the membrane?

Both questions appear to have been answered by the experiments of Braun as early as 1891. Braun [40] employed a glass tube with fine cracks as the membrane. If such a "cracked" tube, for example, were filled with a dilute solution of chloroplatinic acid and immersed in a beaker containing the same solution, a shining mirror would be observed on one side of the glass barrier after passing a direct current of sufficient voltage and duration. The electrodes used by Braun were made of platinum and were placed directly in the solutions across the glass barrier. The side on which the metallic mirror was deposited faced the anode, and on the other side the evolution of a gaseous product was frequently noted.

Braun, and later others, have investigated this highly fascinating phenomenon in detail and termed it "electrostenolysis." More recently, Pope and Kallmann [40a] have reported the use of a thin crystal layer of anthracene as the barrier and observed a number of redox reactions as well as photoconductivity. Among the many interesting observations described in their paper, Pope and Kallmann state that the photocurrent of the system could be as large as the dark current in the presence of strong oxidizing agents such as Ce^{4+} and I_2. In the present-day language of electrochemistry, electrostenolysis simply means that a reduction reaction occurs at the side of the barrier where the positive electrode is situated, and an oxidation reaction takes place on the other surface of the membrane. This explanation implies a movement of electrons through the membrane. Further, it should be mentioned that a very large potential gradient across the barrier is one of the prerequisites for the phenomenon. In other words, the barrier or membrane must possess a <u>high</u> electrical resistance.

Returning to our discussion of black lipid membranes (BLM) formed from chloroplast extracts, such BLM upon illumination can exhibit two interesting photoelectric phenomena: (i) a photovoltaic effect and (ii) photoconductivity. In the case of the photovoltaic effect, an open-circuit photo-emf greater than 100 mV can be generated across the BLM under asymmetrical conditions. This is created by the addition of a modifier to one side or different modifiers to the opposite sides of the bathing solutions. For example, the presence of Fe^{3+} in one side of the BLM separating two aqueous solutions (0.1 M acetate at pH 5) is generally used. The photoeffects in the presence of Fe^{3+} (an electron acceptor) are found to be independent of the direction of light, but the polarity of the photo-emf is determined by the location of Fe^{3+}, being always negative with respect to the iron-free side. The open-circuit photo-emf (E_{op}) as a function of light intensity (I) has been found to follow the simple relationship

$$E_{op} = A \log\left(1 + \frac{I}{B}\right), \tag{11}$$

where A and B are constant for a given BLM at a particular temperature. Under conditions of low light intensity ($B \gg I$), as would be expected, E_{op} is directly proportional to I. Recent work has shown that photoelectric spectra of these membranes are practically identical to the absorption spectrum of the lipid solution used (see Fig. 10). Thus, in an aqueous environment, the existence of semiconduction is not only possible but has been demonstrated in a reconstituted membrane. It is anticipated that the results of further studies on these chloroplast BLM will require interpretations for which the concepts of semiconductor physics will be most useful.

V. SEMICONDUCTION AND RESPIRATION

The reverse of photosynthesis is known as respiration, a process which occurs in all living organisms. The overall reaction for respiration can be written as

$$[CH_2O] + O_2 \rightarrow CO_2 + H_2O + \Delta F , \tag{11a}$$

where ΔF represents a quantity of useful energy for metabolism. The

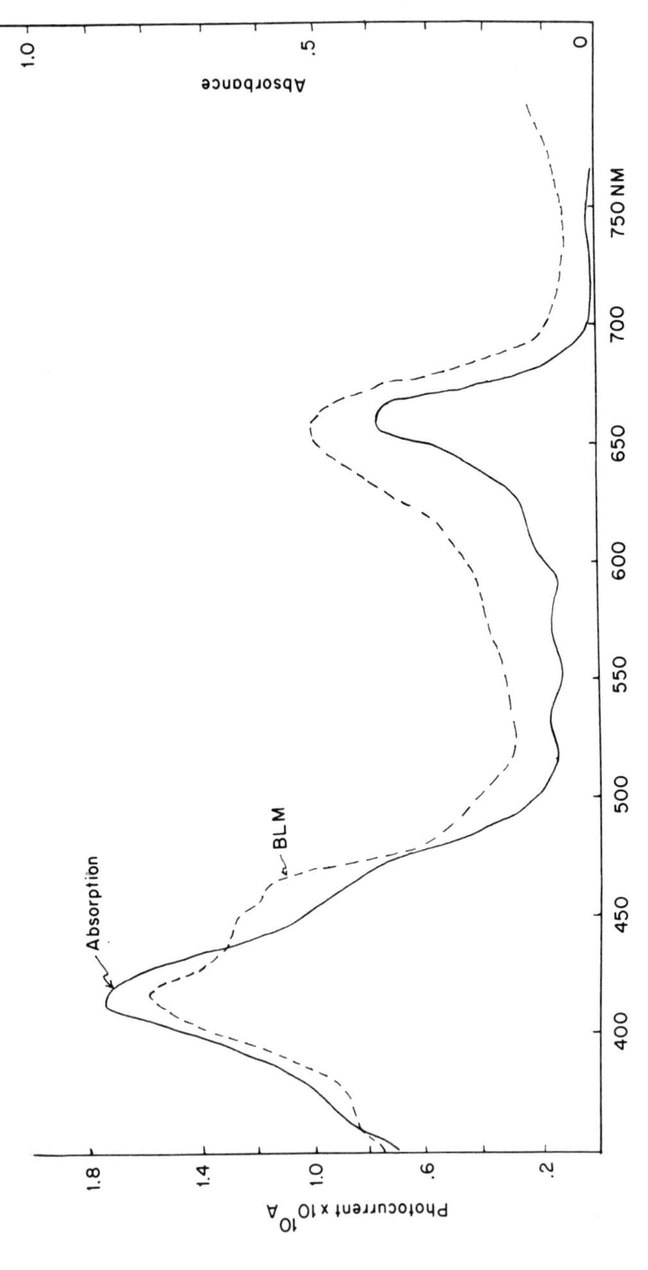

FIG. 10. Absorption and photoredox spectra of chloroplast extract: solid curve, in n-octane solution; dashed curve, in BLM. (Ref. 19 and J. Phys. Chem., 74, 3559, 1970.)

majority of biochemists believe that the process is accomplished in a series of steps, involving the so-called respiratory chain or electron-transport chain, in which a flow of electrons from $[CH_2O]$ through a system of spatially organized cytochromes and dehydrogenases to O_2 takes place. Since living organisms do not operate like heat engines, the ΔF liberated in Eq. (11a) has to be stored in a form that is readily available when required. This is usually "trapped" in the so-called energy-rich chemical bonds in a compound such as ATP (Adenosine Tri-Phosphate). Some of the questions that arise are: (i) Exactly how is energy converted and trapped in a chemical bond? (ii) What is the nature of this energy transducer? There are no definite answers to these questions at the present. Nonetheless we will attempt to discuss the respiration process in terms of semiconduction concepts in the following paragraphs. First, a brief description of the known biochemical and cytological aspects of respiration is in order.

The basic operational unit of respiration resides in the mitochondrion, which is frequently called the powerplant of the cell. The sequences of events taking place in the mitochondrion have been described in considerable detail by biochemists since 1925 [41]. A simplified scheme of the so-called respiratory electron-transport chain adopted from the work of Chance and Williams [42] and others is summarized in Fig. 11. The overall process consists of three separate but closely related steps: (i) the production of ATP during the oxidation of the coenzymes, (ii) the rearrangement and oxidation of carbohydrates to form reduced coenzymes, (iii) the oxidation of these coenzymes by molecular oxygen.

As has been demonstrated by Green and his colleagues [43], the active unit of the electron-transport chain is not the isolated protein but rather a lipoprotein complex. Four such complexes have been identified; each of these complexes represents the smallest unit in which a segment of the electron-transport chain can be isolated without loss of native activities, such as the ability to react with electron

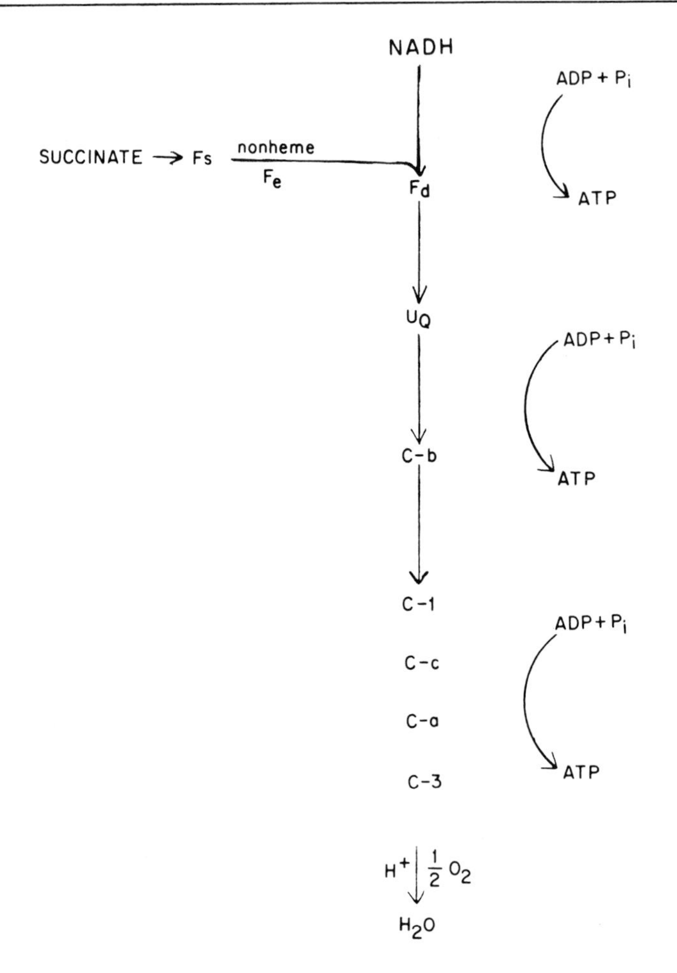

FIG. 11. Diagrammatic representation of the respiratory electron transport chain and the oxidative phosphorylation. NADH — reduced diphosphopyridine nucleotide; Fs — succinic acid dehydrogenase; Fd — NADH dehydrogenase; UQ — ubiquinone or coenzyme Q; C — various cytochromes (b, 1, c, a, and 3).

acceptors, and the appropriate values of redox potentials of the component proteins [44]. To accomplish these activities, the membranous structures are required. In fact, it has been strongly argued [43] that these electron-transport complexes are the membrane-forming elements of the mitochondrial inner membrane. We will now digress for a moment to describe the structural aspects of mitochondria.

The fundamental structure of mitochondria as revealed under the electron microscope appears to consist of two or double membranes [15]. The outer membrane encloses the mitrochondrion, whereas the inner membrane folds voluminously back and forth across the interior giving rise to compartments known as cristae. Hence the inner membrane is often called the cristae membrane. At the present time there is a controversy regarding the structure of mitochondrial membranes owing to uncertainties involved in the interpretation of electron micrographs. Until more unequivocal evidence is available, the generally accepted structure of mitochondria will be used in the present consideration (see Fig. 5).

An average mitochondrion has a sausage-shaped configuration about 3 μ long and 1 μ in diameter. The space between the membranes is filled with aqueous fluid containing perhaps the auxiliary catalysts known as coenzymes. (It will be recalled that enzymes are biological catalysts; they are simple proteins. Coenzymes are small molecules that must combine with the protein moiety of the enzyme in order to perform their assigned tasks.) The outer and cristae membranes serve as the structural framework for the mitochondrion. Without going into further details, both the outer and cristae membranes consist of a lipid bilayer of the Gorter-Grendel type with adsorbed protein layers. An idealized picture showing surrounding double membranes is presented in Fig. 5. The overall thickness of each membrane is again about 100 Å. It is generally recognized that the cristae membranes, as in the case with thylakoid membranes of chloroplasts, are the site of energy transduction. Referring to the electron-transport chain in Fig. 11, it is believed that, in the process of electron flow, the formation of a compound between the reductant (electron donor) and the oxidant (electron acceptor) takes place with the free energy of the electron stored in the chemical bond (e.g., ATP). The most intriguing and still unanswered question which was posed earlier is: "What is the mechanism of electron flow in the electron transport chain as postulated by biochemists?"

From what has been described, it might appear to the reader that semiconduction is hardly involved at all. Yet "electron flow" in the mitochondria, a long cherished belief of biochemists, must somehow occur. This apparent paradox may be explained by the fact that, until very recently, research on respiration has been almost an exclusive domain of biochemists. They tend naturally to describe these processes in terms of classical electrochemistry. For instance, the electron-transport process in their view is basically a series of redox reactions. As mentioned in the introduction (p. 848), redox reactions across biological interfaces are believed to be electronic in nature. As understood by chemists in a typical redox reaction, the transfer of a pair of electrons is involved. Thus, it may be said that biochemists have been interpreting their findings all along in terms of electron flow or semiconduction. Since a movement of electrons is involved in respiration, it is immaterial what language is used to describe it.

It should be mentioned at this point that, in contrast to redox reactions where two electrons are transferred at a time, one-electron transfer is also possible. A compound (or ion) having an "odd" electron is known as a "free" radical, which is usually a highly reactive species. Instead of a complete electron transfer between two molecules, a partial electron transfer is also possible. The process of partial electron transfer is often called "charge transfer" and the resulting species a "charge transfer complex" [45]. The formation of a charge transfer complex is given by

$$D + A \rightleftarrows DA \rightleftarrows D^{+} - A^{-} \rightleftarrows D^{+} + A^{-}, \qquad (12)$$

where D and A denote, respectively, an electron donor and acceptor species. As has been stressed by Szent-Györgyi, the formation of a charge-transfer complex permits transfer of an electron from one substance to another without appreciable loss in energy [21]. Perhaps, the electron-transfer chain involved in respiration (and photosynthesis) operates more efficiently by the use of a charge-transfer mechanism

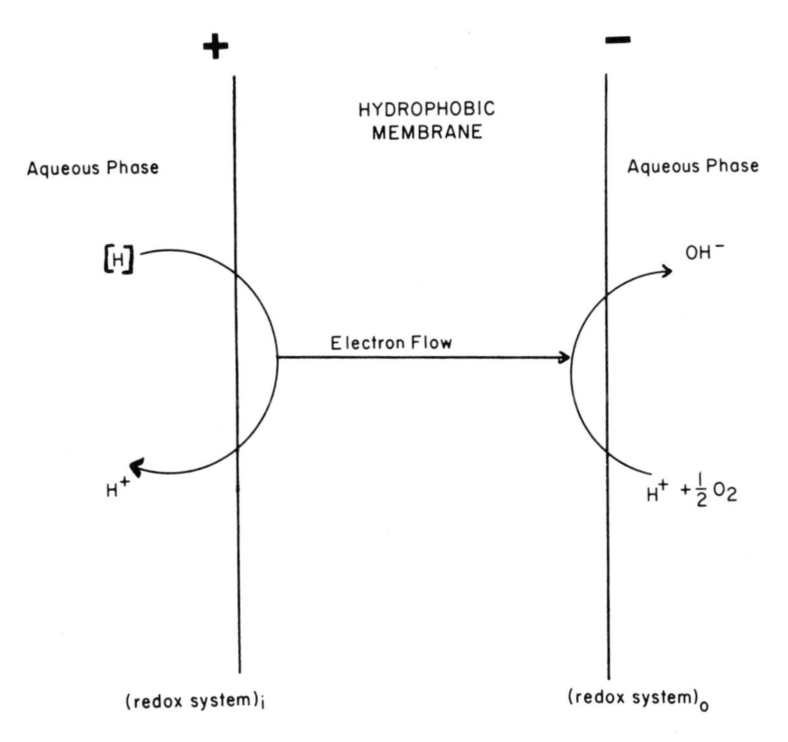

FIG. 12. Principles of operation of a redox membrane system. Electron flow is assumed to take place through the electron-transport chain located in the membrane (see text).

rather than the mechanism of complete charge separation. Further investigations on this question should be fruitful.

To conclude this section on respiration, we will illustrate the operation of a redox mechanism taking place across an ultrathin membrane. The concept of redox membranes in biology has been fully developed by Mitchell [46].

As illustrated in Fig. 12, a redox membrane containing an electron-transport chain separates two aqueous phases. Mitchell's hypothesis is based on the assumption that the membrane-bound redox chain is coupled with the ATPase system. The latter is capable of synthesis and hydrolysis of ATP, depending upon the direction of electron flow [46]. The operation of the redox membrane consists of the oxidation

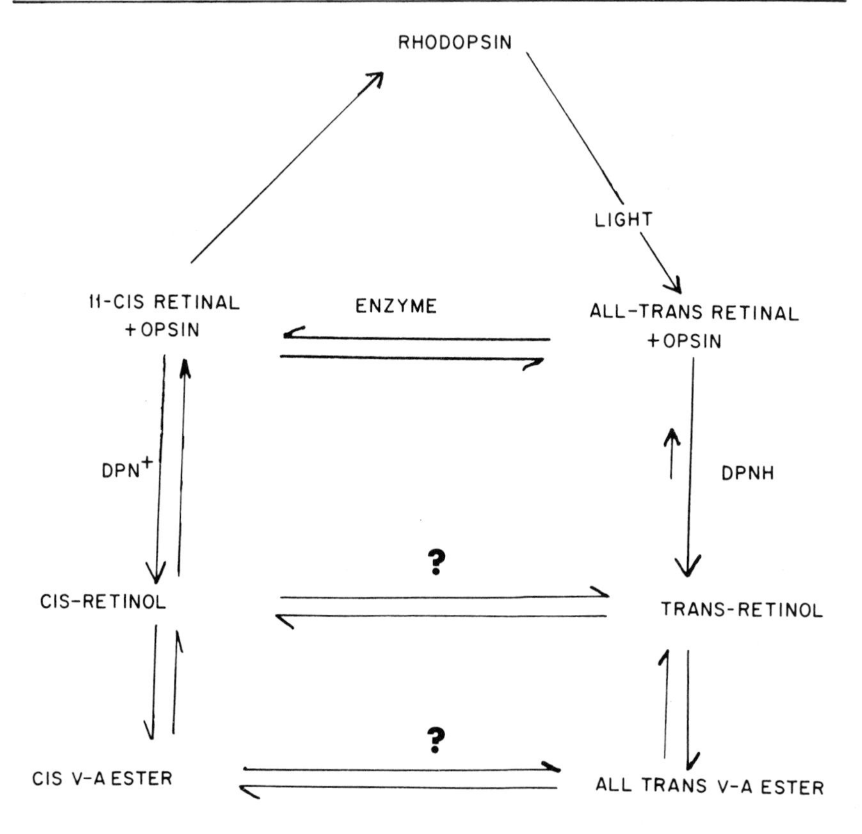

FIG. 13. Biochemistry of rhodopsin. The isomerization cycle of rhodopsin (after Wald [48]).

and reduction at the opposite interfaces across a high resistance hydrophobic membrane (see Fig. 12). The potential drop across the membrane could provide the necessary driving force for the operation. The movement of electrons through the hydrophobic membrane phase could be easily facilitated by the presence of cytochromes and highly conjugated ubiquinones. The operation of a redox membrane can be described as follows: A redox system with its characteristic potential is situated at each of the two interfaces or a biface (a biface is defined as the two coexisting interfaces where mass or electron transport is possible). The two redox systems are assumed to be maintained by unknown processes. Oxidation and reduction will take place across the biface with electrons (or holes) as charge carriers.

Whether the postulated mechanism occurs in the cristae membrane is undertain. But Mitchell's proposal is a useful working hypothesis. At the present time direct tests with the mitochondrial membranes are difficult. Perhaps, experiments with black lipid membranes (BLM) constituted with appropriate constituents would provide us with further understanding. These experimental BLM should prove ideal for investigations of the mechanisms of electronic conduction and active transport [47].

VI. SEMICONDUCTION AND VISION

A. General Considerations

In vision the action of light seems to change the conformation of a visual pigment leading eventually to a nervous impulse. The photochemical aspects of visual excitation have been worked out in detail by Wald [48]. So far as is known, all the visual pigments in vertebrates consist of a complex of retinal and opsin termed rhodopsin. The light-mediated biochemistry of rhodopsin is summarized in Fig. 13. The retinals and retinols are highly conjugated systems with a functional group of aldehyde and alcohol, respectively, at one end of the molecule. Both retinals and retinols possess four double bonds in the side chain, each of which might adopt either cis or trans configuration, producing thereby a number of geometrical isomers (Fig. 2). The most stable and prevalent isomer is the all-trans, whereas the 11-cis isomers are found to be involved in the initial step of light absorption. The various stages in the bleaching of rhodopsin according to Wald are shown in Fig. 14. The initial step in bleaching after absorption of a photon is to isomerize retinal (the chromophore of rhodopsin) from the 11-cis to the all-trans configuration (prelumirhodopsin). Then the structure of opsin opens progressively, ending in the splitting of retinal from opsin. The time lapse in going from metarhodopsin I to II is less than 1 msec. Visual excitation must have taken place by the time metarhodopsin II

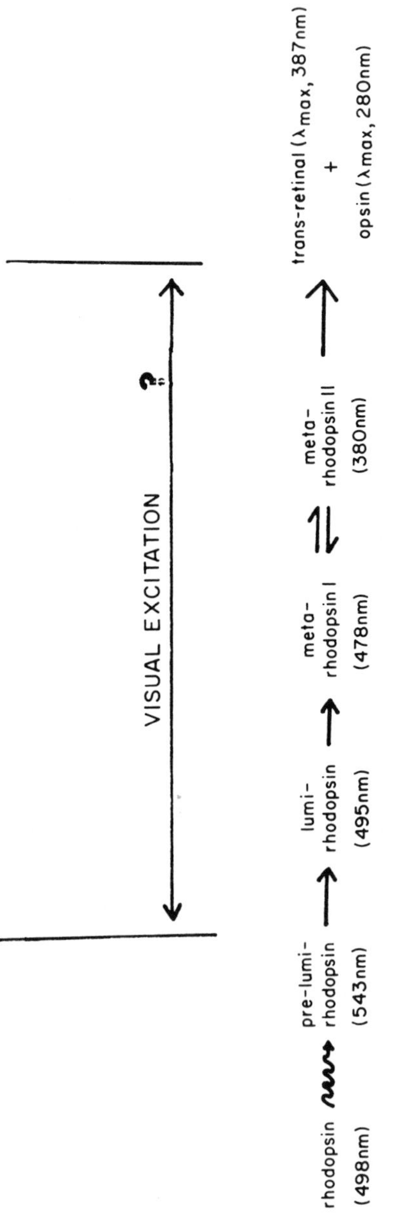

FIG. 14. Stages in the bleaching of rhodopsin (after Wald [48]).

is formed. All subsequent changes are dark reactions. However, under appropriate circumstances, light energy absorbed by any of the all-trans intermediates of bleaching can reisomerize the chromophore to 11-cis, thus generating rhodopsin.

The rhodopsins in vertebrates are located in the highly organized lamellar membranes of the outer segments of rods and cones. From a structural point of view, the photoreceptor consists of three segments: (i) the outer segment membrane where photopigment molecules are located, (ii) the inner segment, which provides the metabolic energy, and (iii) a conducting fiber that takes part in transmitting the excitation to other cells in the retina. Under the electron microscope, the outer segment membranes appear to be a stack of double membrane disks [15]. Each disk membrane, about 120 Å thick, consists of two triple-layered structures which could form a closed sac not unlike those found in the chloroplast. The interpretation of electron micrographs in physical and chemical terms is difficult owing to our lack of knowledge about the reactions that take place during fixing (e.g., OsO_4 treatment) and the precise chemical composition of lipids, proteins, and chromophores present in the outer segments. However, using the findings of electron microscopy and other experiments, Wolken [49] has proposed a schematic model for a rod outer segment membrane. Since the prosthetic groups (e.g., retinol) of the visual pigment molecules are highly surface-active [50], it is postulated that the "retinene" molecules align themselves at the aqueous-solution/membrane interface, with the conjugated chain in the lipid phase. It has been suggested that, if electronic conduction were to occur in the photoreceptor, it would have to be in the outer segment membranes since they are the only "crystalline" (liquid?) structures revealed by the electron microscope [15].

B. Mechanisms of Visual Excitation

At the present time two hypotheses attempting to explain the mechanisms of visual excitation have been suggested. The first approach is the classical view based upon the photoisomerization of retinal mentioned above. As suggested by Wald [48] and others, light absorbed by a rhodopsin molecule in the outer segment membrane isomerizes the retinal chromophore producing metarhodopsin (λ_{max} = 498). This leads to a series of reactions in which a substantial protein conformation change is involved, leading to the formation of metarhodopsin (λ = 380). Since it is postulated that visual pigments are an integral part of a highly organized lamellar structure in the rod outer segment membrane, which could result in significant changes in volume and electrical polarization, this in turn can generate the nerve impulse by a mechanism similar to that proposed for the nerve membrane (see Section VII). It should be mentioned that the participation by an enzyme with a large turnover rate in connection with impulse generation has been suggested [48].

The second approach based upon charge carrier generation by light is of recent origin, although the possibility of electronic conduction in carotenoid pigments had been proposed as early as 1948 [51]. In particular, Jahn [52] has suggested that light, in isomerizing 11-cis retinal to all-trans form, lengthens the conjugated hydrocarbon chain in the lamellar membrane. It is argued that all-trans retinal is a better electron transferring agent than the 11-cis retinal. This is consistent with the view that all-trans retinal is in a planar configuration, hence it is much more "resonant" than the 11-cis isomer, which has a twisted and bent chain (see Fig. 2). Jahn has also suggested that a biological membrane can be considered as a charge condenser with the charge stored in the reactants of the redox system located across the membrane. If these suggestions are correct, it seems that the chromophore of rhodopsin behaves in essence like a light-actuated switch. The photoreceptor is switched on when in the all-trans configuration; the 11-cis isomer existing only in the dark represents the off position. In the Jahn theory the conjugated chain of retinal is depicted as a "conducting wire," the sources of charge carriers are provided by enzymes situated on the two sides of the membrane.

Of interest to our discussion are the investigations of Smith and Brown [53], Becker and Cone [54], Rosenberg [55], and Arden et al. [56]. These workers have observed rapid electrical responses in all types of photosensitive systems. Whether electronic conduction is involved in these biological photoreceptors is not clear; however, recent investigations on BLM containing carotenoid pigments are of interest [57]. Tien and Kobamoto have found that, for example, BLM containing all-trans retinal can generate, upon illumination, temperature-sensitive biphasic photopotentials which are quite similar to the early receptor potentials (ERP) of visual receptors and pigmented animal and plant tissues. The results of BLM containing carotenoid pigments such as all-trans retinal are summarized below.

1. The photoresponses of the membrane consist of the fast component (R1), and the slow component (R2) which showed the sign opposite to R1 which is thought to be the potential determined mainly by hole diffusion from the following observations:

(i) The sign and magnitude of R1 can be influenced only by strong electron acceptors such as ferric ions.

(ii) R1 is independent of pH gradient and buffer capacity.

(iii) The side containing ferric ions is always negative with respect to the iron-free side.

2. Under proper experimental conditions a complete phase shift from R1 to R2 occurs as the temperature is increased from 9°C to 45°C. R2 is identified as the potential which is strongly dependent on protonic diffusion by the following observations:

(i) Changing proton concentration gradient across the membrane can control R2 in such a manner that the magnitude of R2 is linearly proportional to the logarithm of the proton concentration gradient.

(ii) R2 is not observed in buffered bathing solutions.

3. R1 and R2 are not influenced by changes in the ionic strength of K^+, Na^+, and Cl^-, nor by generation of a concentration gradient of these ions across the membrane.

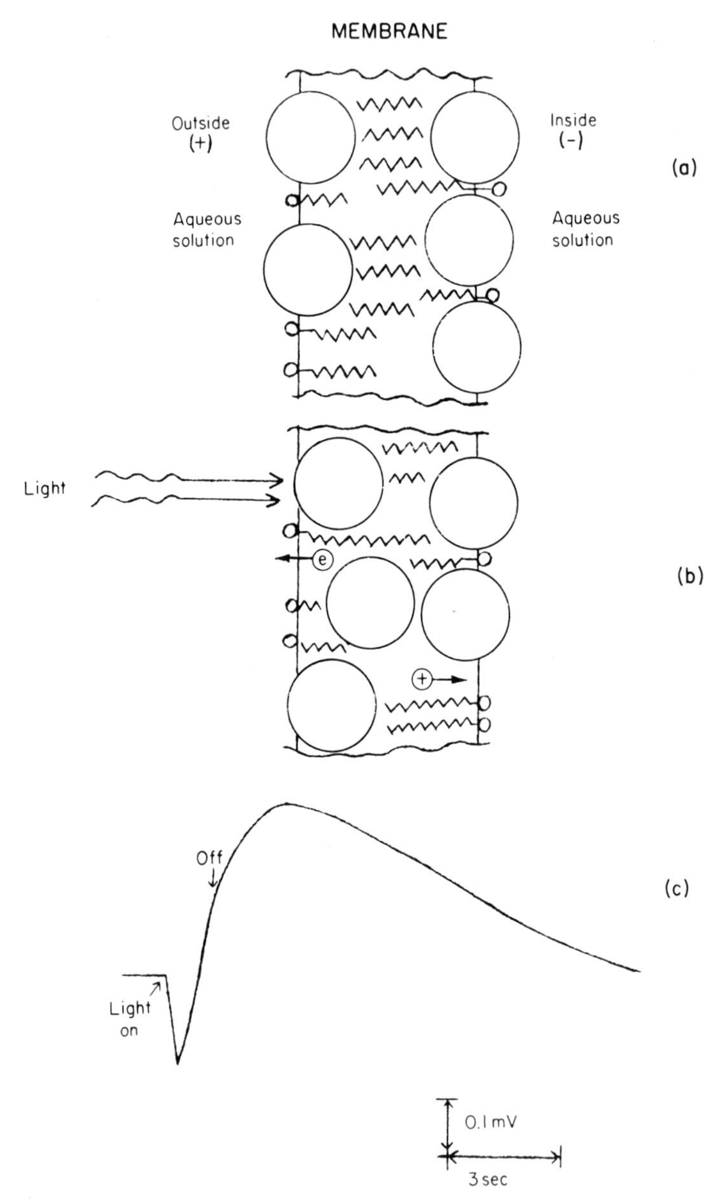

FIG. 15. A photovoltaic hypothesis of visual excitation. (a) A rod outer segment membrane separating two aqueous solutions. The membrane is polarized with inside negative. Circles represent protein moiety (opsin) and zigzag lines represent phospholipids (e.g., phosphatidylethanolamine) and chromophores (e.g., retinal). The hydrocarbon

The effect of increasing temperature can be interpreted as an enhancement not only in R2 but also in dark membrane potentials and membrane conductances. This finding implies that the effect of temperature is to increase proton mobility in the hydrocarbon layer of the membrane. The temperature dependence of the biphasic photopotentials can then be explained satisfactorily as the consequence of an increase in the proton conductance relative to the hole conductance.

In view of these ideas and recent experiments with carotenoid lipid membranes, one can further suggest that the rod outer segment membrane behaves like a photoelectric device operating in conjunction with the redox system across the membrane. The suggested sequence of events is shown in Fig. 15. Also shown in Fig. 15 is a typical time course of light-induced biphasic response in a chromophore-containing black lipid membrane (BLM).

In addition to the references cited in this section, two excellent reviews on the photochemical and macromolecular aspects of vision have been published by Abrahamson and Ostroy [58] and by Arden [59].

VII. SEMICONDUCTION AND NERVE EXCITATION

In the previous sections we have considered subjects of biophysics where application of semiconductor physics has been fruitful in explaining certain physical aspects of biological processes. In this section

---

chain of chromophores are in the cis configuration. Light is assumed to produce two effects: isomerization of chromophore from cis to trans configuration, and generation of charge carriers. These events are shown in (b). It is suggested that charge carriers migrating in the direction of the respective electrodes cause membrane depolarization, thus initiating the nerve excitation. (c) Photoelectric response in a model black lipid membrane (BLM) containing chromophores (all-trans retinal, cis-retinal, or vitamin A). The membrane was studied in an apparatus similar to the one shown in Fig. 4.

we consider the relationship between electricity and physiology, of which nerve excitation is by far the most intensively investigated topic.

Before considering the phenomenon of nerve excitation from the viewpoint of semiconduction, it is perhaps worth mentioning that the development of modern electronics owes much to the discovery of so-called "animal electricity" (or its modern equivalent "bioelectricity") in the eighteenth century. Galvani reported in 1791 that a dog's leg contracted when he touched one of its nerves with metallic wires. In his theory of animal electricity, Galvani formulated five principles, one of which states that the inner substance of the nerve is specialized for conducting electricity whereas the outer oily layer prevents its dispersal and permits its accumulation [60]. Today, more than 150 years later, the description of the outer oily layer of the nerve sounds remarkably modern (see Section III). The discovery of Galvani and his theory of animal electricity initiated a controversy which led his contemporary Volta to construct the famous voltaic pile, thereby providing a convenient source of electricity for experimentation. The fact that the names of Galvani (a biologist) and Volta (a physicist) are practically synonymous with electricity provides an interesting example where interaction between investigators of different disciplines stimulated further development and understanding.

Beginning with Galvani's observation, the phenomenon of nerve excitation has been, and still is, the focal point of research in electrophysiology. This sustained interest is caused by the belief that information processing and transmission in the nervous system are intimately connected with nerve excitation. At present, our understanding of this fascinating subject is far from complete. With this limitation in mind, we will begin the subject by sketching the overall process of nerve excitation.

Similar to other biological processes, the phenomenon of nerve excitation has to be considered in conjunction with a membranous structure. A typical nerve fiber or axon is very much like a long tubing filled with a salt solution of the same total concentration as that in the outer surrounding solution, although the chemical composition of the two solutions are very different from each other. The structure of the axon which separates the inside from the outside is thought to be similar to the membranes described earlier (see Fig. 5). This very thin oily membrane acts mainly as an insulating layer, but with a very small portion of the surface area selectively permeable to ions and other materials. With a pair of microelectrodes placed across the membrane, a potential of the order of 50-100 mV can usually be measured. The inside solution is negative with respect to the outside. This measured potential is termed the "resting" membrane potential. When a region of the membrane is stimulated in some way, such as by a short current pulse, the ionic permeability of the membrane is altered. During the excitation, according to the Hodgkin-Huxley theory [61], the membrane becomes more permeable to Na^+ ions than to other ions. From the resting state to excitation to the resting state, the sequence of events can be summarized as follows: (a) The nerve membrane is permeable to K^+ and Cl^- ions but impermeable to Na^+ ions in the resting state. (b) The membrane becomes highly permeable to Na^+ ions during excitation; the inward flow of Na^+ ions is checked, however, by the buildup of positive membrane potential. (c) The above event causes K^+ ions to diffuse out (in the direction of the electrochemical gradient) and thus polarizes the membrane to its normal resting state.

The sequence of these events is further illustrated in Fig. 16 [61]. It should be noted that in the Hodgkin-Huxley theory the chemical and structural aspects of the membrane are not emphasized. The membrane serves essentially as a diffusion barrier for various ions. In their electrical circuit model, the nerve membrane is thought to be a combination of a dielectric condenser, resistors, and batteries. In addition to the theory of Hodgkin and Huxley, which is generally accepted by physiologists, there are other theories based upon more recent findings [62]. We will describe only one other theory in which the concepts of semiconductor physics have been invoked.

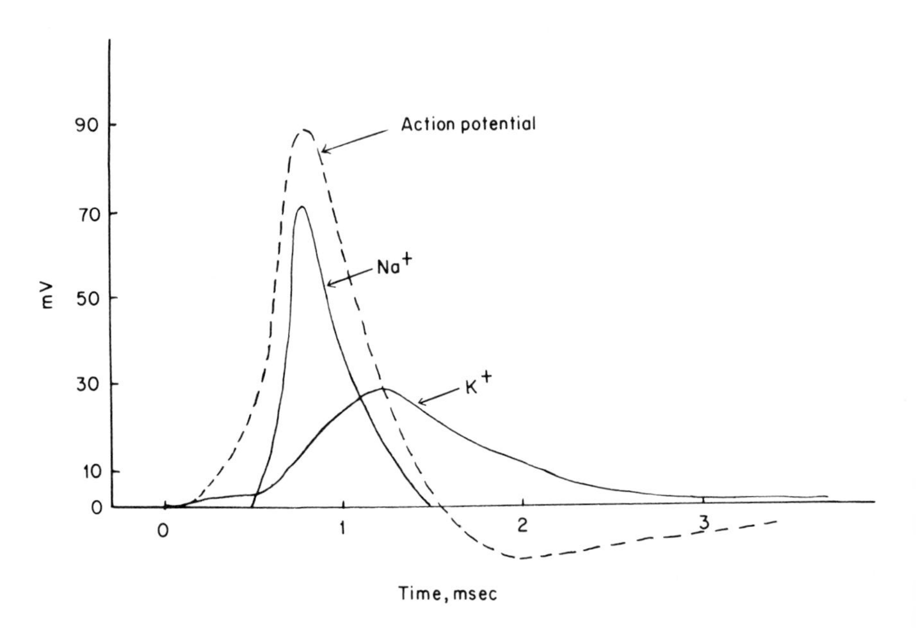

FIG. 16. The time course of the propagated action potential in a nerve axon. To initiate an action potential, the membrane potential must be lowered by 15 mV or more. This lowering of the membrane potential permits the Na^+ current to flow inward. The Na^+ current will further depolarize the membrane until the membrane potential is reversed in sign. At this stage the Na^+ flow stops and the K^+ flow begins, because the internal K^+ concentration is much greater than the outside. The K^+ flow continues outward until the membrane potential is restored to its original resting state (after Hodgkin and Huxley [61]).

Broadly speaking, the "semiconduction" theory proposed recently [63] rests upon two basic considerations: (i) the physics of ions and (ii) the physical structure of the nerve membrane and its adjacent interfaces. As is known, in an electric field the movement of ions constitutes a current, which is given by

$$I = enuE \quad , \tag{13}$$

where e is the electronic charge, n the ion concentration, and u the mobility. In the presence of a concentration gradient, ions will diffuse in the direction of the gradient according to Fick's law:

$$I_d = -eD\frac{dn}{dx} , \qquad (14)$$

where D is the diffusivity. As given by the Nernst-Planck equation, the net current is given by

$$I_{net} = e\left(nuE - D\frac{dn}{dx}\right) . \qquad (15)$$

Further, the generation and recombination of current carriers, i.e., the time rate of the change of concentration, is given by

$$\frac{dn}{dt} = G - R - \nabla I , \qquad (16)$$

where G and R denote, respectively, the rates of generation and recombination. ∇I is the outward flow per unit volume. It is generally recognized that the overall membrane potential (the quantity that is directly measurable) is the sum of the potential drops across the two interfaces and the diffusion potential in the membrane, i.e.,

$$E_t = E_i + E_o + E_m , \qquad (17)$$

where E_i and E_o are the respective potential changes at the inner and outer membrane/solution interfaces and E_m is the potential change due to the membrane itself. Since the membrane is mainly permeable to cations, it must contain negatively charged groups; hence, it is n-type in character. At the inner and outer interfaces, electric double layers are formed and a "p-type" character is acquired. Thus, formally speaking, the nerve membrane may be represented by a p-n-p transistor with ions playing the role of electrons and holes. Since the active carriers of current through the membrane are positive ions, the negative sites on, as well as in, the membrane can interact with the cations during the course of diffusion and drift. Therefore the chemical equivalents of generation and recombination are "dissociation" and "association," i.e.,

$$M^- + C^+ \rightleftarrows M^-{-}C^+ , \qquad (18)$$

where M^- stands for a negative site associated with the membrane and C^+ represents a cation. It should be stressed that this description of nerves in terms of semiconductor language in no sense implies that a nerve membrane is an electronic semiconductor. However, it does suggest that the well-known concepts of semiconductor physics can be extended to provide us with a new approach toward an understanding of nerve excitation.

More recently, the possibility of electron tunneling in biological membranes has also been suggested [64]. This is based on the observation that certain nerve axons exhibit n-shaped current-voltage characteristics. In view of the fact that a nerve or biological membrane is an ultrathin barrier (~100 Å) separating two aqueous solutions, it is within the range for electron tunneling. Further, the solution/membrane interface or junction can act as redox electrodes for charge injection (see Section V). Therefore it seems probable that the observed current through the membrane is the sum of ionic current (J_i) and electron tunneling current (J_e), that is,

$$J_m = J_i + J_e \quad . \tag{19}$$

For a membrane bearing fixed charged groups, J_i is rectified and is directly proportional to the applied potential. If we assume a redox system operating on each side of the membrane, the magnitude of J_e would be controlled by both the membrane and the redox systems. The interplay of these two currents could readily give rise to the observed negative resistance in biological and artificial membranes [18].

VIII. SEMICONDUCTION AND HEARING

The possibility that semiconduction may play a role in the mechanics of the organ of hearing is much less certain and has not been fully explored. Nonetheless, a number of schemes based upon solid state physics have been proposed. Among the various schemes, we will describe the one proposed by Naftalin [65], which is particularly relevant for the present discussion. First, the general aspects of the biophysics of hearing will be considered.

The basic process of hearing involves the transduction of airborne acoustic energy into electrical impulses to be received by the audio centers in the brain. Briefly, longitudinal waves in the air which impinge on the eardrum are transmitted, without significant loss of energy, to the ear fluid located in the inner ear. The acoustic vibrations in the inner ear fluid create a pattern of motion in the adjacent membrane (tectorial membrane) which in turn causes the attached pressure-sensitive nerve ending to fire (nerve impulses). Since individual nerve fibers cannot transmit impulses above 1 kHz, the higher frequencies are believed to be transmitted by the groups of fibers making up the auditory nerve. The mechanism of higher frequency transmission is based upon the so-called "volley principle" suggested by Wever [66]. The theory states that if, in a bundle of three nerve fibers, each of the three nerve fibers generates an impulse every millisecond but with a time delay of 0.33 msec, then the combination of three nerve fibers will generate impulses three times per millisecond, thereby reproducing a frequency of 3 kHz. The mechanism proposed for the hearing process is centered on the tectorial membrane mentioned above. The membrane is depicted as consisting of free ions and bounded electrons. The oscillation of ions induces forced vibrations in the surrounding electrons, thereby creating a lattice potential. If this assumption is granted, the membrane can possess elastic vibrations in the usual acoustic modes or as phonons. Through the electron-phonon interaction, thermal energy can be transferred to raise the electrons to higher energy levels. It was suggested [65] that the tectorial membrane together with its bounded icelike water exhibits crystalline properties. That is, the membrane can possess the dual properties of piezoelectricity and semiconduction, not unlike those of CdS crystals. Since the membrane in the living state is polarized, a dc potential exists across the aqueous-solution interface. If the direction of polarization is in the same direction as sound propagation, an incoming acoustic signal can add additional energy to the "free electrons" generated by thermal or other causes. Therefore, according to the suggestion of Naftalin, the electrons in the membrane not only do not dissipate the acoustic energy but actually could amplify the sound wave. Whether the ideas of semiconduction discussed above are relevant to the hearing process remains an open question [67]. Nevertheless, the proposed scheme is very intriguing and suggests experimental tests. Perhaps a modified BLM system can be used to provide an understanding of the hearing process at the molecular level.

IX. CONCLUDING REMARKS

We have discussed in this chapter only a few of the many interesting biological materials and systems. Fortunately, an extensive list of these topics may be found in the recent monograph by Gutmann and Lyons [6]. Of interest, but not discussed in this chapter, are the papers by Chance and Devault [68] on energy transfer through the cytochromes, on carcinogenesis by B. and A. Pullman [69], and on mechanisms of enzyme action by Cope [70]. For further details on various topics discussed in this chapter, the interested reader is referred to the following sources: on biological semiconductivity previous reviews have been published by Eley [71], Gergely [72], and by Rosenberg and Postow [73]; the area of photosynthesis has been reviewed by Tollin [74] and others [75]; for vision, proceedings of a symposium were published in 1965 (see Ref. [65]).

The aim of this chapter has been to give the reader an introduction to an area of biophysics where solid state concepts have been employed. Obviously there are many differences between a living system and a

TABLE 4

An Analogy Between Biological and Inorganic Semiconductors

	Inorganic semiconductor	Biological semiconductor
Base material	Covalent-bonded, crystalline phase (e.g., germanium crystal)	Hydrophobic hydrocarbon phase (e.g., lipid bilayer)
Electron donor	Group V elements (e.g., As, Sb)	Bioreductants (e.g., cytochromes, ferrous ions, and H_2O)
Electron acceptor	Group III elements (e.g., Ga, In)	Biooxidants (e.g., ferredoxins, quinones, and H_2O)
Electron pathway	Crystal proper	Conjugated hydrocarbon chain and ring systems
Connector	Metallic wire	Electrolyte solution

semiconductor device. Emphasis in this chapter has been on similarities. To complete this line of argument, an analogy between biological semiconductors and inorganic semiconductors is presented in Table 4.

The theme of this chapter on semiconduction and biology has been centered around the membranous structures of living systems as a basis for biological semiconduction. It should be evident that further advances in our understanding of many of the vital biological processes require the combined efforts of solid state physicists and chemists as well as molecular biologists and biophysicists. At the present time, among numerous problems to be solved, the problem of energy conversion in biological systems is the most elusive. Perhaps this is because the problem entails an integrated knowledge of many disciplines. Until we have a new breed of scientists trained in diverse fields, the solution of some of the problems in life sciences awaits investigators who are willing to step outside their own discipline. This step may seem arduous, but as Bentley Glass [76] observed in 1960 at the conclusion of a symposium on Light and Life: ". . . The advancing front of science now lies in the borderlands between the older sciences (biology, chemistry, and physics), and the investigator who would be successful today must explore many fields, learn many skills, and dare to apply to a challenging problem in one area the insight he has gained in studying many. That man is lost who would 'stick to his last,' specializing more and more on less and less. The great problems of life and light will yield only to those whose knowledge of light suffuses their knowledge of life, whose knowledge of life quickens their knowledge of light."

REFERENCES

1. A. Szent-Györgyi, Science, 93, 609 (1941).
2. E.J. Lund, J. Expt. Zool., 51, 327 (1928).
3. R.D. Stiehler and L.B. Flexner, J. Biol. Chem., 126, 603 (1938).
4. H. Lundagärdh, Nature, 143, 203 (1939).
5. G.G.B. Garrett, in Semiconductors (N.B. Hannay, ed.), Reinhold, New York, 1959, pp. 634-675.
6. F. Gutmann and L.E. Lyons, Organic Semiconductors, Wiley, New York, 1967.
7. L.E. Lyons, J. Chem. Soc. (London), 5001 (1957).
8. J. Kommandeur, in Physics and Chemistry of the Organic Solid States (D. Fox, M.M. Labes, and A. Weissberg, eds.), Wiley-Interscience, New York, 1965, p. 1.
9. M. Silver, D. Olness, M. Swicord, and R.C. Jarnagin, Phys. Rev. Lett., 10, 12 (1963).
10. G. Gaines, Insoluble Monolayers at Liquid-Gas Interfaces, Wiley, New York, 1965.
11. E. Gorter and F. Frendel, J. Expt. Med., 41, 439 (1925).
12. H.T. Tien and L.K. James, in Chemistry of Cell Interface, Part A (H.D. Brown, ed.), Academic, New York, 1971, pp. 205-253.
13. A.D. Bangham in Advances in Lipid Research (R. Paoletti and D. Kritchevsky, eds.), Vol. 1, Academic, New York, 1963, pp. 65-104.
14. E.D. Goddard, ed., Molecular Association in Biological and Related Systems, American Chemical Society, Washington, D.C., 1968.

15. F. S. Sjöstrand, Rev. Mod. Phys., 31, 301 (1959).

16. J. D. Robertson, in Electron Microscopy in Anatomy, J. D. Boyd, F. R. Johnson, and J. D. Lever, eds., Arnold, London, 1961.

17. S. K. Malhotra, Progr. Biophys. Mol. Biol., 20, 67 (1970).

18. P. Mueller, D. O. Rudin, H. T. Tien, and W. C. Wescott, Nature, 194, 979 (1962); J. Phys. Chem., 67, 534 (1963).

19. H. T. Tien in The Chemistry of Biosurfaces (M. L. Hair, ed.), Marcel Dekker, New York, pp. 233-348.

20. D. E. Green and Y. Hatefi, Science, 133, 13 (1961).

21. A. Szent-Györgyi, Introduction to a Submolecular Biology, Academic, New York, 1960.

22. M. Calvin and J. A. Bassham, The Photosynthesis of Carbon Compounds, Benjamin, New York, 1962.

23. E. Rabinowitch and Govindjee, Photosynthesis, Wiley, New York, 1969.

24. W. Menke, Brookhaven Sym. Biol., 19, 328 (1967).

25. R. Emerson and W. Arnold, J. Gen. Physiol., 16, 191 (1932).

26. E. Katz in Photosynthesis in Plants (W. E. Loomis and J. Franck, eds.), Iowa State College, Iowa, 1949, p. 287.

27. M. Calvin, Rev. Mod. Phys., 31, 147 (1959); see also Ref. [75].

28. T. L. Jahn, J. Theoret. Biol., 2, 129 (1962).

29. G. Tollin, J. Theoret. Biol., 2, 105 (1962).

30. K. Muhlethaler, in Biochemistry of Chloroplasts, Academic, New York, 1966, p. 49.

31. W. Arnold and H. K. Sherwood, Proc. Nat. Acad. Sci., 43, 105 (1957).

32. R. C. Nelson, J. Chem. Phys., 27, 864 (1957).

33. A. Terenin, E. Putzeiko, and I. Akimov, Dis. Faraday Soc., 27, 83 (1959).

34. V. B. Evstigneev, in Elementary Photoprocesses in Molecules (B. S. Neporent, ed.), Consultants Bureau, New York, 1968, pp. 184-200.

35. G. Tollin, D. R. Kearns, and M. Calvin, J. Chem. Phys., 32, 1013 (1960).

36. B. Rosenberg and J. F. Camiscoli, J. Chem. Phys., 35, 982 (1961).

37. D. D. Eley and R. S. Snart, Biochim. Biophys. Acta, 102, 379 (1965).

38. E. Rabinowitch, Trans. Faraday Soc., 27, 161 (1959).

39. H.T. Tien and S.P. Verma, Nature, 227, 1232 (1970).

40. G. Braun, Wied. Ann., 42, 450; 44, 473 (1891).

40a. M. Pope and H. Kallmann, in Electrical Conductivity in Organic Solids (H. Kallmann and M. Silver, eds.), Wiley, New York, 1961, pp. 83-104.

41. D. Keilin, Proc. Roy. Soc. (London), B98, 312 (1925).

42. B. Chance and G.R. Williams, Adv. Enzymol., 17, 65 (1956).

43. D.E. Green, D.W. Allmann, E. Bachmann, H. Baum, K. Kopaczyk, E.F. Korman, S.H. Lipton, D.H. Maclenman, D.G. McConnell, J.F. Perdue, J.S. Rieske, and A. Tzagoloff, Arch. Biochem. Biophys., 119, 312 (1967).

44. Y. Hatefi, A.G. Haavik, L.R. Fowler, and D.E. Griffiths, J. Biol. Chem., 237, 2661 (1962).

45. L.E. Orgel, Quart. Rev., 8, 422 (1954).

46. P. Mitchell, Biol. Rev., 41, 445 (1966).

47. M.K. Jain, A. Strickholm, and E.H. Cordes, Nature, 222, 871 (1969).

48. G. Wald, Science, 162, 230 (1968).

49. J.J. Wolken, J. Am. Oil Chemists' Soc., 45, 241 (1968).

50. A.D. Bangham, J.T. Dingle, and J.A. Lucy, Biochem. J., 90, 133 (1964).

51. H.J.A. Dartnall, Nature, 162, 122 (1948).

52. T.L. Jahn, Vision Res., 3, 25 (1963).

53. T.G. Smith and J.E. Brown, Nature, 212, 1217 (1966).

54. H.E. Becker and R.A. Cone, Science, 154, 1051 (1966).

55. B. Rosenberg, Adv. Rad. Biol., 2, 193 (1966).

56. G.B. Arden, H. Ikeda, and I.M. Siegel, Vision Res., 6, 373 (1966).

57. H.T. Tien and N. Kobamoto, Nature, 224, 1107 (1969); N. Kobamoto, Ph.D. Thesis, Michigan State University, 1970.

58. E.W. Abrahamson and S.E. Ostroy, Progr. Biophys., 17, 179 (1967).

59. G.B. Arden, Progr. Biophys. Mol. Biol., 19, 371 (1969).

60. T. Shedlovsky, Electrochemistry in Biology and Medicine, Wiley, New York, 1955, pp. 1-5.

61. A.L. Hodgkin, The Conduction of Nervous Impulse, Thomas, Springfield, Illinois, 1964.

62. I. Tasaki, Nerve Excitation, Thomas, Springfield, Illinois, 1968.

63. L. Y. Wei, IEEE Spectrum, 3, 123 (1966).

64. F. Gutmann, Nature, 221, 1359 (1968).

65. L. Naftalin, Sym. Quant. Biol., 30, 169 (1965).

66. E. G. Wever, Theory of Hearing, Wiley, New York, 1949.

67. G. von Bekesy, Experiments in Hearing, McGraw-Hill, New York,

68. B. Chance and D. Devault, Ber. Bunsenges. Phys. Chem., 68, 722 (1964).

69. B. Pullman and A. Pullman, Rev. Mod. Phys., 32, 428 (1960).

70. F. W. Cope, Proc. Natl. Acad. Sci. (U. S.), 51, 809 (1964).

71. D. D. Eley, in Horizons in Biochemistry (M. Kasha and B. Pullman, eds.), Academic, New York, 1962, pp. 341-380.

72. J. Gergely, in Electrical Conductivity in Organic Solids (H. Kallman and M. Silver, eds.), Wiley, 1961, pp. 369-383.

73. B. Rosenberg and E. Postow, Ann. N. Y. Acad. Sci., 158, Art. 1, pp. 161-190 (1969).

74. G. Tollin, Adv. Rad. Biol., 1, 33 (1964).

75. "Energy Conversion by the Photosynthetic Apparatus," Brookhaven Sym. Biol., 19 (1967).

76. B. Glass, in Light and Life (W. D. McElroy and B. Glass, eds.), Johns Hopkins, Baltimore, Maryland, 1961, p. 911.

AUTHOR INDEX

Numbers in parentheses are reference numbers and indicate that an author's work is referred to although his name is not cited in the text. Underlined numbers give the page on which the complete reference is listed.

N

O

P

R

S

SUBJECT INDEX

N

O

Q

R

S

T